Analysis of Climate Variability

Springer-Verlag Berlin Heidelberg GmbH

H. von Storch A. Navarra (Eds.)

Analysis of Climate Variability

Applications of Statistical Techniques

Proceedings of an Autumn School Organized by the Commission of the European Community on Elba from October 30 to November 6, 1993

2nd, Updated and Extended Edition

With 91 Figures
and 13 Tables

Springer

Editors:

Prof. Dr. Hans von Storch
GKSS Research Center
Institute of Hydrophysics
Max-Planck-Straße 1
D-21502 Geesthacht (Germany)

Dr. Antonio Navarra
Istituto ISAO-CNR
Via Gobetti 101
I-40129 Bologna (Italy)

ISBN 978-3-642-08560-4

Library of Congress Cataloging-in-Publication Data
Analysis of climate variability: applications of statistical techniques: proceedings of an autumn school organized by the Commission of the European Community on Elba from October 30 to November 6, 1993/Hans von Storch, Antonio Navarra (eds.). - 2nd. updated and extended ed. p. cm. Includes bibliographical references and index.

ISBN 978-3-642-08560-4 ISBN 978-3-662-03744-7 (eBook)
DOI 10.1007/978-3-662-03744-7

1. Climatic changes - Statistical methods Congresses. I. Storch, H. v. (Hans von), 1949- II. Navarra, A. (Antonio), 1956- . III. Commission of the European Communities. QC981.8.C5A52 1999 551.6-dc21 99-44751 CIP

Originally published by Springer-Verlag Berlin Heidelberg New York in 1999
Softcover reprint of the hardcover 2nd edition 1999

Typesetting: Camera ready by authors/editors
Cover: Erich Kirchner, Heidelberg
SPIN 10741323 32/3136-5 4 3 2 1 0 - Printed on acid-free paper

Preface

This volume has grown from an *Autumn School* about "Analysis of Climate Variability - Applications of Statistical techniques" on Elba in November 1993. We have included those lectures which referred explicitly to ***applications*** of statistical techniques in climate science, since we felt that general descriptions of statistical methods, both at the introductory and at advanced level, are already available. We tried to stress the application side, discussing many examples dealing with the analysis of observed data and with the evaluation of model results (Parts I and II). Some effort is also devoted to the treatment of various techniques of pattern analysis (Part III). Methods like teleconnections, EOF, SSA, CCA and POP are becoming routine tools for the climate researcher and it is probably important for graduate students to be exposed to them early in their academic career in a hopefully clear and concise way.

A short subject index is included at the end of the volume to assist the reader in the search of selected topics. Rather than attempting to reference every possible occurrence of some topic we have preferred to indicate the page where that topic is more extensively discussed.

The Autumn School was part of the training and education activities of the European Programme on Climatology and Natural Hazards (EPOCH), and is continued under the subsequent research programme (ENVIRONMENT 1990–1994). It aimed at students in general, taking first and second year courses at the graduate level. Since then, the idea of organizing week-long schools of this sort, with proceedings published by Springer Verlag, has flourished; one of the editors, Hans von Storch, has set up with his institution, the GKSS Research Centre in Germany, the "GKSS School on Environmental Research" which has dealt by now with "Anthropogenic Climate Change" and "Modelling in Environmental Research".

This collection of lectures created a significant response in the scientific community, so that in late 1998 the first edition was out of stock, and the publisher, Springer Verlag, offered us the publication of a *second, revised edition*. It turned out that in spite of the impressive progress of climate sciences, the lectures of this volume were mostly up-to-date; several chapters were updated, in particular Chapters 3, 4, 8, 10, 12, 13 to 15. For instance, recent developments with respect to detecting climate change are summarized in Chapter 4, an SVD based method for comparing observed and modeled data is presented in Chapter 12, the techniques of Redundancy Analysis and Empirical Orthogonal Teleconnections, and of Analysis-of-Variance are sketched in Chapter 13 and 8. Of course, the extensive literature list has been updated and extended, and fine tuning with formulation was made throughout the book for further improving the clarity.

Hans von Storch and Antonio Navarra, July 1999

Contents

List of Contributors

- **Keith Briffa**
  Climate Research Unit
  University of East Anglia
  GB - NR4 7TJ NORWICH

- **Claude Frankignoul**
  Laboratoire d'Océanographie Dynamique et de Climatologie
  Université Pierre & Marie Curie
  4, Place Jussieu, Tour 14-15, 2èmé étage, F - 75252 PARIX CEDEX 05

- **Phil Jones**
  Climate Research Unit
  University of East Anglia
  GB - NR4 7TJ NORWICH

- **Dennis P. Lettenmaier**
  Department of Civil Engineering
  University of Washington
  164 Wilcox Hall, FX 10, USA 98195 SEATTLE Wa

- **Robert Livezey**
  Climate Analysis Centre
  W/NMC 53, Room 604, World Weather Building
  USA - 20 233 WASHINGTON DC

- **Antonio Navarra**
  Istituto per lo studio delle Metodologie Geofisiche Ambientali (IMGA-CNR)
  Via Gobetti 101, I 40129 Bologna

- **Robert Vautard**
  Laboratoire de Météorologie Dynamique
  24, Rhue Lhomond, F - 75231 PARIS CEDEX 05

- **Hans von Storch**
  Institute of Hydrophysics, GKSS Research Centre
  PO Box, D 21502 Geesthacht

- **Jin-Song von Storch**
  Meteorologisches Institut der Universität Hamburg
  Bundesstrasse 55, D 20146 HAMBURG

- **Neil Ward**
  Hadley Centre, Meteorological Office
  London Road, GB - RG12 2SY BRACKNELL, Berkshire

Part I

Introduction

Chapter 1

The Development of Climate Research

by Antonio Navarra

Il n'y a plus, avec le systeme statistique - de signification profonde des lois - de sens pur.
P. Valery, Cahiers

1.1 The Nature of Climate Studies

The geophysical sciences in general and climate research in particular have always played a special role among the other scientific disciplines for their special relationship with the object of their study. Climate research deals with climate, that is to say the response of the atmosphere, ocean, land and ice systems to the solar forcing over a prescribed time interval. In this respect climatology diverges from the other quantitative "hard" sciences in a simple, but very important detail: the impossibility of performing the "crucial" experiment on which the classic paradigm of investigation of the physical sciences is built.

Since Galileian times, scientific investigation has been based on the interplay between theory and experiment. The job of the theorist is to speculate mathematical relationship between observable quantities to explain the physical phenomena on the basis of some hypothesis. Competing theories, that is mutually excluding explanations for the various facts, have to be sorted out by resolving to the experience. The experimentalists would therefore set up ad hoc physical systems, the experiments, in which the differences between the theories can be artificially stressed as if they were under a magnifying glass. This interplay was the foundation of the success of the natural sciences in the 1800's and the early part of this century. A foremost example of a crucial experiment is the famous interferometry experiment of Michelson and Morley that put an end to the raucous debate on the existence of the ether, the strange medium that was being supported by many physicists to explain the propagation of the electro-magnetic waves. In that case a single experiment, in a small laboratory, was able to settle the dispute once for all.

The situation for climatology is very different. Climatology is a one-experiment science. There is a single experiment for the entire discipline that has been going on for hundreds of million of years, our planet. In this case the mainstream scientific approach is not feasible. We can hardly modify the rotation rate of the Earth or other planetary parameters and then observe the result on the Earth climate. In fact, even the definition of the present state of the Earth, namely the documentation of the present situation of the Earth climate in physical, chemical and biological terms, is a tremendous problem complicated by basic uncertainties and basic sampling problems in space and time for the entire system.

The complexity of the climate system required from the early years the usage of statistical methods to identify relations among the climate quantities that were hidden in the multitude of spatial and temporal scales in play. As the science reached a mature stage, it came to rely more and more on statistical techniques of increasing sophistication. Paul Valery, a French philosopher, in the quote that opens this chapter, expressed some doubts on the "quality" of scientific laws deduced with the help of statistical methods: are statistically-obtained laws less pure than the others ? Do they carry a less profound significance ? In the course of this book it should become increasingly clear that statistical techniques are a powerful and clever tool that can lead to an understanding of nature as profound as other scientific devices. Like other powerful tools it works wonders in competent hands, but it can create havoc otherwise. Power tools can make your work easy, but they can drill you a hole in the arm; it depends on you.

Just to give a feeling on how much climate science has depended on some sort of statistical techniques, I will briefly discuss two cases that clarify how much statistics, or some sort of careful consideration of data, was necessary to understand a physical problem.

1.1.1 The Big Storm Controversy

The development of the atmospheric sciences was greatly accelerated when observations became more reliable and more readily available, around 1860. The great development of merchant marine traffic was a powerful motivation for elaborating new methods for studying and eventually predicting storms. The main problem in those times was to make sense of the complication of the winds and current behavior to try to isolate some generalization, some pattern or some simple dynamical law.

In general terms, these efforts were aiming at identifying and isolating patterns in the observed data. The invention of the synoptic map (1860) contributed to the search for physical phenomena that could be identified as a recurrent feature of the weather. Storms were the most immediate phenomena to be studied and also one whose practical importance was enormous. Great effort was therefore put in trying to define and understand the patterns of wind, pressure and rain, collectively known as "storms".

The main controversy that occupied the XIX century meteorologists had to do with the nature and origin of "storms". In the beginning of the century, before synoptic charts were invented, there were wild disagreements on the nature of storms and particularly intense was the controversy between Americans and European researchers on the rotational nature of the wind in storms. It was well known at that time that storms were characterized by surface low pressure, but the relation between pressure and wind was a little sketchy. Some people, like the famous and powerful German meteorologist H. Dove, refused the idea of rotary storms altogether and theorized that the meteorological events observed as storms were merely the displacement of tropical currents with polar currents and vice versa. Dove's law of gyration of winds ("Drehungsgesetz") that describes how the winds shift during the transition between the two currents was commonly used in Europe until the 1860's. On the other hand, in the United States, observations by W. C. Redfield seemed to support very strongly the rotational character of storms. From a modern point of view the controversy was fueled by the total confusion about the physical phenomena to be explained. There was no attempt to distinguish between storms of different scales and there was no attempt to separate storms between midlatitude cyclones and tropical cyclones. In fact, Dove's observations in central Europe were actually south of the predominant path of the low pressure center, making it difficult to realize the rotational movement of the cyclone, whereas Redfield mostly focused on hurricanes and tornadoes, whose rotational character was immediate.

An alternative explanation of storms was offered by J.P. Espy in 1841. In the late 1870's it became the accepted scientific theory for the scientific understanding of storms. Espy identified the condensation process of water vapor as the main energy source for the storms. In his theory the circulation inside a storm was to be directed towards the low pressure center where the vertical motion would produce enough condensation to sustain energetically the circulation.

Essential for Espy's theory was the vertical profile of temperature in the storm. In Espy's vision, cyclones were cold and anticyclones were warm air vortices with no vertical structure, so when upper air observations at mountain stations in the Alps and in the Rocky Mountain started to show that the vertical temperature profile of a storm was more complicated, the theory was under attack. The evidence brought forward by Julius Hann, director of the newly formed Austrian Weather Service, caused an uproar. Hann was violently attacked by Ferrell and the dust did not settle until F.H. Bigelow suggested that storms were in fact connected to the general circulation. After 1900 a number of studies were able to find out that storms go through a "life cycle" from North America to Europe and their energy budget and dynamical properties are changing through the various stages of their evolution. This concept was finally synthesized in a consistent scheme by the so called polar front theory of cyclones in the 1920. The full story of the understanding of storms in the XIX century is described in detail in a wonderful monograph by G. Kutzbach (1979). But it was not until the late 1940's that the first consistent dynamical interpretation of the polar front cyclones, the baroclinic instability theory, was discovered by Jules Charney. Meanwhile, other studies were gathering evidence of other large scale patterns on an even larger scale.

1.1.2 The Great Planetary Oscillations

The research of the XIX century concentrated on storms for good practical reasons. The motivations for the research at that time were found in the need to provide ship's captains with some tool to make sense of the behavior of the atmosphere and the oceans, with the double aim of avoiding calamitous storms and deftly exploiting swift winds. This necessity led meteorology to recognize the fact that American and European storms were part of the same phenomena, weather patterns connected together on time scales of days and weeks. The interest for this short time scale was so overwhelming that relatively little effort was made to discover connections on different time and spatial scales.

However, in the 1920's and 1930's, Sir Gilbert Walker, the Director of the Indian Meteorological Office, devoted himself to the study of statistical connections between atmospheric variables over large space scales and long periods of time. Sir Gilbert was motivated by the puzzle of the Indian monsoon and its remarkable regularity. A shift of a few weeks in the appearance of the rains would mean the difference between famine and a good crop for the Indian sub-continent. His approach was to try to find relations between time series of monthly means of pressure observations at different places, in the hope that signs that could anticipate the monsoon would be evident in the large scale pressure pattern. Working with his colleague Bliss (Walker and Bliss, 1932), he was able to discover through this analysis that surface pressure was in phase and out of phase in several locations on the globe, resulting in broad oscillatory features. Two of these "oscillations" were remarkable, the Southern Oscillation (SO) and the North Atlantic Oscillation (NAO).

The Southern Oscillation shows up as an out of phase relation between the surface pressure in the Eastern and Western Pacific. It was first noted by Hildebrandsson (1897), in the time series of surface pressure between Sidney and Buenos Aires. It is evident in many places, but traditionally the observation taken at Darwin (in the north of Australia) and the island of Tahiti are used to define it. Unfortunately, the observation taken by Captain Bligh and his Bounty are not available, but the record is sufficiently long that the Oscillation can be established beyond any doubt. It is obviously not an oscillation, i.e. a sinusoidal wave, rather a more irregular quasi-periodic oscillation, but the out of phase relation between Tahiti and Darwin is clear. When the pressure in Tahiti tends to rise, then it falls in Darwin and vice versa. Sir Gilbert's statistical analysis of the observation, in this case a simple correlation analysis, was clearly pointing to some unknown physical process that linked together the Western and Eastern tropical Pacific (Philander, 1989; Troup, 1965). Similarly surprising is the North Atlantic Oscillation, described by Walker as a correspondence in pressure between the Azores and Iceland. The surface pressure in the two sites shows a strong tendency to anticorrelate, with high pressure over Iceland corresponding to low pressure over the Azores.

More modern studies have shown (van Loon and Rogers, 1978) that these local indices are part of larger structures, big circulation patterns stretching over the entire Pacific, in one case, and over the Atlantic and West European sector in the other. At the beginning of the 1980's it was clear that much larger patterns existed in the atmosphere, as will be discussed in Chapter 12.

1.2 The Components of Climate Research

The availability of large scale computers made it possible to overcome the absence of crucial experiments. If crucial experiments could not be done because the Earth was reluctant to be modified according to the scientists' whim, the scientists would just do without the Earth and simulate a new Earth using the same equation and physical systems that are thought to be active in the real Earth. The simulated Earth, a very large ensemble of mathematical equations, can be computed with computers and experiments can be designed to verify hypothesis on the simulated Earth. Numerical experiments became the best approximation to the crucial experiments of the other physical disciplines. A new mode of scientific investigation has been added to theory and experiments: numerical experiments. Numerical simulations blend some features of experimental techniques and the abstraction of theoretical investigation. In latest years, numerical simulations have come to play a predominant role in climate research.

The appearance of large scale computers also made a definite advance on the way to better define the present state of the climate by allowing the statistical treatment of large amounts of data. Statisticians could elaborate more and more sophisticated techniques to extract information from a large ensemble of data. The highly nonlinear nature of the fluid dynamics equations and of the physical processes describing the atmosphere make the interpretation of model results not much easier than investigating the real atmosphere. Sometimes, the subtleties of the interplay at work create a situation not much better than Dove or Redfield faced. The atmosphere is being forced by the sun in a very regular way, the seasonal march of the sun, but the response of the climate system spans many scales. Intuition is a very powerful tool to find out the relations between the various components that are masked by the chaotic behavior of the system, but very often it is necessary to have powerful statistical methods to guide that intuition towards more fruitful directions. In summary, climate research seems therefore to be dominated by three main components whose careful and balanced combination is required to obtain a good scientific result: dynamical theory, numerical experimentation and statistical analysis.

1.2.1 Dynamical Theory

The theory of the atmospheric and oceanic circulation related to the climate problem is based on utilizing basic concepts of hydrodynamics and thermodynamics. It has proven to be an indispensable tool to *understand* climate.

The impact of the baroclinic instability paper by Charney in 1949 can be hardly overestimated. Pedlosky (1990) gives a very vivid analysis of the importance of that paper for the development of meteorology and oceanography. The theory of baroclinic instability has allowed us to put the process of cyclone formation on scientifically firmer ground and it has offered an entirely new interpretative framework, opening the ground to the zonal mean-eddy paradigm that has dominated the atmospheric sciences for the following 30 years. The theory of baroclinic instability has also allowed an energetically consistent treatment of the cyclone formation and evolution process.

Sverdrup (1947) paper was the first work to recognize the role of the wind stress in generating vorticity into the interior of the ocean. The famous relation between the curl of the wind stress and the flow in the ocean interior was formulated for the first time in this work. A related result, though developed in a quite independent way, is the paper on the Gulf Stream boundary intensification by Stommel (1948), in which the crucial role of β, the gradient of the planetary vorticity, is recognized as causing the asymmetry in response to winds between the east and the west coasts of the oceans.

Stommel recognized that the Gulf Stream was generated by a mechanism similar to the generation of boundary layer in flow problems, smoothly joining Sverdrup solution away from the boundary. Afterward, Munk, Groves and Carrier (1950) generalized these models to include nonlinear effects. A beautiful discussion of the intellectual story of the formulation of the theories of the large scale ocean circulation can be found in Veronis (1981).

1.2.2 Numerical Experimentation

Numerical experimentation with large "nature-like" models of the ocean, atmosphere and other climate-relevant systems (such as sea-ice) is the youngest actor on the stage. After the pioneering work in the 1960's (Smagorinsky, 1963; Smagorinsky et al., 1965) with atmospheric models and ocean models (Bryan and Cox, 1969), coupled models were developed (Manabe and Bryan, 1969). Later, numerical experiments were able to show that a significant part of the observed atmospheric variability can be reproduced by an atmospheric GCM forced with 15 years of observed sea-surface temperature (Lau, 1985). And more recently, numerical models were instrumental in revealing the potential danger of anthropogenic modifications of the tropospheric greenhouse gas concentrations (Manabe et al., 1994; Cubasch et al., 1992).

Numerical experimentation is also successful in attracting attention and confidence in the scientific community - one reason is that such models appear as tools which can answer, at least in principle, all questions in a "physically consistent" manner. There is still a lot to do. The hydrodynamic part of the models is based on first principles, such as the conservation of mass, energy and angular momentum, but the discretization in space and time introduces errors that cannot be assessed easily. The irreversible thermodynamic processes are described by *parameterizations* which should account for the *net effect* of processes such as turbulence in the boundary layer, convection or clouds. Discretization and the indeterminacy of our understanding of the physical, biological and hydrological processes affect parameterizations even more severely than the hydrodynamics. The package of these parameterizations is usually called the "physics" of the models - but this "physics" is a mix of equations which have been fitted to satisfy first principles, intuition, results of field campaigns or to improve the overall performance of the model on a large scale (see, for instance, Delsol et al., 1971, Miyakoda and Sirutis, 1977 and more recently, Sirutis and Miyakoda, 1990, Tiedtke, 1986; Roeckner et al., 1992, 1996).

1.2.3 Statistical Analysis

Statistical analysis is required for the interpretation of observed and simulated data sets. The need to use statistical techniques originates from the large phase space of climate (see, e.g., Chapter 13). The advantage of statistical approaches is that they allow one to deal directly with information about the *real* system. While certain crucial processes might be misrepresented in a model, observed data are reflecting the influence of all relevant processes. Unfortunately, observed data reflect also all irrelevant processes, which create the *noise* from which the real signal is to be discriminated.

The danger is that the discrimination between signal and noise may not be successful. Then random features of the noise are mistaken as a true signal - and the literature is full of such cases. A problem specific to the application of statistical analysis in climate research is that we have ***only one realization*** of the climate variability. Therefore almost all studies re-cook the same information (not necessarily the same data, but data from the same limited piece of the trajectory the climate system has moved along in the past hundred "instrumental" years). Not surprisingly, every now and then strange "significant" results are created, which are random events which have been upgraded to "significance" by a type of data screening.

The development of climate research has shown that a successful understanding of climate dynamics requires a clever combination of these three components. Sometimes the dynamics are too complex and the simplification required for the dynamical theory cannot be sensibly done. Then the numerical models have to demonstrate that our dynamical concepts, which are encoded in our basic physical equations, really describe the phenomenon at hand. Furthermore, a statistical analysis of observed data should clarify that the model results are not artifacts of the model, which might stem from processes which are disregarded or inadequately parameterized. Models have advanced so much in the recent past that their behavior is sometimes as difficult to interpret as the real atmosphere. Statistical methods are sometimes the only way to shed light on some intricate processes generated from a numerical experiment. In the following we will try to introduce some of the most useful techniques to analyze climate data, either observed or simulated, with a special emphasis on delicate junctures, where the risk of back firing for the carefree apprentice is sometimes very high.

Chapter 2

Misuses of Statistical Analysis in Climate Research

by Hans von Storch

2.1 Prologue

The history of misuses of statistics is as long as the history of statistics itself. The following is a personal assessment about such misuses in our field, climate research. Some people might find my subjective essay of the matter unfair and not balanced. This might be so, but an effective drug sometimes tastes bitter.

The application of statistical analysis in climate research is methodologically more complicated than in many other sciences, among others because of the following reasons:

- In climate research only very rarely it is possible to perform real *independent experiments* (see Navarra's discussion in Chapter 1). There is more or less only one observational record which is analysed again and again so that the processes of building hypotheses and testing hypotheses are hardly separable. Only with dynamical models can independent

---

Acknowledgments: I thank Bob Livezey for his most helpful critical comments, and Ashwini Kulkarni for responding so positively to my requests to discuss the problem of correlation and trend-tests.

data be created - with the problem that these data are describing the real climate system only to some unknown extent.

- Almost all data in climate research are *interrelated* both in space and time - this spatial and temporal correlation is most useful since it allows the reconstruction of the space-time state of the atmosphere and the ocean from a limited number of observations. However, for statistical inference, i.e., the process of *inferring* from a limited sample robust statements about an hypothetical underlying "true" structure, this correlation causes difficulties since most standard statistical techniques use the basic premise that the data are derived in independent experiments.

Because of these two problems the fundamental question of how much *information* about the examined process is really available can often hardly be answered. Confusion about the amount of information is an excellent hotbed for methodological insufficiencies and even outright errors. Many such insufficiencies and errors arise from

- The *obsession with statistical recipes* in particular *hypothesis testing.* Some people, and sometimes even peer reviewers, react like Pawlow's dogs when they see a hypothesis derived from data and they demand a statistical test of the hypothesis. (See Section 2.2.)

- The use of statistical techniques as a *cook-book like recipe* without a real understanding about the concepts and the limitation arising from unavoidable basic assumptions. Often these basic assumptions are disregarded with the effect that the conclusion of the statistical analysis is void. A standard example is disregard of the serial correlation. (See Sections 2.3 and 9.4.)

- The *misunderstanding of given names.* Sometimes physically meaningful *names* are attributed to mathematically defined objects. These objects, for instance the *Decorrelation Time*, make perfect sense when used as prescribed. However, often the statistical definition is forgotten and the *physical meaning* of the *name* is taken as a definition of the object - which is then *interpreted* in a different and sometimes inadequate manner. (See Section 2.4.)

- The *use of sophisticated techniques.* It happens again and again that some people expect miracle-like results from advanced techniques. The results of such advanced, for a "layman" supposedly non-understandable, techniques are then believed without further doubts. (See Section 2.5.)

2.2 Mandatory Testing and the Mexican Hat

In the desert at the border of Utah and Arizona there is a famous combination of vertically aligned stones named the "Mexican Hat" which looks like a human with a Mexican hat. It is a random product of nature and not man-made ...really? *Can we test the null hypothesis "The Mexican Hat is of natural origin"?* To do so we need a *test statistic* for a pile of stones and a probability distribution for this test statistic under the null hypothesis. Let's take

$$t(p) = \begin{cases} 1 & \text{if } p \text{ forms a Mexican Hat} \\ 0 & \text{otherwise} \end{cases} \tag{2.1}$$

for any pile of stones p. How do we get a probability distribution of $t(p)$ for all piles of stones p not affected by man? - We walk through the desert, examine a large number, say $n = 10^6$, of piles of stones, and count the frequency of $t(p) = 0$ and of $t(p) = 1$. Now, the Mexican Hat is famous for good reasons - there is only one p with $t(p) = 1$, namely the Mexican Hat itself. The other $n - 1 = 10^6 - 1$ samples go with $t(p) = 0$. Therefore the probability distribution for p not affected by man is

$$prob\,(t(p) = k) = \begin{cases} 10^{-6} & \text{for } k = 1 \\ 1 - 10^{-6} & \text{for } k = 0 \end{cases} \tag{2.2}$$

After these preparations everything is ready for the final test. We reject the null hypothesis with a risk of 10^{-6} if t(Mexican hat) $= 1$. This condition is fulfilled and we may conclude: *The Mexican Hat is not of natural origin but man-made.*

Obviously, this argument is pretty absurd - but where is the logical error? The fundamental error is that *the null hypothesis is not independent of the data which are used to conduct the test.* We know a-priori that the Mexican Hat is a rare event, therefore the impossibility of finding such a combination of stones cannot be used as evidence against its natural origin. The same trick can of course be used to "prove" that any rare event is "non-natural", be it a heat wave or a particularly violent storm - the probability of observing a rare event is small.

One might argue that no serious scientist would fall into this trap. However, they do. The hypothesis of a connection between solar activity and the statistics of climate on Earth is old and has been debated heatedly over many decades. The debate had faded away in the last few decades - and has been refueled by a remarkable observation by K. Labitzke. She studied the relationship between the solar activity and the stratospheric temperature at the North Pole. There was no obvious relationship - but she *saw* that during years in which the Quasibiennial Oscillation (QBO) was in its West Phase, there was an excellent *positive* correlation between solar activity and North Pole temperature whereas during years with the QBO in its East Phase

Figure 2.1: *Labitzke' and van Loon's relationship between solar flux and the temperature at 30 hPa at the North Pole for all winters during which the QBO is in its West Phase and in its East Phase. The correlations are 0.1, 0.8 and -0.5. (From Labitzke and van Loon, 1988).*

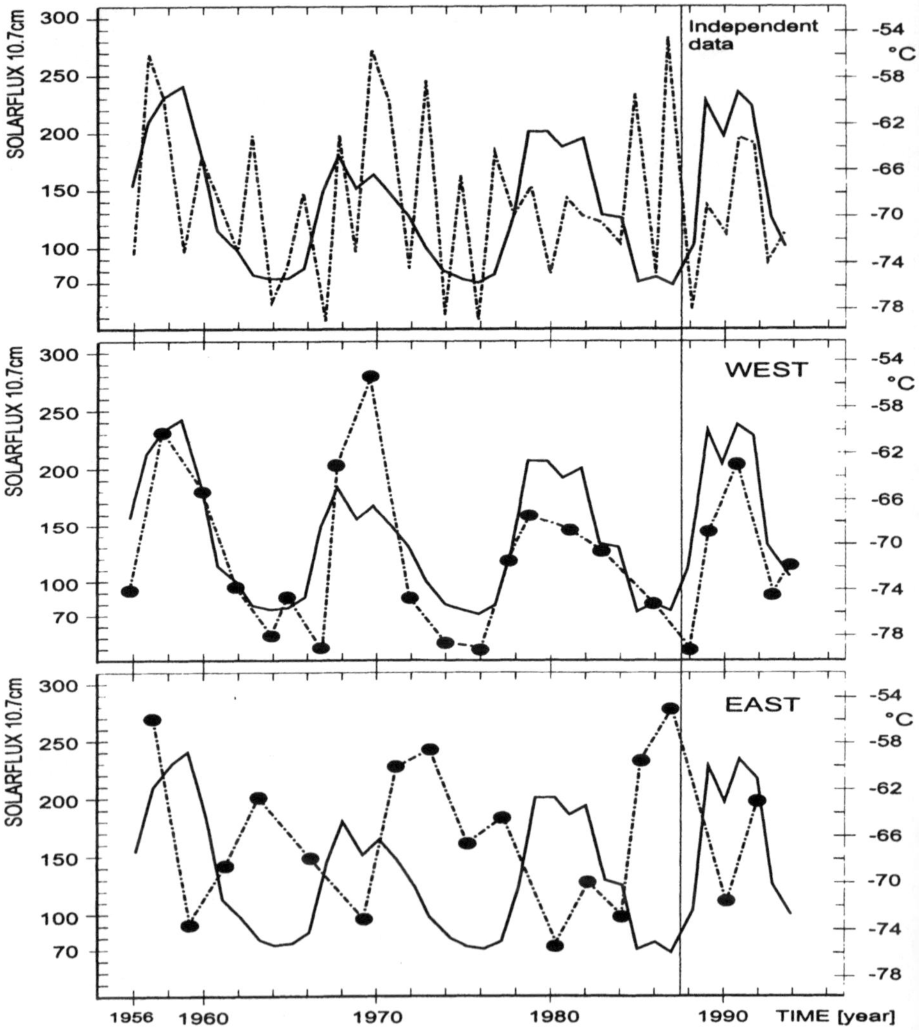

there was a good *negative* correlation (Labitzke, 1987; Labitzke and van Loon, 1988).

Labitzke's finding was and is spectacular - and obviously right for the data from the time interval at her disposal (see Figure 2.1). Of course it could be that the result was a coincidence as unlikely as the formation of a Mexican Hat. Or it could represent a real on-going signal. Unfortunately, the data which were used by Labitzke to formulate her hypothesis can no longer be used for the assessment of whether we deal with a signal or a coincidence. Therefore an answer to this question requires information unrelated to the data as for instance dynamical arguments or GCM experiments. However, physical hypotheses on the nature of the solar-weather link were not available and are possibly developing right now - so that nothing was left but to wait for more data and better understanding. (The data which have become available since Labitzke's discovery in 1987 support the hypothesis.)

In spite of this fundamental problem an intense debate about the "statistical significance" broke out. The reviewers of the first comprehensive paper on that matter by Labitzke and van Loon (1988) demanded a test. Reluctantly the authors did what they were asked for and found of course an extremely little risk for the rejection of the null hypothesis "The solar-weather link is zero". After the publication various other papers were published dealing with technical aspects of the test - while the basic problem *that the data to conduct the test had been used to formulate the null hypothesis* remained.

When hypotheses are to be derived from limited data, I suggest two alternative routes to go. If the time scale of the considered process is short compared to the available data, then split the full data set into two parts. Derive the hypothesis (for instance a statistical model) from the first half of the data and examine the hypothesis with the remaining part of the data.[1] *If the time scale of the considered process is long compared to the time series such that a split into two parts is impossible, then I recommend using all data to build a model optimally fitting the data. Check the fitted model whether it is consistent with all known physical features and state explicitly that it is impossible to make statements about the reliability of the model because of limited evidence.*

2.3 Neglecting Serial Correlation

Most standard statistical techniques are derived with explicit need for statistically independent data. However, almost all climatic data are somehow correlated in time. The resulting problems for testing null hypotheses is discussed in some detail in Section 9.4. In case of the t-test the problem is nowadays often acknowledged - and as a cure people try to determine the "equivalent sample size" (see Section 2.4). When done properly, the t-test

[1] An example of this approach is offered by Wallace and Gutzler (1981).

Figure 2.2: *Rejection rates of the Mann-Kendall test of the null hypothesis "no trend" when applied to 1000 time series of length* n *generated by an AR(1)-process (2.3) with prescribed* α. *The adopted nominal risk of the test is 5%.*
Top: results for unprocessed serially correlated data.
Bottom: results after pre-whitening the data with (2.4). (From Kulkarni and von Storch, 1995)

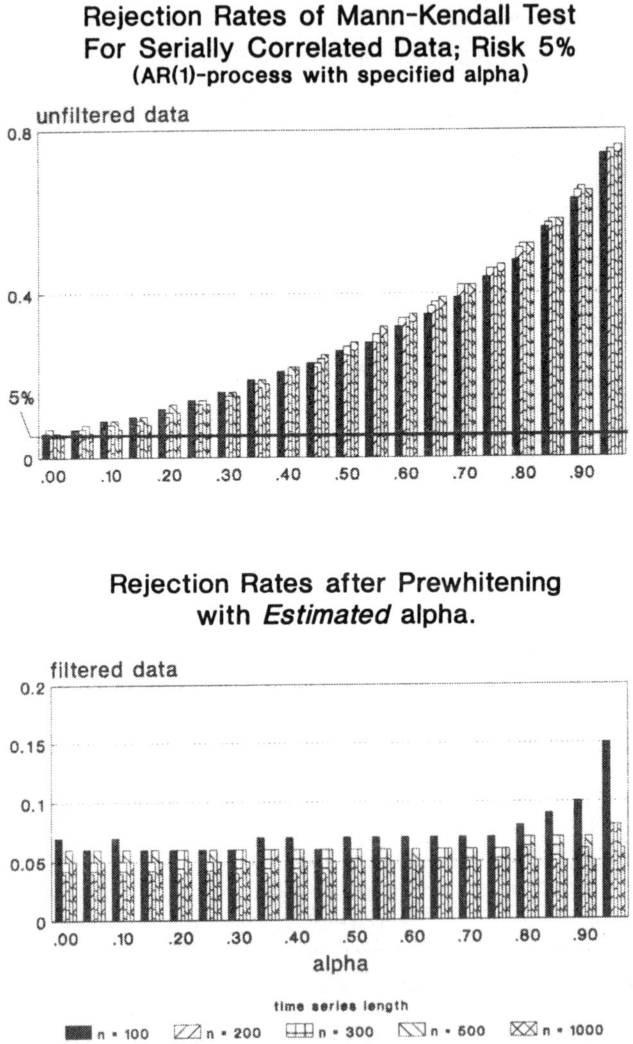

becomes conservative - and when the "equivalent sample size" is "optimized" the test becomes liberal[2]. We discuss this case in detail in Section 2.4.

There are, however, again and again cases in which people simply ignore this condition, in particular when dealing with more exotic tests such as the Mann-Kendall test, which is used to reject the null hypothesis of "no trends". To demonstrate that the result of such a test really depends ***strongly*** on the autocorrelation, Kulkarni and von Storch (1995) made a series of Monte Carlo experiments with AR(1)-processes with different values of the parameter α.

$$\mathbf{X}_t = \alpha \mathbf{X}_{t-1} + \mathbf{N}_t \tag{2.3}$$

with Gaussian "white noise" $\mathbf{N}_t$, which is neither auto-correlated nor correlated with $\mathbf{X}_{t-k}$ for $k \geq 1$. α is the lag-1 autocorrelation of $\mathbf{X}_t$. 1000 iid[3] time series of different lengths, varying form $n = 100$ to $n = 1000$ were generated and a Mann-Kendall test was performed. Since the time series have no trends, we expect a (false) rejection rate of 5% if we adopt a risk of 5%, i.e., 50 out of the 1000 tests should return the result "reject null hypothesis". The actual rejection rate is much higher (see Figure 2.2). For autocorrelations $\alpha \leq 0.10$ the actual rejection rate is about the nominal rate of 5%, but for $\alpha = 0.3$ the rate is already 0.15, and for $\alpha = 0.6$ the rate > 0.30. If we test a data field with a lag-1 autocorrelation of 0.3, we must expect that on average at 15% of all points a "statistically significant trend" is found even though there is no trend but only "red noise". This finding is mostly independent of the time series length.

When we have physical reasons to assume that the considered time series is a sum of a trend and stochastic fluctuations generated by an AR(1) process, and this assumption is sometimes reasonable, then there is a simple cure, the success of which is demonstrated in the lower panel of Figure 2.2. Before conducting the Mann-Kenndall test, the time series is "pre-whitened" by first estimating the lag-autocorrelation $\hat{\alpha}$ at lag-1, and by replacing the original time series $\mathbf{X}_t$ by the series

$$\mathbf{Y}_t = \mathbf{X}_t - \hat{\alpha} \mathbf{X}_{t-1} \tag{2.4}$$

The "pre-whitened" time series is considerably less plagued by serial correlation, and the same Monte Carlo test as above returns actual rejections rates close to the nominal one, at least for moderate autocorrelations and not too short time series. The filter operation (2.4) affects also any trend; however, other Monte Carlo experiments have revealed that the power of the test is reduced only weakly as long as α is not too large.

A word of caution is, however, required: If the process is *not* AR(1) but of higher order or of a different model type, then the pre-whitening (2.4)

---

[2] A test is named "liberal" if it rejects the null hypothesis more often than specified by the significance level. A "conservative" rejects less often than specified by the significance level.

[3] "iid" stands for "independent identically distributed".

is insufficient and the Mann-Kenndall test rejects still more null hypotheses than specified by the significance level.

Another possible cure is to "prune" the data, i.e., to form a subset of observations which are temporally well separated so that any two consecutive samples in the reduced data set are no longer autocorrelated (see Section 9.4.3).

When you use a technique which assumes independent data and you believe that serial correlation might be prevalent in your data, I suggest the following "Monte Carlo" diagnostic: Generate synthetic time series with a prescribed serial correlation, for instance by means of an AR(1)-process (2.3). Create time series without correlation ($\alpha = 0$) and with correlation ($0 < \alpha < 1$) and try out if the analysis, which is made with the real data, returns different results for the cases with and without serial correlation. In the case that they are different, you cannot use the chosen technique.

2.4 Misleading Names: The Case of the Decorrelation Time

The concept of "the" *Decorrelation Time* is based on the following reasoning:[4] The variance of the mean $\bar{\mathbf{X}}^n = \frac{1}{n}\sum_{k=1}^{n}\mathbf{X}_k$ of n identically distributed and independent random variables $\mathbf{X}_k = \mathbf{X}$ is

$$\text{VAR}(\bar{\mathbf{X}}^n) = \frac{1}{n}\text{VAR}(\mathbf{X}) \tag{2.5}$$

If the $\mathbf{X}_k$ are autocorrelated then (2.5) is no longer valid but we may *define* a number, named the *equivalent sample size* n' such that

$$\text{VAR}(\bar{\mathbf{X}}^n) = \frac{1}{n'}\text{VAR}(\mathbf{X}) \tag{2.6}$$

The *decorrelation time* is then *defined* as

$$\tau_D = \lim_{n\to\infty}\frac{n}{n'}\cdot\Delta t = \left[1 + 2\sum_{\Delta=1}^{\infty}\rho(\Delta)\right]\Delta t \tag{2.7}$$

with the autocorrelation function ρ of $\mathbf{X}_t$.

The decorrelation times for an AR(1) process (2.3) is

$$\tau_D = \frac{1+\alpha}{1-\alpha}\Delta t \tag{2.8}$$

There are several conceptual problems with "the" Decorrelation Time:

[4]This section is entirely based on the paper by Zwiers and von Storch (1995). See also Section 9.4.3.

- The definition (2.7) of a decorrelation time makes sense ***when dealing with the problem of the mean of n-consecutive serially correlated observations.*** However, its arbitrariness in defining a ***characteristic time scale*** becomes obvious when we reformulate our problem by replacing the ***mean*** in (2.6) by, for instance, the ***variance***. Then, the characteristic time scale is (Trenberth, 1984):

$$\tau = \left[1 + 2\sum_{k=1}^{\infty} \rho^2(k)\right] \Delta t$$

Thus characteristic time scales τ depends markedly on the ***statistical problem*** under consideration. ***These numbers are, in general, not physically defined numbers.***

- For an AR(1)-process we have to distinguish between the physically meaningful processes with positive memory ($\alpha > 0$) and the physically meaningless processes with negative memory ($\alpha < 0$). If $\alpha > 0$ then formula (2.8) gives a time $\tau_D > \Delta t$ representative of the decay of the auto-correlation function. Thus, in this case, τ_D may be seen as a physically useful time scale, namely a "persistence time scale" (but see the dependency on the time step discussed below). If $\alpha < 0$ then (2.8) returns times $\tau_D < \Delta t$, even though probability statements for any two states with an even time lag are identical to probabilities of an AR(p) process with an AR-coefficient $|\alpha|$.

  Thus the number τ_D makes sense as a characteristic time scale when dealing with red noise processes. But for many higher order AR(p)-processes the number τ_D does not reflect a useful physical information.

- The Decorrelation Time depends on the time increment Δt: To demonstrate this dependency we consider again the AR(1)-process (2.3) with a time increment of $\Delta t = 1$ and $\alpha \geq 0$. Then we may construct other AR(1) processes with time increments k by noting that

$$\mathbf{X}_t = \alpha^k \mathbf{X}_{t-k} + \mathbf{N}'_t \tag{2.9}$$

with some noise term $\mathbf{N}'_t$ which is a function of $\mathbf{N}_t \ldots \mathbf{N}_{t-k+1}$. The decorrelation times τ_D of the two processes (2.3,2.9) are because of $\alpha < 1$:

$$\tau_{D,1} = \frac{1+\alpha}{1-\alpha} \cdot 1 \geq 1 \quad \text{and} \quad \tau_{D,k} = \frac{1+\alpha^k}{1-\alpha^k} \cdot k \geq k \tag{2.10}$$

so that

$$\lim_{k\to\infty} \frac{\tau_{D,k}}{k} = 1 \tag{2.11}$$

Figure 2.3: *The dependency of the decorrelation time $\tau_{D,k}$ (2.10) on the time increment k (horizontal axis) and on the coefficient α (0.95, 0.90, 0.80, 0.70 and 0.50; see labels). (From von Storch and Zwiers, 1999).*

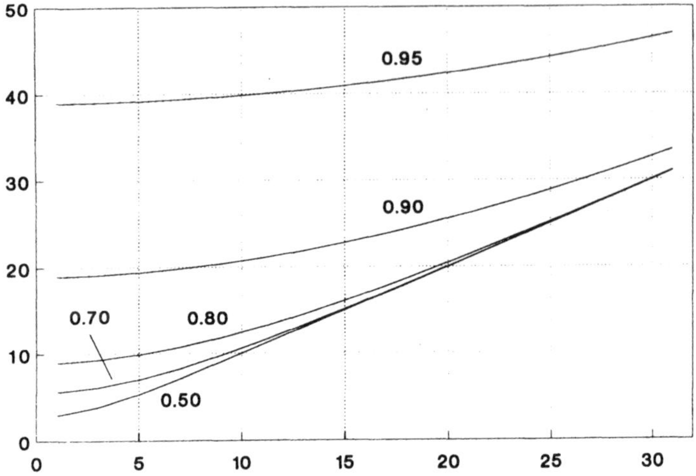

That means that the decorrelation time is as least as long as the time increment; in case of "white noise", with $\alpha = 0$, the decorrelation time is always equal to the time increment. In Figure 2.3 the dimensional decorrelation times are plotted for different α-values and different time increments k. The longer the time increment, the larger the decorrelation time. For sufficiently large time increments we have $\tau_{D,k} = k$. For small α-values, such as $\alpha = 0.5$, we have virtually $\tau_{D,k} = k$ already after $k = 5$. If $\alpha = 0.8$ then $\tau_{D,1} = 9$, $\tau_{D,11} = 13.1$ and $\tau_{D,21} = 21.4$. If the time increment is 1 day, then the decorrelation time of an $\alpha = 0.8$-process is 9 days or 21 days - if we sample the process once a day or once every 21 days.

We conclude that the absolute value of the decorrelation time is of questionable informational value. However, the relative values obtained from several time series sampled with the same time increment are useful to infer whether the system has in some components a longer memory than in others. If the decorrelation time is well above the time increment, as in case of the $\alpha = 0.95$-curve in Figure 2.3, then the number has some informational value whereas decorrelation times close to the time increment, as in case of the $\alpha = 0.5$-curve, are mostly useless.

We have seen that the name "Decorrelation Time" is not based on physical reasoning but on strictly mathematical grounds. Nevertheless the number is often incorrectly ***interpreted*** as the minimum time so that two consecutive observations $\mathbf{X}_t$ and $\mathbf{X}_{t+\tau_D}$ are independent. If used as a vague estimate with the reservations mentioned above, such a use is in order. However, the number is often introduced as crucial parameter in test routines. Probably the most frequent victim of this misuse is the conventional t-test.

We illustrate this case by a simple example from Zwiers and von Storch (1995): We want to answer the question whether the long-term mean winter temperatures in Hamburg and Victoria are equal. To answer this question, we have at our disposal daily observations for one winter from both locations. We treat the winter temperatures at both locations as random variables, say $\mathbf{T}_H$ and $\mathbf{T}_V$. The "long term mean" winter temperatures at the two locations, denoted as μ_H and μ_V respectively, are parameters of the probability distributions of these random variables. In the statistical nomenclature the question we pose is: do the samples of temperature observations contain sufficient evidence to reject the ***null hypothesis*** $H_0 : \mu_H - \mu_V = 0$.

The standard approach to this problem is to use to the ***Student's t-test.*** The test is conducted by imposing a statistical model upon the processes which resulted in the temperature samples and then, within the confines of this model, measuring the degree to which the data agree with H_0. An essential part of the model which is implicit in the t-test is the assumption that the data which enter the test represent a set of statistically independent observations. In our case, and in many other applications in climate research, this assumption is not satisfied. The Student's t-test usually becomes "liberal" in these circumstances. That is, it tends to reject that null hypothesis on weaker evidence than is implied by the significance level[5] which is specified for the test. One manifestation of this problem is that the Student's t-test will reject the null hypothesis more frequently than expected when the null hypothesis is true.

A relatively clean and simple solution to this problem is to form subsamples of approximately independent observations from the observations. In the case of daily temperature data, one might use physical insight to argue that observations which are, say, 5 days apart, are effectively independent of each other. If the number of samples, the sample means and standard deviations from these reduced data sets are denoted by n^*, $\tilde{\mathbf{T}}^*_H$, $\tilde{\mathbf{T}}^*_V$, $\tilde{\sigma}^*_H$ and $\tilde{\sigma}^*_V$ respectively, then the test statistic

$$t = \frac{\tilde{\mathbf{T}}^*_H - \tilde{\mathbf{T}}^*_V}{\sqrt{(\tilde{\sigma}^{*2}_H + \tilde{\sigma}^{*2}_V)/n^*}} \tag{2.12}$$

has a Student's t-distribution with n^* degrees of freedom provided that the null hypothesis is true[6] and a test can be conducted at the chosen signif-

[5]The significance level indicates the probability with which the null hypothesis will be rejected when it is true.

[6]Strictly speaking, this is true only if the standard deviations of $\mathbf{T}_H$ and $\mathbf{T}_V$ are equal.

icance level by comparing the value of (2.12) with the percentiles of the $t(n^*)$-distribution.

The advantage of (2.12) is that this test operates as specified by the user provided that the interval between successive observations is long enough. The disadvantage is that a reduced amount of data is utilized in the analysis. Therefore, the following concept was developed in the 1970s to overcome this disadvantage: The numerator in (2.12) is a random variable because it differs from one pair of temperature samples to the next. When the observations which comprise the samples are serially uncorrelated the denominator in (2.12) is an estimate of the standard deviation of the numerator and the ratio can be thought of as an expression of the difference of means in units of estimated standard deviations. For serially correlated data, with sample means $\tilde{\mathbf{T}}$ and sample standard deviations $\tilde{\sigma}$ derived from all available observations, the standard deviation of $\tilde{\mathbf{T}}_H - \tilde{\mathbf{T}}_V$ is $\sqrt{(\tilde{\sigma}_H^2 + \tilde{\sigma}_V^2)/n'}$ with the *equivalent sample size* n' as defined in (2.6). For sufficiently large samples sizes the ratio

$$t = \frac{\tilde{\mathbf{T}}_H - \tilde{\mathbf{T}}_V}{\sqrt{(\tilde{\sigma}_H^2 + \tilde{\sigma}_V^2)/n'}} \tag{2.13}$$

has a standard Gaussian distribution with zero mean and standard deviation one. Thus one can conduct a test by comparing (2.13) to the percentiles of the standard Gaussian distribution.

So far everything is fine.

Since $t(n')$ is approximately equal to the Gaussian distribution for $n' \geq 30$, one may compare the test statistic (2.13) also with the percentiles of the $t(n')$-distribution. The incorrect step is the heuristic assumption that this prescription - "compare with the percentiles of the $t(n')$, or $t(n'-1)$ distribution" - would be right for small ($n' < 30$) equivalent samples sizes. The rationale of doing so is the tacitly assumed fact that the statistic (2.13) would be $t(n')$ or $t(n'-1)$-distributed under the null hypothesis. However, *this assumption is simply wrong.* The distribution (2.13) is not $t(k)$-distributed for any k, be it the equivalent sample size n' or any other number. This result has been published by several authors (Katz (1982), Thiébaux and Zwiers (1984) and Zwiers and von Storch (1995)) but has stubbornly been ignored by most of the atmospheric sciences community.

A justification for the small sample size test would be that its behaviour under the null hypothesis is well approximated by the t-test with the equivalent sample size representing the degrees of freedom. But this is not so, as is demonstrated by the following example with an AR(1)-process (2.3) with $\alpha = .60$. The exact equivalent sample size $n' = \frac{1}{4}n$ is known for the process since its parameters are completely known. One hundred independent samples of variable length n were randomly generated. Each sample was used to test the null hypothesis $H_o : \mathrm{E}(\mathbf{X}_t) = 0$ with the t-statistic (2.13) at the 5% significance level. If the test operates correctly the null hypothesis should be (incorrectly) rejected 5% of the time. The actual rejection rate (Figure 2.4)

Figure 2.4: *The rate of erroneous rejections of the null hypothesis of equal means for the case of auto correlated data in a Monte Carlo experiment. The "equivalent sample size" n' (in the diagram labeled n_e) is either the correct number, derived from the true parameters of the considered AR(1)-process or estimated from the best technique identified by Zwiers and von Storch (1995). (From von Storch and Zwiers, 1995).*

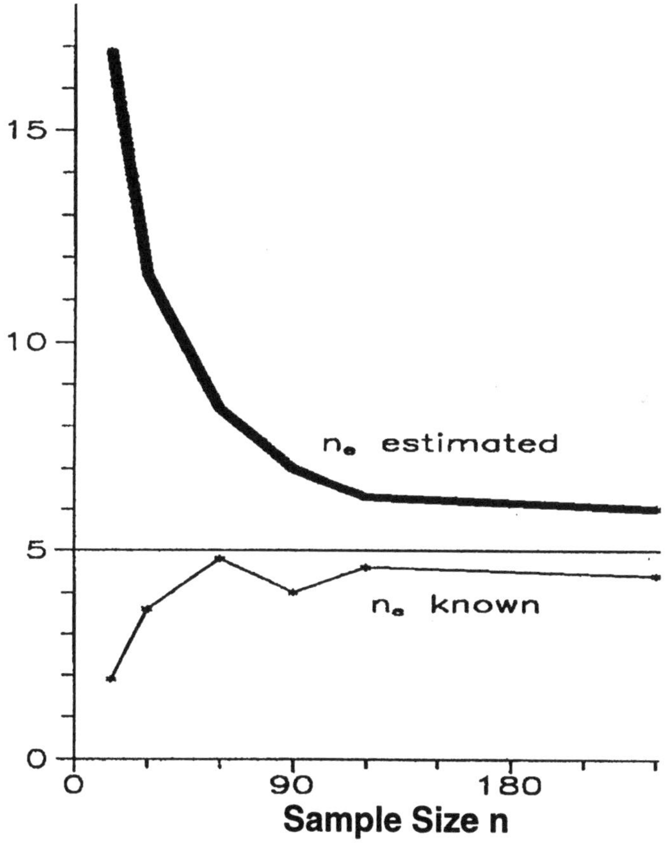

is notably *smaller* than the expected rate of 5% for $4n' = n \leq 30$. Thus, the t-test operating with the true equivalent sample size is *conservative* and thus *wrong*.

More problems show up when the equivalent sample is unknown. In this case it may be possible to specify n' on the basis of physical reasoning. Assuming that conservative practices are used, this should result in underestimated values of n' and consequently even more conservative tests. In most applications, however, an attempt is made to estimate n' from the same data that are used to compute the sample mean and variance. Monte Carlo experiments show that the actual rejection rate of the t-test tends to be greater than the nominal rate when n' is estimated. Also this case has been simulated in a series of Monte Carlo experiments with the same AR(1)-process. The resulting rate of erroneous rejections is shown in Figure 2.4 - for small ratio sample sizes the actual significance level can be several times *greater* than the nominal significance level. Thus, the t-test operating with the estimated equivalent sample size is *liberal* and thus *wrong*.

Zwiers and von Storch (1995) offer a "table look-up" test as a useful alternative to the inadequate "t-test with equivalent sample size" for situations with serial correlations similar to red noise processes.

2.5 Use of Advanced Techniques

The following case is an educational example which demonstrates how easily an otherwise careful analysis can be damaged by an inconsistency hidden in a seemingly unimportant detail of an advanced technique. When people have experience with the advanced technique for a while then such errors are often found mainly by instinct ("This result cannot be true - I must have made an error.") - but when it is new then the researcher is somewhat defenseless against such errors.

The background of the present case was the search for evidence of bifurcations and other fingerprints of truly nonlinear behaviour of the dynamical system "atmosphere". Even though the nonlinearity of the dynamics of the planetary-scale atmospheric circulation was accepted as an obvious fact by the meteorological community, atmospheric scientists only began to discuss the possibility of two or more stable states in the late 1970's. If such multiple stable states exist, it should be possible to find bi- or multi-modal distributions in the observed data (if these states are well separated).

Hansen and Sutera (1986) identified a bi-modal distribution in a variable characterizing the energy of the planetary-scale waves in the Northern Hemisphere winter. Daily amplitudes for the zonal wavenumbers $k = 2$ to 4 for 500 hPa height were averaged for midlatitudes. A "wave-amplitude indicator" $\mathbf{Z}$ was finally obtained by subtracting the annual cycle and by filtering out all variability on time scales shorter than 5 days. The probability density function f_Z was estimated by applying a technique called the *maximum penalty technique* to 16 winters of daily data. The resulting f_Z had two maxima separated by a minor minimum. This bimodality was taken as proof of the existence of two stable states of the atmospheric general circulation: A "zonal regime", with $\mathbf{Z} < 0$, exhibiting small amplitudes of the planetary waves and a "wavy regime", with $\mathbf{Z} > 0$, with amplified planetary-scale zonal disturbances.

Hansen and Sutera performed a "Monte Carlo" experiment to evaluate the likelihood of fitting a bimodal distribution to the data with the maximum penalty technique even if the generating distribution is unimodal. The authors concluded that this likelihood is small. On the basis of this statistical check, the found bimodality was taken for granted by many scientists for almost a decade.

When I read the paper, I had never heard about the "maximum penalty method" but had no doubts that everything would have been done properly in the analysis. The importance of the question prompted other scientists to perform the same analysis to further refine and verify the results. Nitsche et al. (1994) reanalysed step-by-step the same data set which had been used in the original analysis and came to the conclusion that the purportedly small probability for a misfit was *large*. The error in the original analysis was not at all obvious. Only by carefully scrutinizing the pitfalls of the maximum penalty technique did Nitsche and coworkers find the inconsistency between the Monte Carlo experiments and the analysis of the observational data.

Nitsche et al. reproduced the original estimation, but showed that something like 150 years of daily data would be required to exclude with sufficient certainty the possibility that the underlying distribution would be unimodal. What this boils down to is, that the null hypothesis according to which the distribution would be unimodal, is *not rejected by the available data* - and the published test was *wrong* . However, since the failure to reject the null hypothesis does not imply the acceptance of the null hypothesis (but merely the lack of *enough* evidence to reject it), the present situation is that the (alternative) hypothesis "The sample distribution does *not* originate from a unimodal distribution" is not falsified but still open for discussion.

I have learned the following rule to be useful when dealing with advanced methods: Such methods are often needed to find a signal in a vast noisy phase space, i.e., the needle in the haystack - but after having the needle in our hand, we should be able to identify the needle as a needle by simply looking at it.[7] *Whenever you are unable to do so there is a good chance that something is rotten in the analysis.*

[7] See again Wallace's and Gutzler's study who identified their teleconnection patterns first by examining correlation maps - and then by simple weighted means of few grid point values - see Section 12.1.

2.6 Epilogue

I have chosen the examples of this Chapter to advise users of statistical concepts to be aware of the sometimes hidden assumptions of these concepts. Statistical Analysis is not a ***Wunderwaffe***[8] to extract a wealth of information from a limited sample of observations. More results require more assumptions, i.e., information given by theories and other insights ***unrelated to the data under consideration.***

But, even if it is not a ***Wunderwaffe*** Statistical Analysis is an indispensable tool in the evaluation of limited empirical evidence. The results of Statistical Analysis are not miracle-like enlightenment but sound and understandable assessments of the consistency of concepts and data.

[8]Magic bullet.

Part II

Analyzing The Observed Climate

Chapter 3

Climate Spectra and Stochastic Climate Models

by Claude Frankignoul

3.1 Introduction

As the power spectra of extratropical atmospheric variables are essentially white on time scales between about 10 days and a few years, many climatic fluctuations can be understood as the response of the slow climate variables to stochastic forcing by the fast atmospheric fluxes (Hasselmann, 1976). The stochastic climate model explains the statistical properties of mid-latitude sea surface temperature anomalies on time scales of up to a few years, and has been applied to long term climate changes, quasi-geostrophic ocean fluctuations, sea ice variability, soil moisture fluctuations, and ocean circulation variability, providing a simple but powerful framework to investigate climate variability. After briefly describing the spectral characteristics of the atmospheric variables, the concept of the stochastic climate model is introduced in this chapter. Its application to climate variations is then illustrated in the context of the midlatitude sea surface temperature anomalies. The other applications are then briefly discussed.

Figure 3.1: *Spectra of atmospheric pressure (p), east (τ_1) and north (τ_2) wind stress, and wind stress magnitude at Ocean Weather Station C ($52.5^o N, 35.5^o W$). (From Willebrand, 1978).*

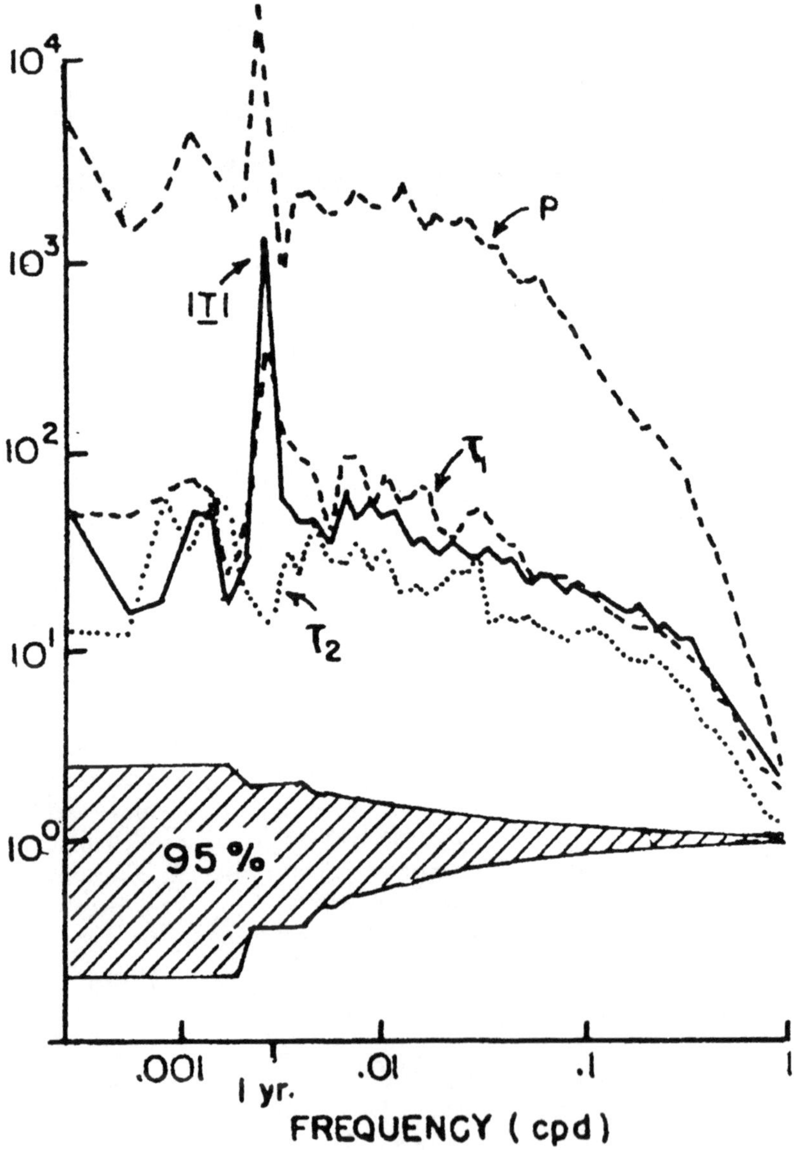

3.2 Spectral Characteristics of Atmospheric Variables

Since the weather changes irregularly and is largely unpredictable beyond a few days, the frequency spectra of extratropical atmospheric variables are approximately white on time scales between about 10 days and a few years, except for the seasonal peaks, and the variance is dominated by the *white noise extension* of the daily weather fluctuations. This is shown in Figure 3.1 in a "log - log" presentation which emphasizes spectral slopes, and in Figure 3.2 (top) in a "frequency × spectrum - log frequency" plot which conserves variance, therefore indicating more clearly the dominant time scales. The level of the frequency spectra varies with both position and season, and high resolution analysis may reveal weak spectral peaks superimposed on the white background (e.g., Chave et al., 1991). As one approaches the equator, the frequency spectra become increasingly red and the level of high frequency variability decreases.

Wavenumber-frequency analysis indicates that the time scale of the atmospheric fluctuations is larger at large scales, so the dominant period at the planetary scale is of the order of 20-30 days while it is only a few days at large wavenumbers (Figure 3.3). A more complete representation of the spectra of atmospheric variables at a given level is thus given by the frequency-wavenumber spectrum $\Gamma(\vec{k}, f)$, which is related to the space-time covariance function $\gamma(\vec{r}, u)$ by

$$\Gamma(\vec{k}, f) = (2\pi)^{-3} \int_{-\infty}^{\infty} \gamma(r, u) e^{-i(\vec{k}\cdot\vec{r} - fu)} d\vec{r} du \tag{3.1}$$

and to the frequency spectrum by

$$\Gamma(f) = \int_{-\infty}^{\infty} \Gamma(\vec{k}, f) d\vec{k} \tag{3.2}$$

Relation (3.2) indicates that the frequency spectra of the atmospheric fields which are dominated by the largest scales (e.g. geopotential height, sea level pressure) start falling off at larger periods than those which are dominated by small scales (e.g. wind stress curl). The turbulent air-sea fluxes of momentum and energy in the extratropical ocean, which are quadratic or higher-order functions of atmospheric variables, fall between these extremes (see Figure 3.1) and are dominated at non-seasonal frequencies by short time scale weather changes. At low frequencies, the exponential in (3.1) can then be set to 1 and the spectrum is white, $\Gamma(\vec{k}, f) = \Gamma(\vec{k}, 0)$. From the reality condition $\Gamma(\vec{k}, f) = \Gamma^*(\vec{k}, f) = \Gamma(-\vec{k}, -f)$, where the asterisk indicates complex conjugate, it then follows that the spectra are symmetric, i.e. there is no preferred propagation direction. At higher frequencies, however, the extratropical fluctuations are dominated by eastward propagation, reflecting the propagation of the storm systems (Figure 3.4).

Figure 3.2: *Top: Spectrum of sensible-plus-latent heat flux at Ocean Weather Station P ($50^o N, 145^o W$) over the period 1958-1967. The error bars represent 95% confidence intervals; the continuous line shows the white noise level. Bottom: corresponding sea-surface temperature spectrum. The continuous curve is the stochastic model prediction. (From Frankignoul, 1979).*

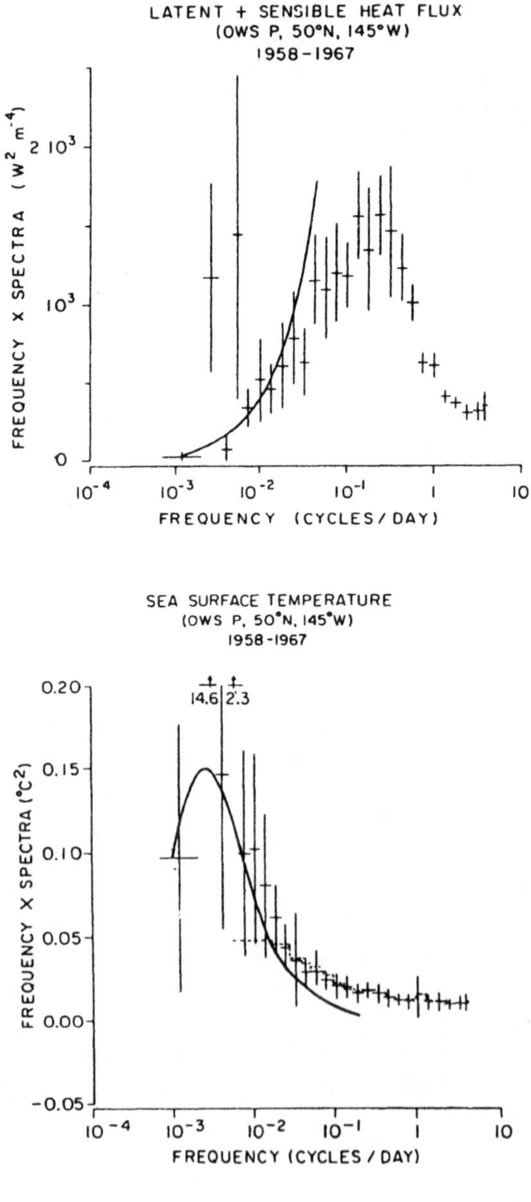

Figure 3.3: *Normalized spectra, with approximately 19 degrees of freedom, of the 500 hPa geopotential height at 45°N, wavenumbers 1 to 12. Below 10-1 cpd, the unsmoothed spectra (with 1.6 degrees of freedom) are also plotted. (From Wilson, 1975).*

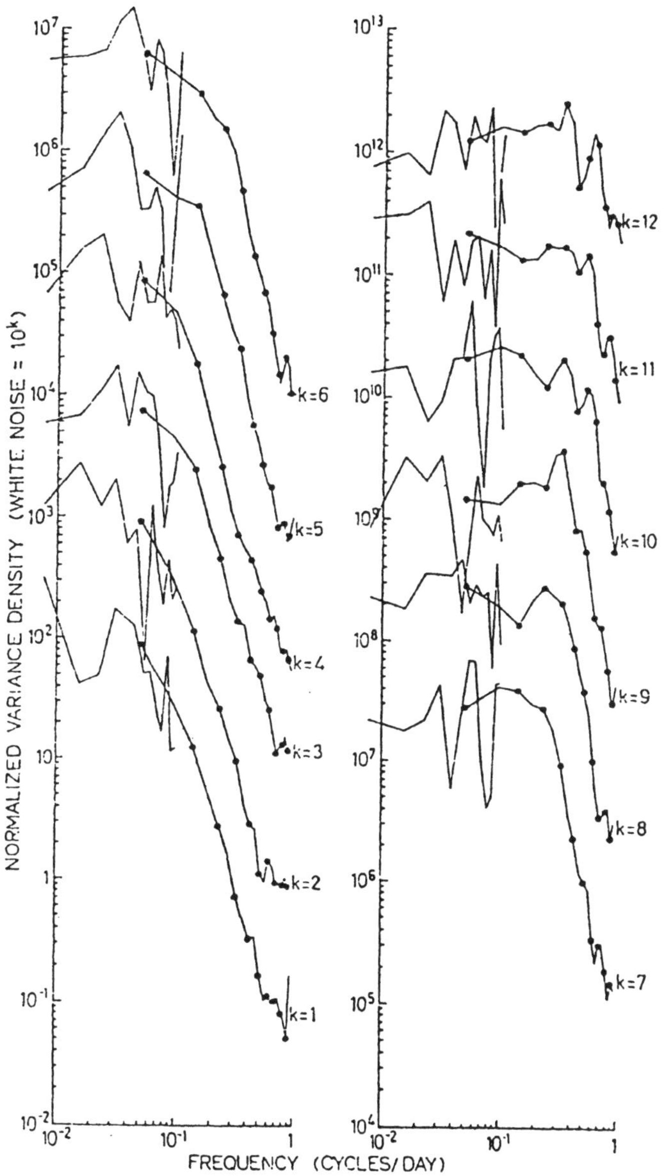

Figure 3.4: *Zonal wavenumber spectrum for the interval July 1985 through June 1988 for the North Pacific site 41°N, 162°W. It is computed using the maximum likelihood method, with a constant bandwidth at 0.027 cpd. Power is expressed in decibels down from the peak at each frequency at the peak meridional wavenumber. (From Chave et al. 1991)*

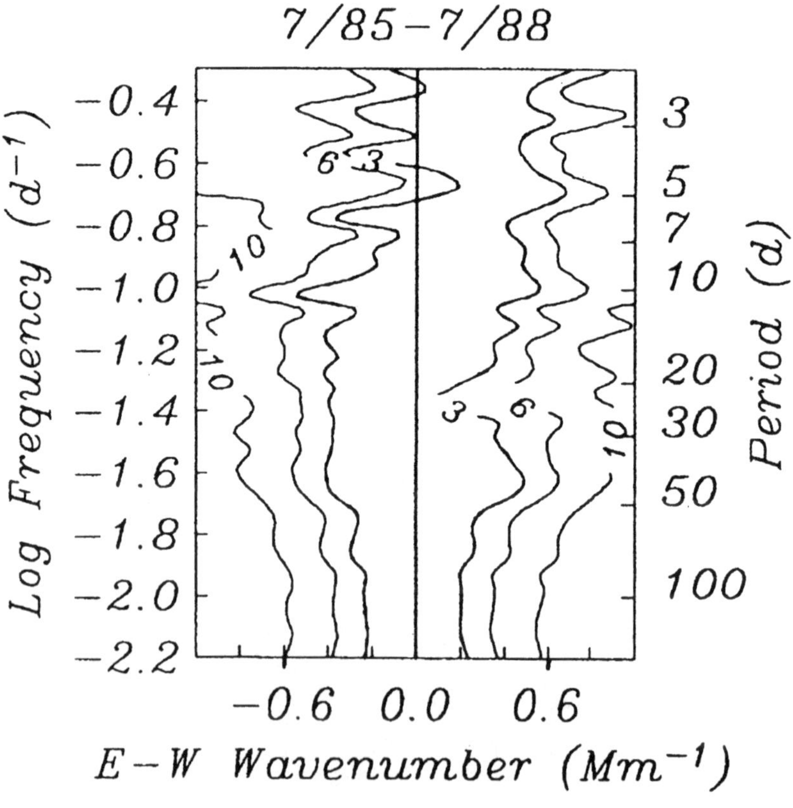

Wavenumber spectra of tropospheric variables have been primarily estimated from hemispheric or global data derived from operational products. Some spectra have been calculated for surface variables and fluxes, but they are difficult to interpret in view of the spatial heterogeneity of the fields and their limited spatial resolution, so that idealized representations have been constructed for air-sea interaction studies (Frankignoul and Müller, 1979). Note that Freilich and Chelton (1986) have analyzed surface winds measured by satellite, showing that an approximately k^{-2} behavior holds for their spectrum in the Pacific ocean over wavelengths from 200 to 2200 km. As discussed in Chave et al. (1991), there is also spatial and interannual variability in the atmospheric frequency-wavenumber spectra.

When atmospheric spectra are estimated from long time series, some redness is found at very low frequencies, in particular for the dominant large scale atmospheric patterns. This is illustrated in Figure 3.5 for the area-weighted sea level pressure over the north Pacific region 30° to 65^oN, 160^oE to 140^0W, which depicts changes in the Aleutian low in winter and is well-correlated with the Pacific North American (PNA) teleconnection pattern, a preferred mode of variability in the Northern Hemisphere winter. Although no spectral peak is significant, there is enhanced variance between 2 and 6 years, and a marked redness at periods > 20 years (interdecadal climate variability). This variability is found throughout the troposphere and is associated with large scale changes in sea surface temperature. It appears to be associated to a small extent with the El Niño-Southern Oscillation (ENSO) phenomenon, with changes in the tropical Pacific slightly leading the extra-tropical ones (Trenberth and Hurrell, 1994; Zhang et al., 1996).

3.3 Stochastic Climate Model

At a time when most climate researchers were trying to link climatic changes to (sometimes far-fetched) variable external factors and hypothetical positive feedback within the climate system, Hasselmann (1976) pointed out that climate variability might be explained more simply as the integral response of the slowly varying parts of the climate system to internal random forcing by the always present short time scale weather fluctuations. The resulting climate fluctuations would have a random walk character, in agreement with the observed redness of the climate spectra, and the challenge was to find the positive and negative feedback mechanisms which enhance or damp this continual generation of climate fluctuations.

Because there is a separation of time scale between the fast (atmosphere) and the slow (ocean, cryosphere, soil, etc.) components of the climate system, its evolution can be described by two subsystems: a system for the fast "weather" variables $\vec{\mathbf{X}}$ of short time scale t_x (geopotential height, wind stress,...) where the slow "climate" variables $\vec{\mathbf{Y}}$ can be regarded as constant, as in most weather prediction models, and a system for the slow climate vari-

Figure 3.5: *Spectrum of atmospheric variability in the North Pacific (details see text) for November through March averages for the 67 years from 1924 to 1990. Also shown is the corresponding red spectrum with the same lag one autocorrelation coefficient (0.1) and the 5 and 95% confidence limits. (From Trenberth and Hurrell, 1994).*

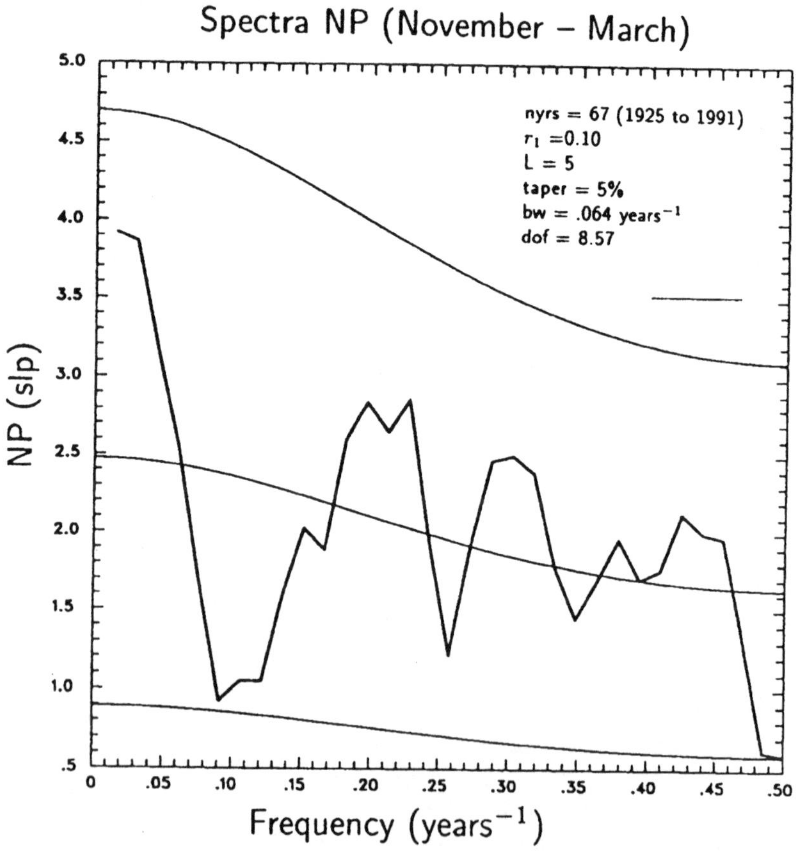

ables $\vec{\mathbf{Y}}$ of time scale t_y (ocean temperature, ice thickness,...), with $t_x \ll t_y$:

$$\frac{d\mathbf{X}_i}{dt} = u_i(\vec{\mathbf{X}}, \vec{\mathbf{Y}}) \tag{3.3}$$

$$\frac{d\mathbf{Y}_i}{dt} = v_i(\vec{\mathbf{X}}, \vec{\mathbf{Y}}) \tag{3.4}$$

where u_i and v_i are, in general, nonlinear functions. External forcing factors can be added. In most investigations of climate variability, in particular those based on Statistical Dynamical Models, it had been argued that, on the climatic time scale, the rapidly varying components could be ignored, so that by averaging over the time interval t_i in the range $t_x \ll t_i \ll t_y$, Eq. (3.4) could be replaced by

$$\frac{d\mathbf{Y}_i}{dt} = \langle v_i(\vec{\mathbf{X}}, \vec{\mathbf{Y}})\rangle \tag{3.5}$$

where the averaged rate of change $\langle v_i \rangle$ of $\mathbf{Y}_i$ would only depend on the averaged statistical properties of $\vec{\mathbf{X}}$, which would then be expressed as a function of $\vec{\mathbf{Y}}$ only. Except in cases of chaotic behavior, the reduced Eq. (3.5) would then be deterministic and external forcing necessary to produce climate changes. Hasselmann (1976) pointed out that, even though (3.5) is valid in an ensemble average sense (over a set of realisations of $\vec{\mathbf{X}}$ for given $\vec{\mathbf{Y}}$), it is not appropriate for a particular "realization" of the climate evolution, which should rather obey an equation of the form

$$\frac{d\mathbf{Y}_i}{dt} = \langle v_i(\vec{\mathbf{X}}, \vec{\mathbf{Y}})\rangle + v_i'(\vec{\mathbf{X}}, \vec{\mathbf{Y}}), \tag{3.6}$$

where the stochastic forcing term $v'(\vec{\mathbf{X}}, \vec{\mathbf{Y}})$ has zero ensemble mean (defined as above). The implication is that climate evolution is a statistical rather than a deterministic phenomenon.

The climate change from an initial state may be divided into a mean and a fluctuating term, where the latter, denoted by $\mathbf{Y}_i'$, is given by

$$\frac{d\mathbf{Y}_i'}{dt} = v_i'(\vec{\mathbf{X}}, \vec{\mathbf{Y}}). \tag{3.7}$$

For short integration time $t \ll t_y$, $\vec{\mathbf{Y}}$ can be regarded as constant in (3.7). Then, a statistically steady atmospheric forcing creates a non-stationary climate response, whose covariance increases linearly with time

$$\langle \mathbf{Y}_i' \mathbf{Y}_j' \rangle = 2D_{ij}t \tag{3.8}$$

where D_{ij} is called, by analogy with Brownian motion, a diffusion coefficient, and is given by the integral of the covariance function of v_i' and v_j'.

In the frequency domain, (3.7) predicts that, for f such that ${t_y}^{-1} \ll f$, climate spectra are red and given by

$$\mathbf{\Gamma}_{ij}^{Y}(f) = \frac{\mathbf{\Gamma}_{ij}^{X}(f)}{f^2}, \tag{3.9}$$

where $\mathbf{\Gamma}_{ij}^{X}(f)$ is the frequency cross-spectrum of the atmospheric forcing. In the range $t_y^{-1} \ll f \ll t_x^{-1}$, the latter is white and (3.9) takes the simple form

$$\mathbf{\Gamma}_{ij}^{Y}(f) = \frac{\mathbf{\Gamma}_{ij}^{X}(0)}{f^2}. \tag{3.10}$$

On time scales of the order of t_y, $\vec{\mathbf{Y}}$ cannot be considered as constant in (3.7), and the dynamics of the climate subsystem must be taken into account, introducing the effects of dissipation, feedback, resonance ... into the evolution equation for $\mathbf{Y}_i'$. Furthermore, the level $\mathbf{\Gamma}_{ij}^{X}(0)$ of stochastic forcing may also change as the climate evolves, introducing additional feedback. Provided the two-scale approximation remains valid, the problem can be investigated by writing a Fokker-Planck equation for the evolution of the probability distribution of climate state in the climatic phase space, in which the random weather excitation is represented by diffusion terms, even though in the general case the equation can only be solved numerically by constructing solutions with the Monte Carlo method (see Hasselmann, 1976, for details).

For small changes (denoted by a prime) about a climate equilibrium state $\vec{\mathbf{Y}}_0$, the $\vec{\mathbf{Y}}$-dependence in Eq. (3.7) can in some cases be linearized, yielding

$$\frac{d\mathbf{Y}_i'}{dt} = \sum_j V_{ij}\mathbf{Y}_j' + v_i'(\vec{\mathbf{Y}}_0) \tag{3.11}$$

which provides explicit solutions on the t_y time scale. For a stable solution, the matrix V_{ij} must be negative definite. Since for $t \gg t_x$, v_i' acts as a white noise generator, Eq. (3.11) represents a multivariate first-order Markov process. For $f \ll t_x{}^{-1}$, one has

$$\mathbf{\Gamma}_{ij}^{Y}(f) = \sum_{k,l} T_{ik}T_{jl}^{*}\mathbf{\Gamma}_{kl}^{X}(0) \tag{3.12}$$

with the matrix $T \doteq (ifI - V)^{-1}$ (I is the unit matrix). In the univariate case, (3.11) reduces to

$$\frac{d\mathbf{Y}'}{dt} = v'(\mathbf{Y}_0) - \lambda\mathbf{Y}', \tag{3.13}$$

and (3.12) yields for the frequency spectrum

$$\mathbf{\Gamma}^{Y}(f) = \frac{\mathbf{\Gamma}^{X}(0)}{f^2 + \lambda^2}, \tag{3.14}$$

where λ is the feedback. Relation (3.14) and the corresponding covariance function, which decays exponentially, provide simple statistical signatures which can even be tested when no simultaneous information is available on the weather forcing.

The correlation and the cross-spectrum between forcing and response can also be calculated from (3.11), providing more stringent signatures than the power spectra. In the univariate case, the covariance $\gamma_{yv}(u)$ between atmospheric forcing and climate response obeys

$$\frac{d\gamma_{yv}}{du} = \gamma_{vv} - \lambda\gamma_{yv} \tag{3.15}$$

which yields the cross correlation between **X** and **Y**. When **Y** leads, the correlation is negligible, while when **Y** lags, the correlation has a maximum at small lags. The coherence

$$Coh_{xy}(f) = \frac{|\mathbf{\Gamma}^{XY}(f)|}{\sqrt{\mathbf{\Gamma}^{X}(f)\mathbf{\Gamma}^{Y}(f)}} \leq 1 \tag{3.16}$$

and the phase lead of $\mathbf{X}(t)$ over $\mathbf{Y}(t)$, given by $\Theta^{XY}(f) = -tan^{-1}[Q^{XY}(f)/C^{XY}(f)]$, can be obtained similarly from the cross-spectrum

$$\mathbf{\Gamma}^{XY}(f) = C^{XY}(f) - iQ^{XY}(f), \tag{3.17}$$

where $C^{XY}(f) = Re\mathbf{\Gamma}^{XY}(f)$ is the co-spectrum and $Q^{XY}(f) = -Im\mathbf{\Gamma}^{XY}(f)$ the quadrature spectrum. For $f \ll t_x^{-1}$, the coherence between stochastic forcing and climate response is unity and the phase given by $-arctan(f/\lambda)$.

As illustrated below, the model (3.13) is consistent with the statistical properties of the observed anomalies of mid-latitude sea surface temperature, soil moisture and sea ice extent, on the monthly to yearly time scales. However, these are particular cases where the dynamics of the climate subsystem play little role. In other cases, the climate subsystem may exhibit nonlinearities, resonances and more complex feedback mechanisms. Then, the response spectra will differ from (3.12) and (3.14), reflecting primarily the internal dynamics of the system. Stochastic climate models remain testable, however, by focusing on energy levels and, if atmospheric observations are available, cause-to-effect relationships (Müller and Frankignoul, 1981). Of course, not all the climate changes can be explained by stochastic forcing, and some climate variations are forced deterministically (e.g. by the changes in the orbital parameters of the earth), or reflect the chaotic nature of the climate subsystem.

3.4 Sea Surface Temperature Anomalies

Frankignoul and Hasselmann (1977) have shown that the stochastic climate model successfully explains the main statistical properties of sea surface temperature (SST) anomalies in the midlatitudes, as they mainly reflect the response of the oceanic mixed layer to the day-to-day changes in the air-sea

Figure 3.6: *SST anomaly spectrum at Ocean Weather Station I (59°N, 19°W) for the period 1949-1964, with 95% confidence interval. The smooth curve is the stochastic model prediction. (From Frankignoul and Hasselmann, 1976).*

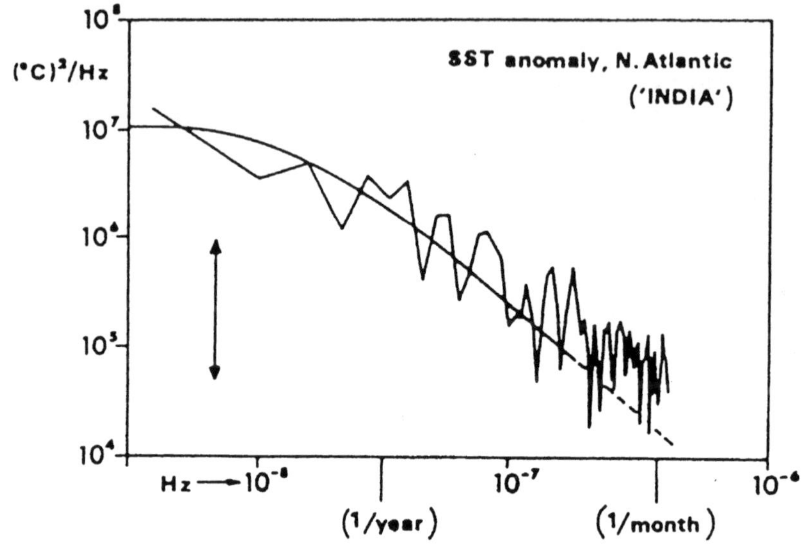

fluxes. In a simple one-dimensional mixed-layer model, the heat content equation is

$$h\frac{dT}{dt} + w_e(T - T_d) = \kappa h \nabla_h^2 T + \frac{Q - Q_d}{\rho c_p} \tag{3.18}$$

where T denotes temperature, h the mixed-layer depth, w_e the entrainment velocity, κ the horizontal mixing coefficient, Q the surface heat flux into the ocean, ρ the water density, c_p the specific heat at constant pressure and ∇_h^2 the horizontal Laplacian, and the index $_d$ indicates values just below the mixed layer base. Denoting anomalies by a prime and means by an overbar, and neglecting the advection by the mean current, the SST anomaly equation can be approximately written (Frankignoul, 1985)

$$\begin{aligned} \partial_t T' \quad = \quad & \frac{Q'}{\rho c_p \bar{h}} - \frac{(h\vec{u})' \cdot \nabla_h T}{\bar{h}} + \frac{h'}{\bar{h}}\partial_t \bar{T} \\ & - \frac{(T' - T_d')(\bar{w}_e + w_e')}{\bar{h}} - \frac{\bar{T} - \bar{T}_d}{\bar{h}} w_e' + \kappa \nabla_h^2 T' \end{aligned} \tag{3.19}$$

It can be shown that this equation has the form (3.13), where v' includes the terms involving both the short time scale atmospheric forcing and the mixed layer depth response, and λ represents the (linearized) dissipation and feedback processes. The model explains most mid-latitude SST anomaly

Figure 3.7: *(Top) Correlation between the dominant EOF of SST and sea level pressure anomalies over the North Pacific during 1947-1974; (Bottom) Theoretical correlation assuming that the weather forcing can be modeled as a first-order Markov process with 8.5-day decay time, and using* $\lambda = (6\ month)^{-1}$. *The correlation is given without smoothing (dashed line) and as estimated from monthly averages (continuous line). (After Frankignoul and Hasselmann, 1977).*

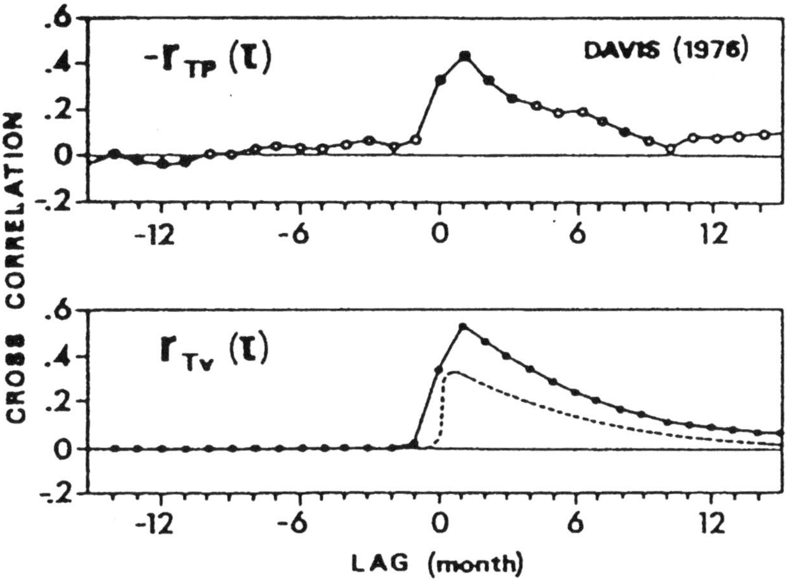

spectra on time scales of up to a few years (Figures 3.2 and 3.6), as well as their lagged correlation (Figure 3.7) or cross-spectrum (Figure 3.8) with the atmospheric variables.

The stochastic forcing model can be refined by including the advection by the mean ocean currents (Frankignòul and Reynolds, 1983; Herterich and Hasselmann, 1987), and by taking into account the seasonal modulation of the atmospheric forcing and the feedback (Ruiz de Elvira and Lemke, 1982; Ortiz and Ruiz de Elvira, 1985). The covariance function then depends on the phase of the annual cycle, consistently with the observations.

As reviewed by Frankignoul (1985), the dominant forcing mechanisms in (3.19) are the synoptic weather fluctuations in surface heat flux and wind stirring. The SST anomaly decay time λ^{-1} is of the order of 3 months, but the larger spatial scales are more persistent (Reynolds, 1978). This is due in part to a small El Niño-related persistence in the North Pacific large scale forcing (e.g. Luksch and von Storch 1992), and also to larger advection effects at small scales. Identification of the main feedback mechanisms has

Figure 3.8: *Phase and coherence between SST and turbulent heat flux at Ocean Weather Station P (see Figure 3.2). The dashed line represents the 95% confidence level for non-zero confidence and the dotted line the stochastic model prediction. (From Frankignoul, 1979).*

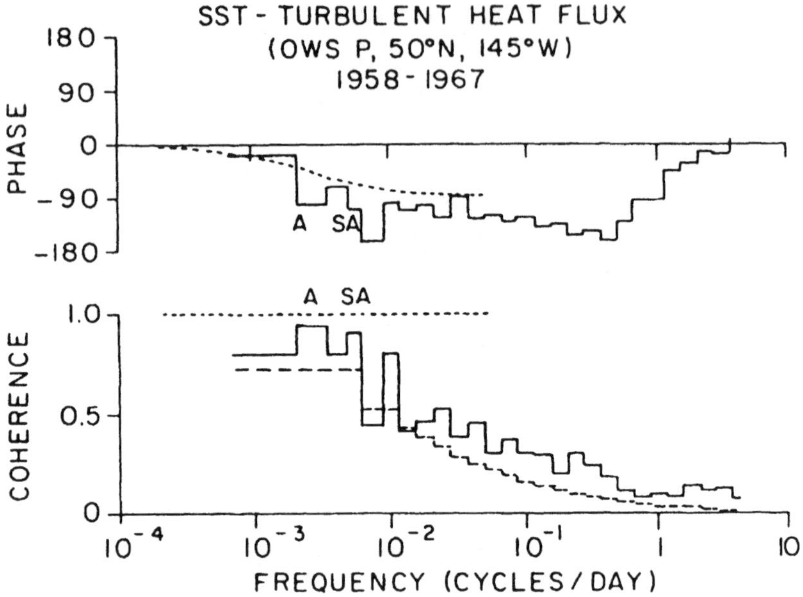

been difficult: oceanic entrainment, vertical mixing and subduction are likely to contribute to SST anomaly damping (Frankignoul, 1985), and also the effective diffusion by surface current fluctuations (Molchanov et al., 1987). A negative feedback may also be caused by the atmosphere, via the turbulent heat exchange. However, as the latter is a function of the atmospheric adjustment to the SST anomalies, its response is not solely local. Atmospheric GCMs indicate that this (somewhat model-dependent) back interaction heat flux strongly depends on the geographical location and the sign of the SST anomalies (e.g., Kushnir and Lau, 1991).

Much of the interest in studying midlatitude SST anomalies has been stimulated by their possible effects on the mean atmospheric circulation, even though they are small and difficult to distinguish from the "natural" variability of the atmosphere. The difficulty in demonstrating the SST influence using observations can be understood in part by the high correlation which is predicted at zero lag, even in the case of a purely passive ocean (see Figure 3.7), so that a correlated behavior in ocean and atmospheric variables does not necessarily indicate an oceanic influence onto the atmosphere.

To better understand the observed air-sea interactions, the model (3.13) can be refined. Frankignoul (1985) has considered the case where only part

Figure 3.9: *Predicted correlation between H' and T' in (3.20) for different atmospheric feedbacks. It was assumed that q' and m' can be modeled as first-order Markov processes with 4-day decay time, that q' (or, to first order, H') has twice the variance of m' and is uncorrelated with it, and that $\lambda = (80\ day)^{-1}$. (From Frankignoul, 1985).*

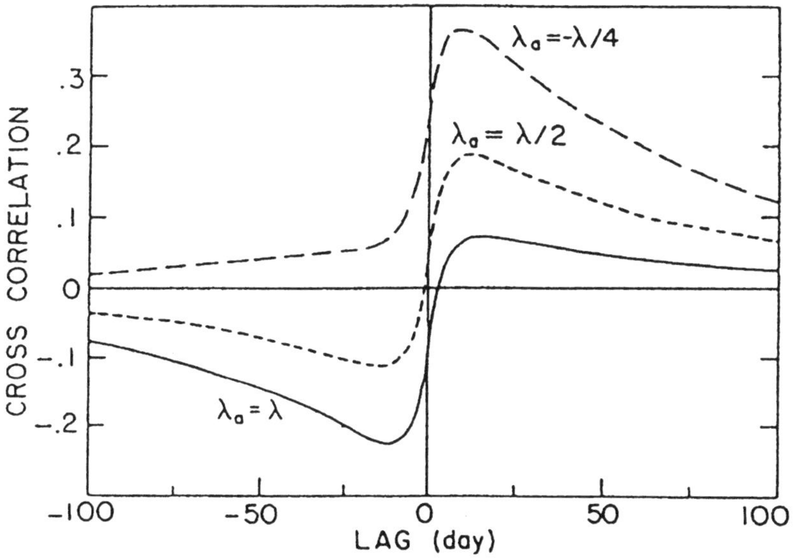

of the atmospheric forcing is measured, namely the surface heat flux which itself includes the atmospheric feedback. The SST anomaly equation is then written

$$\partial_t T' = H' + m' - \lambda_0 T', \tag{3.20}$$

where

$$H' = q' - \lambda_a T' \tag{3.21}$$

is the heat flux term in (3.19), which contributes a feedback λ_a, q' and m' are white noise processes, and λ_0 is the oceanic contribution to the total feedback λ ($\lambda = \lambda_a + \lambda_0$). The shape of the cross-correlation function between T' and H' then depends primarily on the atmospheric feedback (Figure 3.9). When $\lambda_a = 0$, the curve is as in Figure 3.7 (dashed line). When $\lambda_a < 0$ (negative feedback, as expected if the heat flux contributes to SST damping), the cross-correlation function takes an antisymmetric appearance, with zero crossing near zero lag. When $\lambda_a > 0$ (positive feedback), it peaks when SST lags but has the same positive sign for all lags (the same signature would occur if the atmospheric forcing had a slow component, however). As in Figure 3.7, smoothing would increase the correlation and shift the maxima toward lags of plus or minus one (Frankignoul et al., 1998).

These signatures can be used to interpret the correlations in Figure 3.10 between monthly anomalies in zonally-averaged SST and turbulent heat flux in the Atlantic during the period 1949-1979. The anomalies were derived from the COADS data set, and a second order trend removed from the anomaly time series to remove some of the artificial trends introduced by the changes in the wind measurement methods. In the extratropical regions, the atmospheric anomalies lead the oceanic ones, but the slightly anti-symmetric shape of the cross-correlation functions and the smaller peak when the ocean leads by one month suggests that the turbulent heat flux not only contributes to generating the SST anomalies, but also acts as a negative feedback. An estimate of this feedback and its seasonal variation is given in Frankignoul et al. (1998). Below $10^o N$, however, the progressive change in the shape of the cross-correlations suggests an enhancement of the negative feedback, and at the equator, the curve simply peaks at zero lag, indicating that heat flux and SST anomalies vary in phase: the turbulent heat exchanges play no role in generating the SST anomalies, but contribute to their damping. This occurs because SST fluctuations are in part remotely forced by the wind, because of equatorial wave propagation, and because cumulus convection creates an intense air-sea coupling.

In the equatorial Pacific, the SST anomalies are predominantly associated with ENSO and generated through large scale ocean-atmosphere feedbacks involving remote atmospheric forcing, oceanic adjustment and unstable ocean-atmosphere oscillating modes. These SST anomalies have a longer time scale than in midlatitudes, and a stronger influence on the global climate.

Figure 3.10: *Correlation between zonally averaged monthly SST and turbulent heat flux anomalies in various latitude bands in the Atlantic Ocean for the period 1949-1979. Ocean leads for negative lags.*

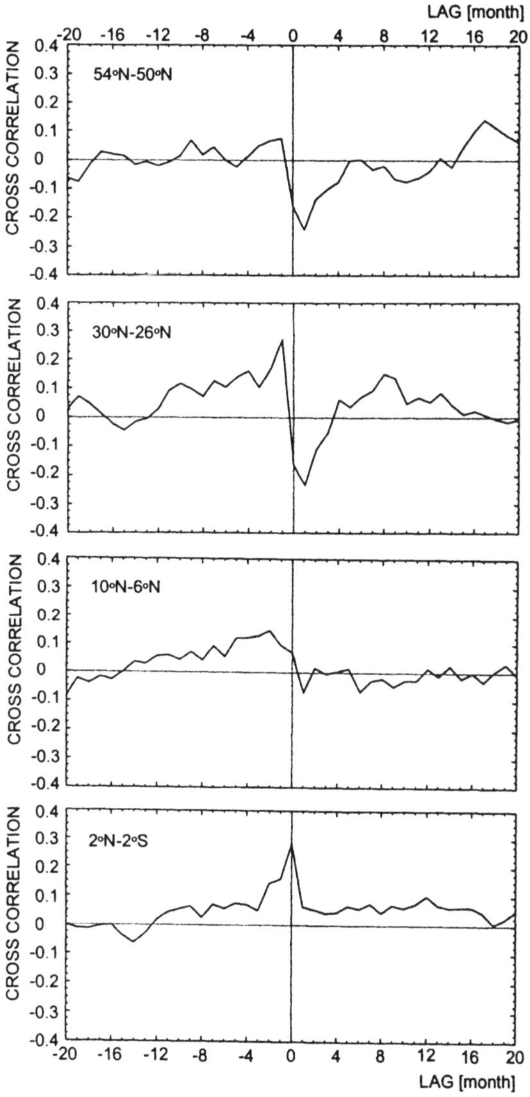

However, modeling studies show that the natural variability of the surface wind stress significantly contributes to the irregularity ENSO by acting as a stochastic forcing (e.g., Blanke et al., 1997). Correspondingly, the triggering effect of equatorial the triggering effects of equatorial Kelvin wave generated by westerly wind bursts have been documented in the observations.

On longer time scales, the SST anomaly spectrum remains slightly (3.15), Barsugli and Battisti (1998) have suggested that at low frequencies the oceanic mixed layer would be adjusted to the surface air temperature, thereby attenuating the damping of the atmospheric fluctuations by surface heat exchanges. This would increase the variance and persistence of the anomalies in the two media, while decreasing the heat flux variance at very low frequencies. As discussed by Deser and Blackmon (1993), the SST fluctuations observed in the North Atlantic are irregular and have a weak spectral peak at about 10-15 yr period, and their relationship to the atmospheric anomalies in the wintertime is broadly similar to that found on shorter time scales (Figure 3.11). Hence, the decadal SST anomalies may also be largely generated by the local air-sea fluxes. Halliwell and Mayer (1996) have shown that in the westerlies local anomalous turbulent heat flux is effective in forcing the wintertime SST anomalies down to the decadal periods, and that the SST anomalies propagate to the east and northeast. In addition, basin-scale dynamical interactions between the oceanic circulation and the atmosphere contribute to the decadal SST variability.

3.5 Variability of Other Surface Variables

Delworth and Manabe (1988) have shown that soil moisture anomalies in an atmospheric GCM having a simple soil representation had the red spectrum (3.14) and were well-represented by the stochastic climate model. The soil acted as an integrator of the white noise forcing by rainfall and snow melt, and the feedback time increased with latitude, except in regions of frequent runoff. This was explained by considering the dissipation processes that limit soil wetness: as the energy available for evaporation decreases with increasing latitude, potential evaporation, hence soil moisture damping decreases. However, if precipitation exceeds potential evaporation, the excess precipitation is removed by runoff when the soil is saturated, and λ increases. Although other factors (vegetation, varying soil characteristics, subsurface water flow) need to be considered, Delworth and Manabe's (1988) interpretation seems sturdy: soil moisture anomaly observations in the Soviet Union are indeed well-modeled by a first-order Markov process, with a damping time approximately equal to the ratio of field capacity to potential evaporation (Vinnikov and Yeserkepova, 1991).

On monthly to yearly time scales, sea ice anomalies in the Arctic and Antarctic have been shown by Lemke et al. (1980) to be reasonably well-represented by an advected linearly damped model forced by stochastic

Figure 3.11: *Top: EOF 2 of North Atlantic SST anomalies based on unnormalized winter (November-March) means, for the period 1900-1989. This mode accounts for 12% of the variance. Also shown is the climatological SST distribution (thin contours).*
Bottom: Spectrum of EOF 2. The thin line represents a background red noise spectrum and its 95% confidence limit. (From Deser and Blackmon, 1993).

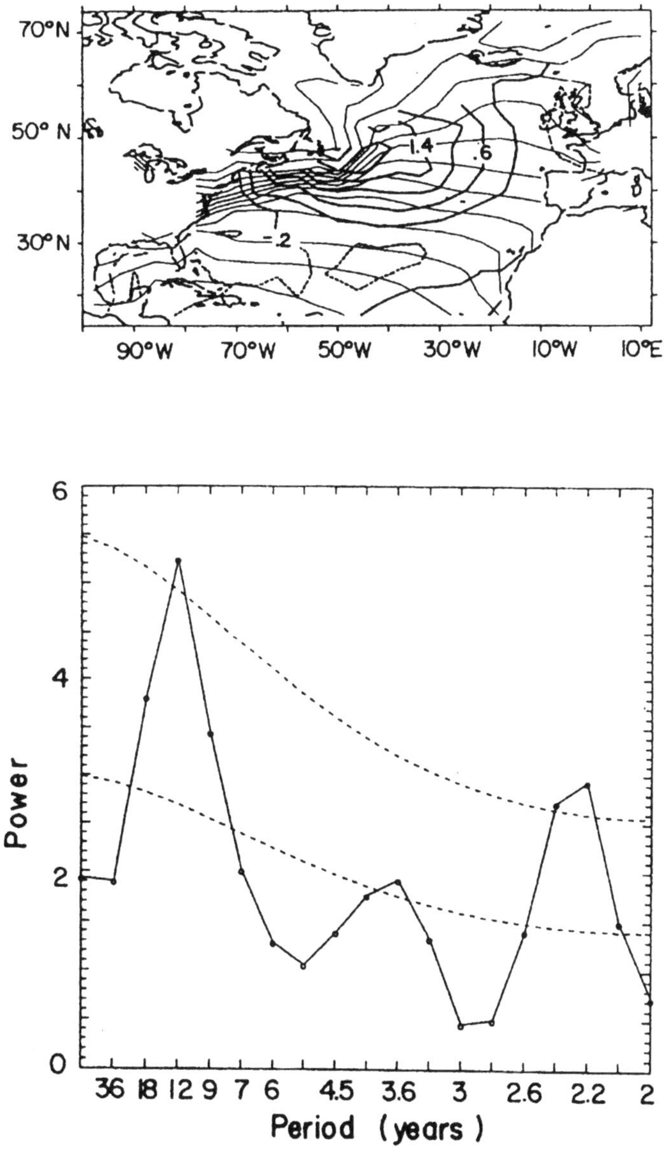

Figure 3.12: *Frequency spectrum of 500-m temperature at Bermuda (32°N, 64°W). Thick line: Panulirus data (the spectrum was estimated with the Tukey method and the 95% confidence interval is indicated). Thin line: stochastic model prediction for the low frequencies (continuous line) and the high frequency slope (dashed line). (From Frankignoul et al., 1997).*

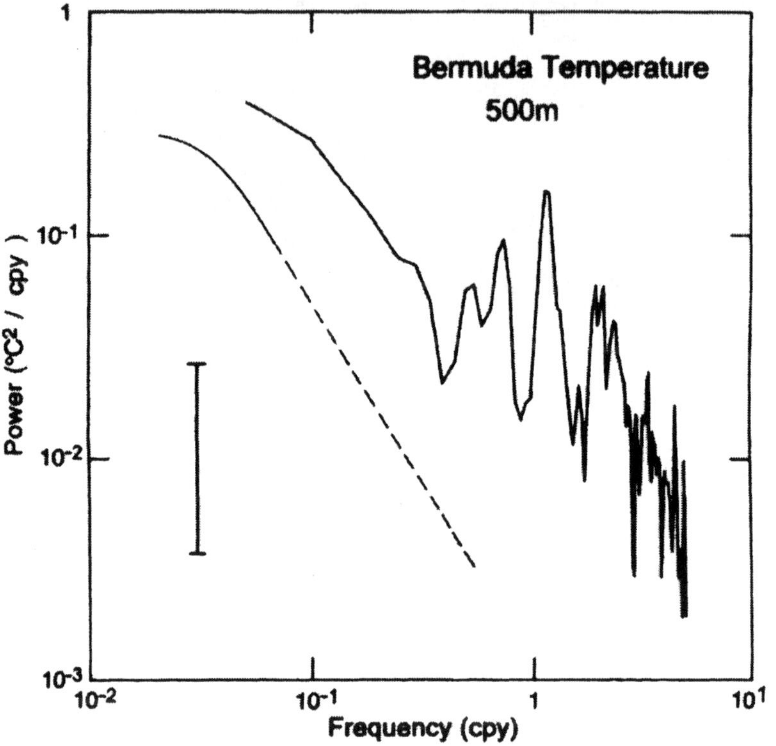

weather fluctuations. On longer time scales, larger sea ice fluctuations occur in the Arctic (e.g., Stocker and Mysak, 1992). Decadal sea ice fluctuations in the Greenland and Labrador seas are related to high latitude surface salinity variations (cf. the "Great Salinity Anomaly" in the northern North Atlantic), and appear to be lagging long term changes in the Northern Hemisphere atmospheric circulation (Walsh and Chapman, 1990). Mysak et al. (1990) have shown that the sea ice anomalies mainly lag the sea surface salinity anomalies and suggested that the latter had been caused by earlier runoffs from northern Canada into the western Arctic Ocean. They then postulated the existence of a complex climate cycle in the Arctic involving advection along the subpolar gyre and changes in the convective overturning in the Greenland sea, but the role of the stochastic forcing by the atmosphere was not assessed.

3.6 Variability in the Ocean Interior

Few long time series are available in the ocean interior, so that the spectral characteristics are virtually unknown below the surface at periods longer than a few years. The longest series of hydrographic stations were made at station "S" off Bermuda ($32^o N$, $64^o W$). At periods of a few years, the temperature spectra are dominated by mesoscale eddies because of the proximity of the Gulf Stream (Frankignoul 1981), but they remain red down to the decadal period (Figure 3.12). At low frequencies, the temperature and salinity are largely independent in the upper layers, but highly correlated in the thermocline, as would be expected from vertical advection; a long-term trend is also seen in the deep layer (Joyce and Robbins, 1996).

The stochastic atmospheric forcing primarily affects the ocean interior via Ekman pumping, surface heat exchanges and freshwater fluxes. On the oceanic mesoscale, the oceanic response to Ekman pumping fluctuations is dominated by resonant Rossby waves, with an energy input rate that may be sufficient to maintain the observed eddy field in regions of weak eddy activity (Frankignoul and Müller, 1979). The predicted response spectra are sensitive to friction and other dynamical effects, however, so that the signatures from the interaction are best seen in the seasonal modulation and the coherence between oceanic and atmospheric variables (Müller and Frankignoul, 1981; Brink, 1989). The observations have only confirmed the importance of the stochastic forcing at high frequencies (Luther et al., 1990). For periods of up to a few months, much of the deep ocean subinertial variability is consistent with the barotropic response to stochastic wind stress forcing, except near the western boundary currents and their large open-ocean extension, where instability dominates and the energy level of the mesoscale eddies is much higher. Using observed SST anomaly spectra to specify the buoyancy forcing by short time scale atmospheric fluctuations, Frankignoul and Müller (1979b) found negligible effects on time scales of up to 102 years.

On the basin scale, the presence of an eastern boundary must be taken into account. The oceanic response to fluctuating Ekman pumping consists of a directly forced part to which a free Rossby wave is added to satisfy the eastern boundary condition. At low frequencies, the response to stochastic forcing is dominated by the first baroclinic mode, and its amplitude limited by the finite width of the ocean basin. Using a simple linear model, Frankignoul et al. (1997) have shown that the thermocline response spectrum was red and its dominant period increased with the distance to the eastern boundary, in broad agreement with the observations in the western North Atlantic thermocline (Figure 3.12). The model is also consistent with the thermocline variability found in long-term integrations with coupled ocean-atmosphere GCMs, although the ocean interior may not only respond passively to the atmospheric forcing but also participate in complex two-way interactions with the atmosphere (e.g., Latif, 1998). In the ventilated regions of the subtropical gyre, large-scale stochastic buoyancy forcing plays a role as well and may explain in part the observed decadal variability of the properties of subducted waters (Czaja and Frankignoul, 1999).

On longer time scales, the fresh water flux contribution to buoyancy forcing may become dominant since surface salinity anomalies have a longer lifetime than the SST anomalies. Mikolajewicz and Maier-Reimer (1990) have driven a global oceanic GCM with stochastic perturbations in the freshwater flux, using the simplifying so-called "mixed boundary conditions" (strong SST relaxation onto a prescribed reference temperature, no salinity feedback). The GCM primarily acted as an integrator to the white-noise forcing, and a decadal mode was identified by Weisse et al. (1994) in the upper level salinity field of the Labrador sea and the northern North Atlantic (Figure 3.13). At low frequencies, the internal dynamics of the ocean model determined the response, causing strong and irregular oscillations and a pronounced 320-year spectral peak well above the f^{-2} level (Figure 3.14). This variability primarily appeared in the form of an intermittent "eigenmode" of oscillation of the Atlantic thermohaline overturning cell which connects the polar regions, involving a positive feedback in deep-water formation rate and large changes in the surface heat fluxes.

However, the mixed boundary conditions are artificial, neglecting the atmospheric and sea ice dynamics. Their use with fixed forcing may also lead to instabilities of the oceanic thermohaline circulation in global ocean models and to large self-sustained oscillations on decadal and century scales; this variability persists when stochastic forcing is added (e.g., Weaver et al. 1993). As the self-sustained oscillations disappear when "restoring" boundary conditions (relaxation for both SST and surface salinity) are used, coupled ocean-sea ice-atmosphere models with more realistic boundary conditions are needed to better understand the origin of the oceanic variability on long time scales. As discussed by Timmermann et al. (1998), stochastic atmospheric forcing and two-way air-sea coupling may also explain the pluridecadal variability found in long-term integrations with coupled models.

3.7 Long Term Climate Changes

The general features of the spectrum of climatic variability on longer time scales have been described by Kutzbach and Bryson (1974), Stocker and Mysak (1992), Shackleton and Imbrie (1990), and others. The spectral density mostly increases with decreasing frequency, down to the large peaks at the orbital frequencies. Early studies with zonally averaged energy balance climate models have suggested that stochastic fluctuations in the meridional atmospheric eddy heat flux could generate a climate variability consistent with the observed spectra over periods of 10^2 to perhaps as much as 10^4 years (Lemke, 1977; Robock, 1978), but the models were very simple and the results have been much debated and refined. The effect of stochastic fluctuations on the periodic forcing at the orbital frequencies has also been

Figure 3.13: *Salinity anomalies in psu in the second model layer (75m) as identified with the POP technique. Areas with negative values are shaded. Shown are the imaginary (top) and the real (middle) parts of the mode, and their power spectra (bottom). The bar on the left side represents the 95% confidence interval. (From Weisse et al., 1994).*

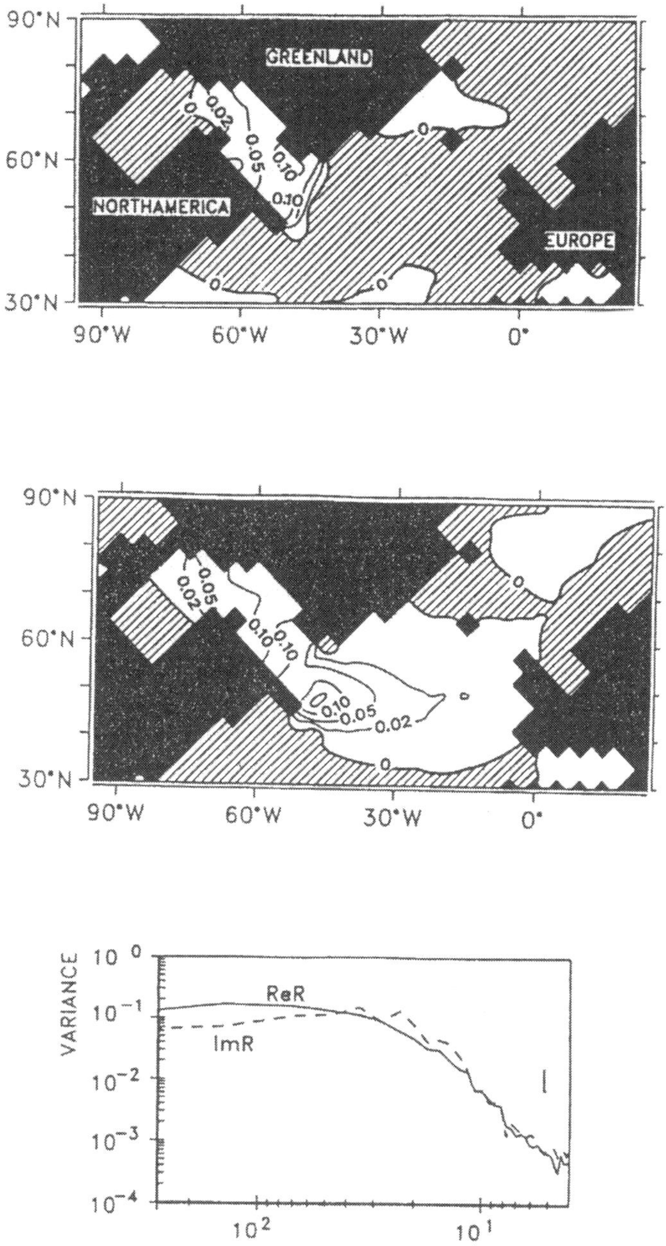

Figure 3.14: *Spectrum of the mass transport through the Drake Passage and the net freshwater flux of the Southern Ocean (dashed line). The thin lines represents the results to be expected from (3.14) with feedback with time constants of 50 years, 100 years, and infinite. The bar represents the 95% confidence interval. (From Mikolajewicz and Maier-Reimer, 1990).*

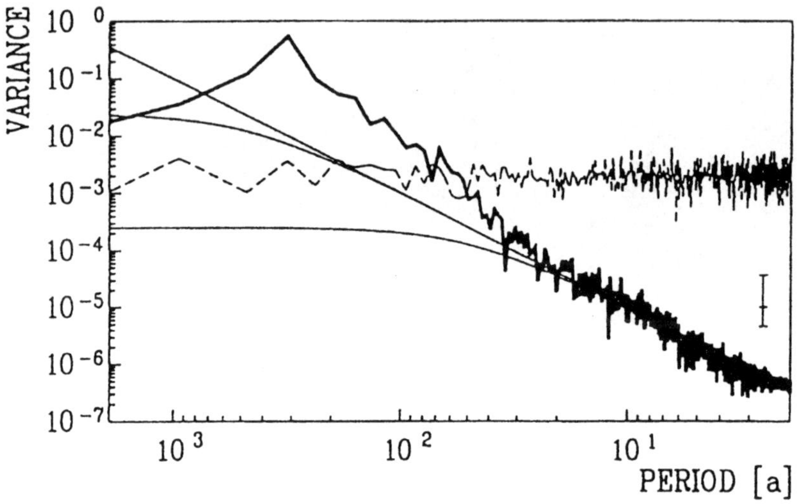

studied (e.g., Sutera, 1981; Saltzman and Maasch, 1991) but, as models with and without stochastic perturbations may show the same level of agreement with the limited observations that are available, the importance of stochastic forcing remains to be assessed.

Chapter 4

The Instrumental Data Record: Its Accuracy and Use in Attempts to Identify the "CO_2 Signal"

by Phil Jones

4.1 Introduction

Although meteorological recording began earlier, the period since the middle of the nineteenth century is that traditionally associated with instrumental records. This review examines in detail the reliability of temperature, precipitation and sea-level pressure data, concentrating on the construction of hemispheric and global average temperature series. The key piece of observational evidence in the "global warming debate" is the "global" temperature series (Jones and Wigley, 1990). How much has the temperature risen? In the last section, pattern correlation statistics are used to search for the

Acknowledgements: Mike Hulme provided the precipitation data used in Figures 4.4 and 4.5 and Ben Santer provided the diagrams 4.8 to 4.11. The assistance of W.L. Gates, M.E. Schlesinger, J.F.B. Mitchell, J.E. Hansen and S. Manabe in providing model results is gratefully acknowledged. Gabi Hegerl provided comments on the this updated version of the chapter. Much of the work and the data sets and time series discussed in the chapter have been developed with the support of the United States Department of Energy, Atmospheric and Climate Research Division and the U.K. Department of the Environment, Transport and the Regions.

greenhouse warming signals predicted by five different General Circulation Models(GCMs) in the observed record of land and ocean surface temperature changes. No trends in the time series of the statistic, $R(t)$, were found using just greenhouse warming signals. Anthropogenic influences were detected, however, using the GCM response patterns from combined greenhouse and sulphate aerosol forcing perturbation experiments.

4.2 Homogeneity

In this section we define homogeneity according to Conrad and Pollak (1962). In this they state that a numerical series representing the variations of a climatological element is called homogeneous if the variations are caused only by variations of weather and climate. Although observers may take readings with meticulous care, non-climatic influences can easily affect the readings. The following sections discuss the most important causes of non-homogeneity.

- Changes in instrument, exposure and measurement technique
- Changes in station location
- Changes in observation times and the methods used to calculate monthly averages
- Changes in the station environment with particular reference to urbanization.

4.2.1 Changes in Instrumentation, Exposure and Measuring Techniques

For land-based temperature, the effects of changes in thermometers have been slight so far this century. More important to the long-term homogeneity is the exposure of the thermometers. Readings now are generally from thermometers located in louvred screened enclosures, which are generally painted white. Earlier readings often were from shaded wall locations, sometimes under roof awnings. These need corrections which can only be estimated by parallel measurements or by reconstructing past locations (Chenoweth, 1992).

A potential discontinuity in temperature time series is currently being induced in first order stations in the United States. Here mercury-in-glass thermometers have been replaced by a thermister installed in a small louvred housing (model HO-83). The change was made by the National Weather Service in the USA at most first order sites during early 1982. There appears to have been little thought given concerning the climatic homogeneity of the record. Parallel measurements were not made, as is generally recommended.

Marine temperature data are seriously affected by homogeneity problems. SST data were generally taken before World War II by collecting some sea water in an uninsulated canvas bucket and measuring the temperature. As there was a short time between sampling and reading, the water in the bucket cooled slightly by evaporative means. Since WWII most readings are made in the intake pipes which take water on to ships to cool the engines. The change appears to have been quite abrupt, occurring within a month during December 1941. The sources of ship's observations in the two marine data banks (COADS and UKMO) changed from principally European to mainly American at this time (for a discussion of marine data, see Bottomley et al., 1990 and Woodruff et al., 1987).

Studies of the differences between the two methods indicate that bucket temperatures are cooler than intake ones by 0.3 to 0.7^oC (James and Fox, 1972). A correction technique for SST bucket measurements has been derived based on the physical principles related to the causes of the cooling (Bottomley et al. 1990; Folland and Parker, 1995). The cooling depends on the prevailing meteorological conditions, and so depends on the time of year and on location. Although a day-to-day phenomenon, the various influences on the bucket are basically linear, so cooling amounts can be estimated on a monthly basis. Assuming that there has been no major change in the seasonal cycle of SSTs over the last 100 years, the time between sampling and reading can be estimated as that which best minimises the residual seasonal cycle in the pre-WWII data. The estimate of between 3 and 4 minutes is both quite consistent spatially and agrees with instructions given to marine observers (Bottomley et al. 1990; Folland and Parker, 1991, 1995).

4.2.2 Changes in Station Locations

It is rare that a temperature recording site will have remained in exactly the same position throughout its lifetime. Information concerning station moves is of primary importance to homogenization. Such information about a station's history is referred to as *metadata*. Assessment of station homogeneity requires nearby station data and appropriate metadata.

4.2.3 Changes in Observation Time and the Methods Used to Calculate Monthly Averages

In the late 19th and early 20th centuries, there was considerable discussion in the climatological literature concerning the best method of calculating true daily and, hence, monthly average temperatures (e.g. Ellis, 1890; Donnell, 1912; Hartzell, 1919; Rumbaugh, 1934). Many schools of thought existed, and unfortunately, no one system prevailed in all regions. At present, there is no uniform system, although the majority of nations use the average of the daily maximum and minimum temperatures to estimate monthly mean temperatures. The remainder use a variety of formulae based on observations at fixed hours, some of which are designed to simulate the true daily mean that would be estimated from readings every hour. There would be no need to correct readings to a common standard, however, provided the observing systems had remained internally consistent. Unfortunately, only in a few countries has the methodology remained consistent since the nineteenth century (Bradley et al. 1985).

In the routine exchange of data, countries exchange monthly mean temperatures, using whatever practice is applied in the country. Since November 1994 the number of meteorological variables routinely exchanged has increased. Now maximum and minimum temperatures are exchanged and it is likely that these will become the preferred method of calculating mean temperature. Maximum and minimum temperatures are extremely useful when considering both the causes and impacts of climate changes. Using maximum and minimum temperature, however, to calculate monthly means is extremely susceptible to the time the observation is made (Baker, 1975; Blackburn, 1983). In the United States, Karl et al. (1986) have developed a model to correct observations made in the morning and afternoon to the 24 hour day ending at midnight. The corrections to the monthly means are called *time-of-observation biases.*

4.2.4 Changes in the Station Environment

The growth of towns and cities over the last 200 years around observing sites can seriously impair the temperature record. Affected sites will appear warmer than adjacent rural sites. The problem is serious at individual sites but appears relatively small in averages calculated for large continental-scale areas (Jones et al. 1990). This conclusion has been confirmed by a recent study (Peterson et al. 1999) defining urban sites using nightlights observed from satellites in addition to population. The influence of urban warming is an order of magnitude smaller than the temperature rise evident over the last 150 years.

An earlier correction procedure has been developed for the contiguous United States by Karl et al. (1988) and depends on population size. Population is generally used as a ready surrogate for urbanization but other indices such as paved area (Oke, 1974), although more useful, are only available for specific cases. The individual nature of the effect means that the corrections applied to U.S. data by Karl et al. (1988) will only be reliable on a regional rather than a site basis. The corrections are also specific to North America and would not apply to Europe, for example, where urban development has taken place over a much longer period of time. The problem is likely to become important in the future in the developing countries as cities expand rapidly (Oke, 1986).

4.2.5 Precipitation and Pressure Homogeneity

Both of these variables are also affected by severe homogeneity problems. For precipitation, there are problems associated with the gauge size, shielding, height above the ground and the growth of vegetation near the gauge. All can impair the performance of the gauge, affecting the efficiency of the catch. The biggest problem concerns the measurement of snowfall, where continual attempts to improve gauge performance affects the long-term homogeneity of the station records. Rodda (1969) has reviewed the possible errors and inhomogeneities that can result in precipitation records.

The total of these problems, coupled with the greater spatial variability of precipitation data, makes the problem of homogeneity much more difficult to deal with than for temperature. Precipitation networks are rarely dense enough to allow for the homogeneity checks discussed in the next section to be undertaken. Instead of studying individual records, analyses of regional precipitation series have been made (Bradley et al. 1987; Diaz et al. 1989; Folland et al. 1990, 1992; Nicholls et al. 1996; Jones and Hulme, 1996).

Pressure data have an advantage over the temperature and precipitation data bases because they are routinely analysed on to regular grid networks for weather forecasting purposes. Monthly-mean data for the Northern Hemisphere (north of 15^oN) extend back to 1873 and for the Southern Hemisphere routine analyses began in the early 1950s. Neither analyses are complete for the entire period, particularly during the earliest years. The sources of the data are discussed and their quality assessed in a number of papers (Northern Hemisphere: Williams and van Loon, 1976; Trenberth and Paolino, 1980 and Jones, 1987, Southern Hemisphere: van Loon, 1972; Karoly 1984; Jones, 1991). Global analyses have been produced at the major operational weather centres since 1979. The most widely available analyses are those from the European Centre for Medium-Range Weather Forecasts (ECMWF) and the National Meteorological Center (NMC) of NOAA. However because of changes to the model and to data assimilation schemes, the homogeneity of the analyses for many climate applications is questionable (Trenberth et al., 1992). To account for the numerous model changes Reanalysis (Kalnay et al. 1996 and Gibson et al. 1997) has been performed at both NCEP (National Centers for Environmental Prediction - the new name for NMC in the USA) and ECMWF for periods up to the last 40 years.

4.2.6 Data Homogenization Techniques

Many methods have been proposed for testing the homogeneity of station records relative to those at adjacent stations (Conrad and Pollak, 1962 and WMO, 1966). Generally, tests involve the null hypothesis that a time series of differences (ratios for precipitation) between neighbouring observations will exhibit the characteristics of a random series without serial correlation (e.g. Mitchell, 1961; Craddock, 1977, Jones et al. 1986a,c, Alexandersson, 1986, Potter, 1981, Young, 1993). The methods work with or without metadata. The scope of the method and the ability to explain and correct inhomogeneities is dramatically improved, however, with detailed metadata. The most comprehensive homogeneity exercise performed produced the United

States Historic Climate Network (Karl and Williams, 1987). Peterson et al. (1998) review many of the recently developed techniques. It is likely that no single technique will be widely applied and continued proliferation of techniques is expected.

4.3 Surface Climate Analysis

4.3.1 Temperature

Various research groups have synthesized the individual site temperature series into gridded and/or hemispheric average time series. The three main groups are the Climatic Research Unit/U.K. Meteorological Office (CRU/UKMO) (Jones, 1988; Jones et al. 1991; Jones et al. 1999a), the Goddard Institute for Space Studies (Hansen and Lebedeff, 1987, 1988) and the State Hydrological Institute in St. Petersburg (Vinnikov et al. 1990). The differences between the groups arise both from the data used and from the method of interpolation. Intercomparison of the results of the three groups can be found in various publications (e.g. Elsaesser et al. 1986; Folland et al. 1990, 1992; Jones et al. 1991; Elsner and Tsonis, 1991 and Nicholls et al. 1996). Here we consider only CRU/UKMO because they are the only group to incorporate marine data.

Time series of the mean hemispheric annual and seasonal temperature anomalies are shown in Figures 5.1 and 5.2. Both series have been widely discussed (Wigley et al. 1985, 1986; Jones and Wigley, 1990; Folland et al. 1990, 1992; Nicholls et al. 1996). The accuracy of the individual seasonal and annual estimates undoubtedly varies with time. The most reliable period is from about 1950. Various techniques have tried to assess error estimates, particularly for years prior to the 1920s. Frozen grid analyses (Jones et al. 1985, 1986a,b, 1991; Bottomley et al. 1990) indicate that hemispheric estimates can be well estimated from the sparse data available during the late nineteenth and early twentieth centuries. Individual annual estimates appear to have about twice the variability of the post-1950 data but the level of temperatures during the late nineteenth century is well approximated by the sparse network. This result has been confirmed by more sophisticated techniques which also use GCM results to estimate the additional uncertainty associated with the lack of data for the Southern Oceans (Trenberth et al. 1992; Madden et al. 1993; Jones et al. 1997a).

In Figures 4.1 and 4.2 warming begins in the hemispheric series around the turn of the century in all seasons. In all seasons over both hemispheres the 1990s are the warmest decade. In the Northern Hemisphere, interannual variability is greater in winter compared to summer. No such seasonal contrast is evident in the Southern Hemisphere. The hemispheric annual time series and their average (the global series), Figure 4.3, represent the major observational evidence in the "global warming" debate (Jones and Wigley,

1990).

Though hemispheric and global series are heavily relied upon as important indicators of past climatic change, they mask the spatial details of temperature variability across the globe. On *a priori* grounds there are only weak arguments to suggest why any regional series should be more indicative of global scale change than any other of comparable size. The issue has been addressed by a number of workers showing spatial patterns of temperature change (see e.g. Barnett, 1978; Jones and Kelly 1983; Folland et al. 1990, 1992; Jones and Briffa 1992; Briffa and Jones, 1993 and Jones et al., 1999a).

Most of the focus of past and future temperature change has been concerned with mean temperature. For past changes, this has resulted from a lack of readily-available databases of monthly mean maximum and minimum temperatures. Recently developed data sets (Karl et al., 1993) have enabled analyses of maximum and minimum temperatures to be made for 37% of the global land mass (encompassing the contiguous United States, Canada, Alaska, the former Soviet Union, China, Japan, the Sudan, South Africa and eastern Australia). The analyses indicate that over the 1952-89 period, minimum temperatures have risen at a rate three times that of maximum temperatures. The reduction in the diurnal temperature range is approximately equal to the temperature increase. The changes are detectable in all of the regions studied in all seasons. Easterling et al. (1997) have confirmed these findings with an enhanced dataset covering the 1951-93 period and 54% of the global land area.

4.3.2 Precipitation

Analyses of precipitation data have tended to focus on regional rather than hemispheric scales. The reasons for this stem from two factors; precipitation data have much greater spatial variability compared to temperature and precipitation data from oceanic areas are virtually non-existent. For most areas of the world, precipitation variability, both in space and time, is so large that it will be generally impossible to detect changes until significant impacts will have been felt in the agricultural and water resource sectors.

Various regional time series were developed for the IPCC Scientific Assessments (Folland et al. 1990, 1992; Nicholls et al. 1996). All of the studied regions in North America, Eurasia, Africa and Australia show large year-to-year variability in either annual or growing-season precipitation totals. All show marked decadal-scale variation but with little long-term change on the century timescale. There are two major exceptions to this. Average annual series for the former Soviet Union show a gradual increase in precipitation since the beginning of this century (Figure 4.4). Most of the increase has occurred in the non-summer season. Some of the increase may be related to the underestimation of snowfall during winter but even allowing for this a significant increase over the USSR south of 70^oN is still evident (Groisman et al. 1991).

Figure 4.1: *Northern Hemisphere surface air temperatures for land and marine areas by season, 1856-1998. Standard meteorological seasons for the Northern Hemisphere are used: winter is December to February, dated by the year of the January. Data are expressed as anomalies from 1961-90. Time series in this and subsequent plots have been smoothed with a ten year Gaussian filter.*

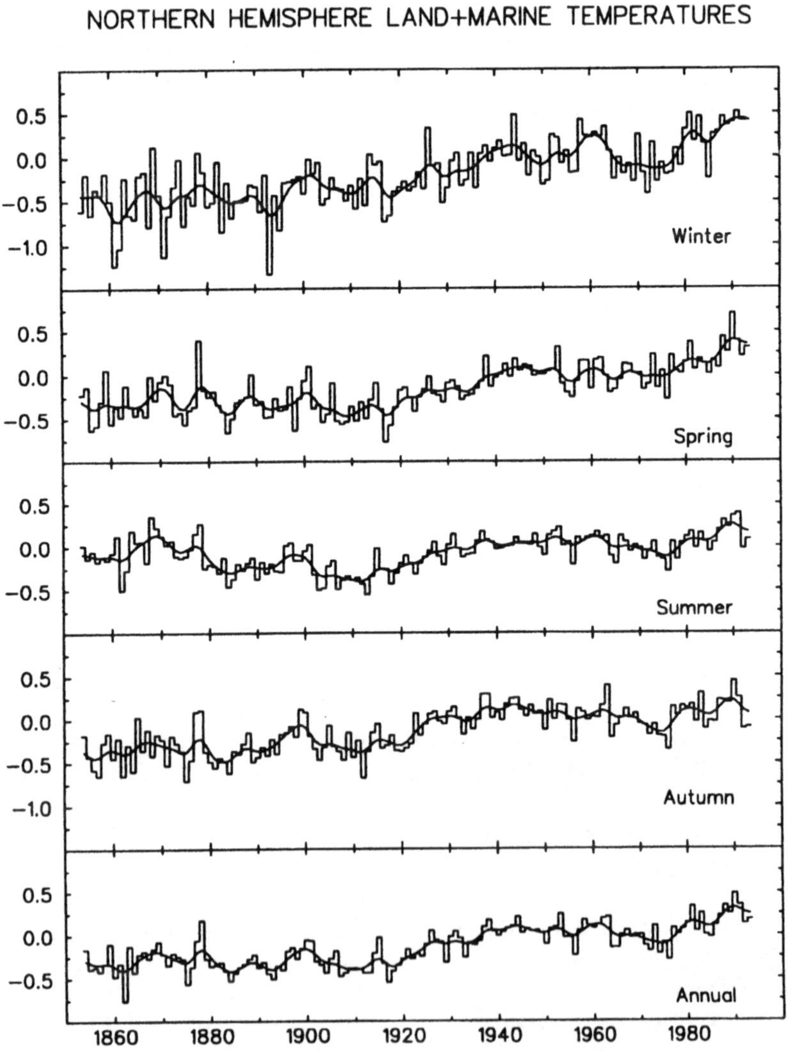

Figure 4.2: *Southern Hemisphere surface air temperatures for land and marine areas by season, 1856-1998. Standard meteoroloigcal seasons for the Southern Hemisphere are used: summer is December to February, dated by the year of the January. Data are expressed as anomalies from 1961-90.*

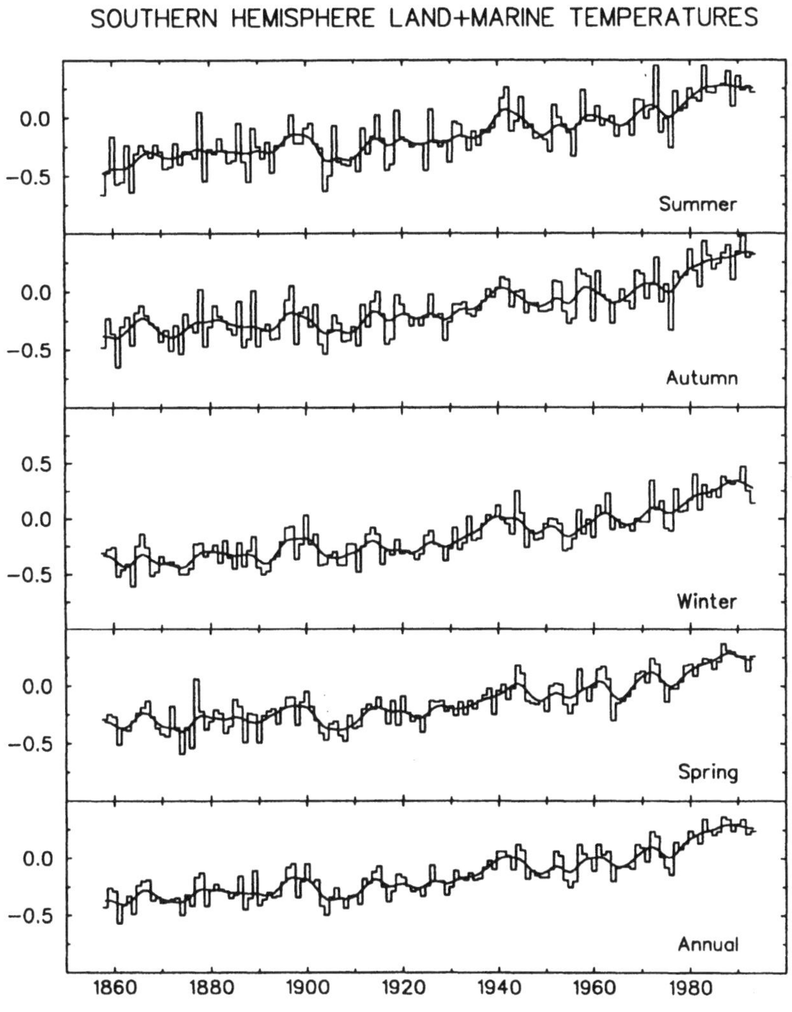

Figure 4.3: *Global and hemispheric annual surface air temperatures for land and marine areas, 1856-1998. Data are expressed as anomalies from 1961-90.*

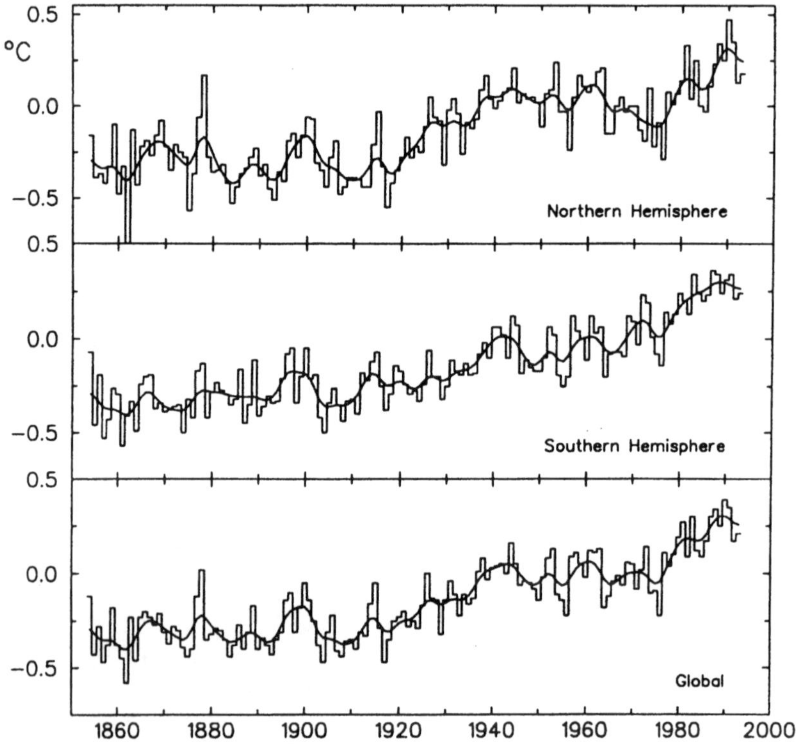

Figure 4.4: *Standardised regional annual precipitation anomaly time series for the territory of the former Soviet Union. Standardization of each station time series is achieved by subtracting the reference period precipitation and dividing by the standard deviation. The regional series is the average of all available station series. No form of areal weighting is used. The reference period used for calculating the annual means and standard deviations was 1951-80.*

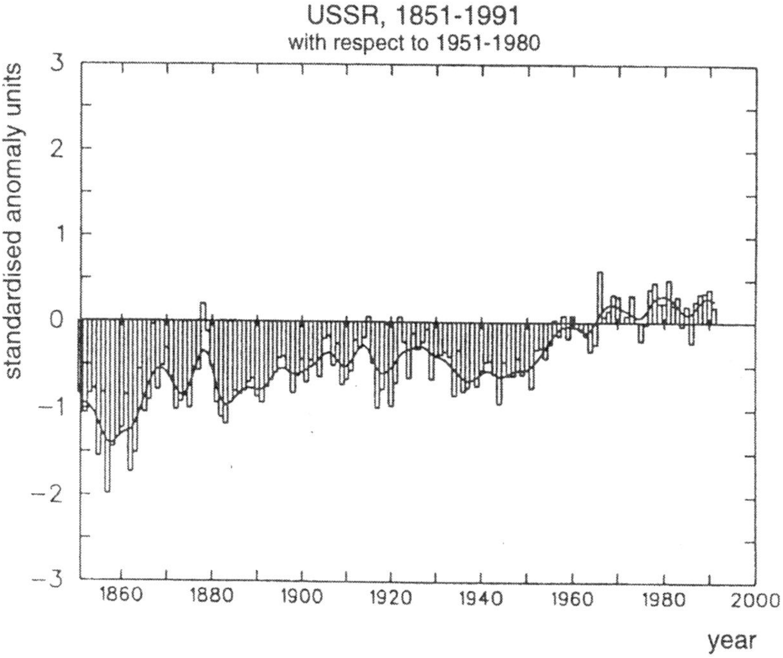

Figure 4.5: *Standardised regional (annual) precipitation anomaly time series for the Sahel regions (10°-15°N, 20°W-35°E). Two reference periods 1931-60 and 1961-90 are used for comparison of the results.*

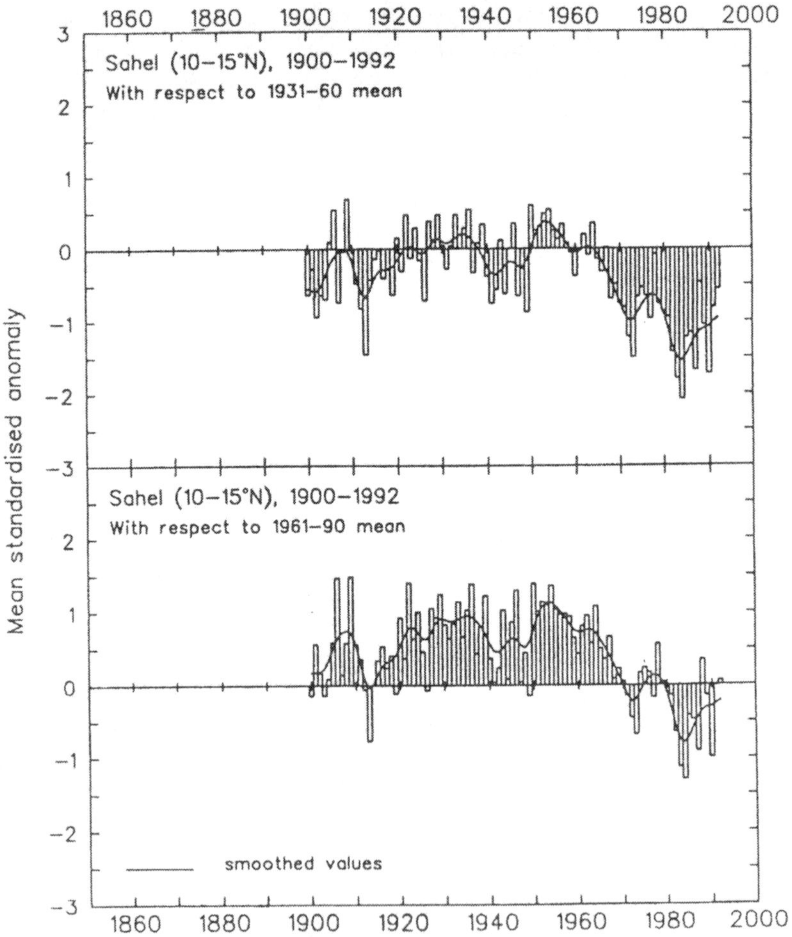

The most significant and dramatic changes in precipitation have occurred over the Sahel region of Subsaharan Africa. Here the reduction of precipitation during the rainy season in recent years has been highly statistically significant. At some sites in the region there has been a decline of about 30% between averages for the two standard WMO periods of 1931-60 and 1961-90. The steep change would be even greater if the periods 1941-70 and 1971-90 were compared as the 1960s were relatively wet. Figure 4.5 plots the average time series for region using the two standard WMO reference periods. Using the most recent period changes the character of the time series from the Sahel drought of the last 20 years (relative to 1931-60) to the Sahel wet period of the 1920s to the 1960s (relative to 1961-90). The difference between the two analyses is not a step change because of the use of standardized anomalies. Interannual variability is also greater at most stations during the 1961-90 period compared to 1931-60

4.3.3 Pressure

The interrelationships of the climate system mean that past changes in the patterns of temperature and precipitation patterns will undoubtedly also be accompanied by changes in circulation throughout the troposphere. Changes that have been noted range from local-scale studies such as the decline in westerlies over the British Isles (Lamb, 1972) to much larger regional scale indices such as the North Atlantic Oscillation (van Loon and Rogers, 1978; Hurrell, 1995 and Jones et al. 1997b), North Pacific winter pressure (Trenberth, 1990) and various Southern Hemisphere indices (including the Southern Oscillation) (Karoly, 1996; Konnen et al. 1998 and Jones et al. 1999b).

The North Atlantic/European region has probably been studied more than other parts of the world, almost certainly because of the availability of long (> 150 years) records. The well-known out-of-phase relationship between winter temperatures over northern Europe (especially Fennoscandia) and Greenland has long been recognized (van Loon and Rogers, 1978). The link between surface temperature anomalies in the region and the circulation of the North Atlantic region is evident in indices such as the North Atlantic Oscillation (NAO). Figure 4.6 shows seasonal (November to March) values of the NAO (difference in normalized pressure between Gibraltar and Reykjavik). Stronger westerlies prevailed during the first three decades of the twentieth century and again since the 1970s. The weakest westerlies occurred during part of the nineteenth century and particularly during the 1960s.

In the North Pacific region a major change in the strength and position of the Aleutian Low took place during the mid-1970s. After about 1977 the strength of the low pressure centre increased and moved slightly south particularly in winter (Figure 4.7). In many respects the change may have been a return to the conditons that occurred during the 1920s, 1930s and the early 1940s. Some of the change since the 1970s has reversed in the most recent years. The effects of the change in the mid-1970s in the climate of the

Figure 4.6: *Winter (November-March) normalized pressure difference between Gibraltar and Reykjavik, Iceland. Years dated by the January. Higher values indicate stronger westerlies over the North Atlantic.*

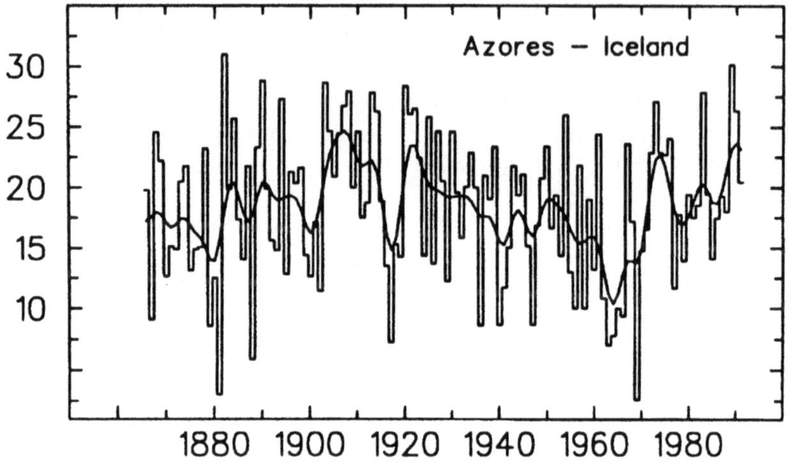

Figure 4.7: *Winter (November-March) pressure average (mb) for the North Pacific 30°-70°N, 150°E-150°W. Year dated by the January.*

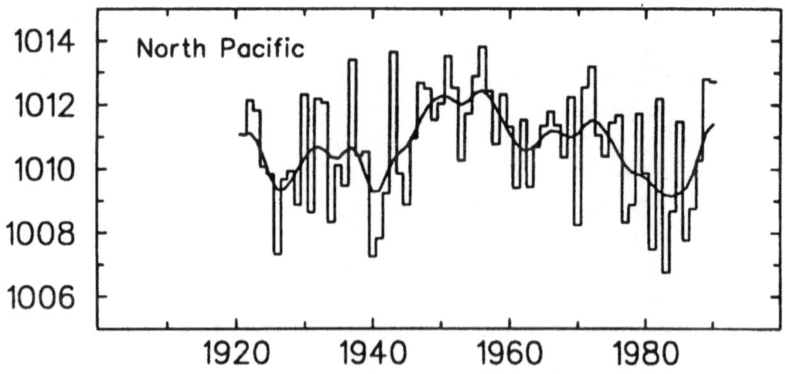

whole North Pacific region have been discussed by numerous workers (e.g. Douglas et al. 1982; Nitta and Yamada, 1989 and Trenberth, 1990). The decreased pressures led to greater advection of warm maritime air on to the northwest of North America leading to enhanced precipitation and warmer temperatures. Alaska has experienced the greatest regional warmth over the Northern Hemisphere with average decadal temperatures 1^oC warmer than the 1951-80 period during the 1980s. In contrast, sea temperatures have been much cooler than normal in the central North Pacific and cooler air temperatures have been experienced over Japan and the Okhotsk/Kamchatka region. Although the region only explains part of the hemispheric increase, the change in climate of the region, particularly in the winter, has been abrupt.

4.4 The Greenhouse Detection Problem

In detecting the enhanced greenhouse effect, the key step is to be able to attribute an observed change in climate to this specific cause. Attribution almost certainly requires the identification in the observational record of a multivariate signal characteristic of (and, ideally, unique to) greenhouse-gas-induced climatic change. This type of detection approach has been called the *fingerprint method* (Madden and Ramanathan, 1980; MacCracken and Moses, 1982). Pre-1995 fingerprint studies have been inconclusive either because the signal has been obscured by the noise of regional-scale natural climatic variability (Barnett, 1986; Barnett and Schlesinger, 1987; Santer et al., 1991; Barnett, 1991), or because of uncertainties regarding the level and structure of both the signal and natural variability (Madden and Ramanathan, 1980; Karoly, 1987, 1989; Wigley and Barnett, 1990).

A fingerprint detection variable may be considered to be a vector whose components are either different scalar variables (e.g., temperature, precipitation, etc.) and/or the same variable measured at different points or averaged over different regions (e.g., temperatures at different locations on the Earth's surface or at different levels in the atmosphere) (Wigley and Barnett, 1990). The basic detection strategy is to compare the observed time series of this vector either with an estimate of the equilibrium greenhouse-gas signal (as may be inferred by differencing the results of equilibrium model experiments for $1\times CO_2$ and $2\times CO_2$), or with a time-evolving signal (inferred from a model simulation with time-dependent greenhouse-gas forcing).

4.4.1 Definition of Detection Vector and Data Used

Before performing any analysis, some consideration should be given to the choice of the detection vector. There are at least four criteria which can be used to evaluate the usefulness of a given vector for detection studies.

- First, the individual components should have high *signal-to-noise ratios* (SNR). Results from climate model experiments indicate that temperature and atmospheric moisture content are relatively good choices in this regard, while precipitation and sea level pressure have low SNR (Barnett and Schlesinger, 1987; Santer et al., 1991).

- Second, the signal vector should not be model-specific. If different models gave different signal vectors, then this would lower confidence in the signal from any one model.

- Third, the signal vector should be easily distinguished from (i.e., near orthogonal to) both the signals due to other forcing factors and the noise of natural internal variability on the 10-100 year time scale relevant to the enhanced greenhouse effect (Barnett and Schlesinger, 1987; Wigley and Barnett, 1990). A poor choice in this regard might be the troposphere/lower stratosphere temperature change contrast (but see also Karoly et al., 1994), where the expected greenhouse signal is apparently similar to observed natural variability (Liu and Schuurmanns, 1990). Most post-1995 detection studies are using upper air data increasingly (see later section). The spatial pattern of near-surface temperature changes, which is used here, is also less than ideal, since the feedback processes that influence the spatial character of the enhanced greenhouse signal must also operate with other external forcing mechanisms, such as changes in solar irradiance (Wigley and Jones, 1981).

- Fourth, suitable observational data must exist. Because it is the 10-100 year time scale that is of concern, long data records are needed. Only surface-based data have records exceeding ~40 years in length, and of these, only temperature data satisfy the high signal-to-noise ratio condition.

Part of the reason why detection is so difficult is because the intersection of the sets satisfying the above criteria is effectively empty. The observed data we use satisfy these criteria better than any other data set, viz. $5^o \times 5^o$ gridded monthly-mean, land-based surface air temperatures and sea surface temperatures (SST). These data come from the combined land-ocean data set described by Jones et al. (1991,1999a). Here they are expressed as anomalies relative to 1950-79 and span the 90-year interval, 1900-89. Only data from 1900 are used because of the reduced coverage and decreased reliability of earlier data. Data are available for the area shaded in Figure 4.8. Other possible detection variables such as upper air data have been considered by Karoly et al. (1994).

The data we use for defining a greenhouse-gas signal come from five separate climate model experiments. The experiments differ in terms of the models employed and the forcings applied. All experiments use an atmospheric general circulation model (AGCM) coupled to an ocean model. The

Figure 4.8: *Observed coverage of land-based surface air temperature and sea surface temperature in the Jones et al. (1991) combined land-ocean data set. Annual mean coverage is shown for 1900-39, the period used for calculating the observed standard deviation used in pointwise normalization.*

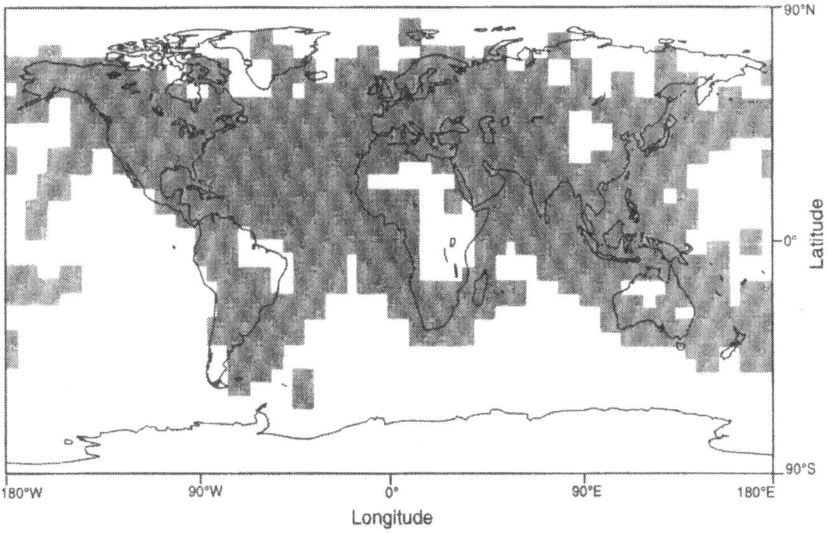

ocean models vary from simple mixed-layer models in which horizontal heat transport is prescribed (viz., OSU [Schlesinger and Zhao, 1989], Geophysical Fluid Dynamics Laboratory [GFDL; Wetherald and Manabe, 1990] and United Kingdom Meteorological Office [UKMO; Wilson and Mitchell, 1987]) to a model in which heat transport below the mixed-layer is parameterized as a diffusive process (Goddard Institute for Space Studies [GISS; Hansen et al., 1988]) and an ocean model intermediate in complexity between the mixed-layer models and fully-dynamic ocean general circulation models (Lawrence Livermore National Laboratory [LLNL; Gates and Potter, personal communication], which uses a two-layer ocean [Pollard, 1982]).

Results are taken from experiments with two different types of forcing: step-function and time-dependent. In the former (OSU, GFDL, UKMO and LLNL), the steady-state response to a step function doubling of CO_2 is considered. In the latter (the GISS *Transient A* experiment; see Hansen et al., 1988), greenhouse forcing commences in 1958 and increases with time in a manner similar to the IPCC "Business as Usual" scenario (Houghton et al., 1990). Control run data and information from years 81-90 of the GISS perturbed run are used.

4.4.2 Spatial Correlation Methods

Let us consider two observed data sets D_1 and D_2 (D = Data), and two model data sets M_1 and M_2 (M = Model). Each are two-dimensional (x, t) arrays where x and t are independent discrete variables running over space (grid-points) and time (years), with $x = 1 \ldots m$ and $t = 1 \ldots n$ respectively. Here, D_1 and D_2 represent near-surface temperatures averaged over two different periods of time. M_1 and M_2 are corresponding temperatures simulated in a control run and a greenhouse-gas experiment, respectively. If M_1 and M_2 are derived from equilibrium experiments, we are concerned with a signal pattern which has no time dependence viz. the equilibrium surface temperature change between a $2 \times CO_2$ experiment and a $1 \times CO_2$ control run

$$M(x) = M_2(x) - M_1(x) \tag{4.1}$$

Our detection strategy, which follows that used by Barnett (1986) and Barnett and Schlesinger (1987), is to compare the signal pattern, $M(x)$, with a series of time-evolving observed patterns of change

$$D(x,t) = D_2(x,t) - D_1(x) \tag{4.2}$$

where the index t runs over the total number of observed mean change fields. Here, $D_1(x)$ is the time-average over some fixed reference period, defined as

$$D_1(x) = \frac{1}{p} \sum_{u=t_0}^{t_0+p-1} D(x,u) \tag{4.3}$$

where $p < n$ and t_0 is the initial year of the reference period. Similarly, $D_2(x,t)$ is defined by

$$D_2(x,t) = \frac{1}{p} \sum_{u=t}^{t_0+p-1} D(x,u) \tag{4.4}$$

where $t > t_0 + p$. Here we take $p = 10$ (years). The use of decadal averages and overlapping observed decades yields a smoother estimate of the observed changes. In later examples we assume that $t_0 = 0$, corresponding to the year 1900, and therefore compare a single model-derived signal pattern with 71 observed mean fields of change relative to 1900-09 for the overlapping decades from 1910-19 to 1980-89.

If the enhanced greenhouse effect is actually becoming stronger in the observed data, and if the model-based estimate of the signal is reasonable, $M(x)$ and $D(x,t)$ should become increasingly similar - i.e., there should be a positive trend in the statistic used to measure similarity. Detection then becomes a problem of identifying a significant trend in the similarity statistic. The next choice, therefore, is to select an appropriate indicator of pattern similarity. In the past, some inappropriate choices have been made.

As a similarity indicator, Barnett and Schlesinger (1987) use a statistic, $C(t)$, which involves the uncentered cross-moment between the $M(x)$ and $D(x,t)$ fields

$$C(t) = \frac{\sum_{x=1}^{m} D(x,t)M(x)}{\sum_{x=1}^{m} M(x)^2} \tag{4.5}$$

(Note that in their original usage Barnett and Schlesinger (1987) used a fixed single year to define the reference period rather than a decadal-average as used here). More conventionally, pattern similarities are quantified using a centered statistic such as a correlation coefficient (i.e., where $D(x,t)$ and $M(x)$ are centered on their respective spatial means). This is the approach which we adopt here. In its general form, the centered detection statistic, $R(t)$, is defined as the pattern correlation between the observed and simulated fields.

$$R(t) = \frac{\sum_{x=1}^{m} (D(x,t) - \overline{D}(t))(M(x) - \overline{M})}{m S_D(t) S_M} \tag{4.6}$$

where S_D^2 and S_M^2 are the spatial variances defined as

$$S_D^2(t) = \frac{1}{m} \sum_{x=1}^{m} \left(D(x,t) - \overline{D}(t)\right)^2 \tag{4.7}$$

(S_M^2 similarly) and $\overline{D}$ and $\overline{M}$ are spatial means defined as

$$\overline{D}(t) = \frac{1}{m} \sum_{x=1}^{m} D(x,t)$$

($\overline{M}$ similarly). It turns out that $C(t)$ is *not* an appropriate statistic. To see this, we need to determine the relationship between $C(t)$ and $R(t)$.

If we define

$$Z^2 = \frac{1}{m}\sum_{x=1}^{m} M(x)^2 = S_M^2 + \overline{M}^2$$

then $C(t)$ can be written as

$$\begin{aligned} C(t) &= \frac{\sum_{x=1}^{m}(D(x,t) - \overline{D}(t))(M(x) - \overline{M}) + m\overline{D}(t)\overline{M}}{mZ^2} \\ &= \frac{S_D(t)S_M R(t) + \overline{D}(t)\overline{M}}{Z^2} \end{aligned}$$

Hence

$$C(t) = aR(t) + b\overline{D}(t) \tag{4.8}$$

where

$$a = \frac{S_D(t)S_M}{Z^2} \tag{4.9}$$

$$b = \frac{\overline{M}}{Z^2} \tag{4.10}$$

Equation (4.8) shows that $C(t)$ has an $R(t)$ component and a component determined by the spatial-mean of the D field. If, therefore, $\overline{D}(t)$ has a trend, this will produce a trend in $C(t)$ independent of the behaviour of $R(t)$. It may be, therefore, that $C(t)$ tells us little more than $\overline{D}(t)$, in which case $C(t)$ would be of little value as a fingerprint test statistic.

The reason $C(t)$ (or, probably, any uncentered statistic) does not succeed as a multivariate test statistic is because it does not separate the effect of the changing mean (which, for $\overline{D}(t)$, is a strongly time-varying quantity) from the true pattern similarity. This will be particularly true for spatially uniform unless attempts are made at optimal detection strategies (Hegerl et al. 1996 and Tett et al. 1997)

$R(t)$ is now calculated using the observed data and the five sets of model data discussed earlier. Pattern correlations were calculated using the grid-point data (after excluding grid-points with missing observed data and the corresponding model grid-points), defined by (4.1-4.4). The results for the five models are shown in Figures 4.9 (annual), 4.10 (seasonal for the UKMO model) and 4.11 (seasonal for the GFDL model). In none of the five models is there any increase of $R(t)$ with time. In other words, there appears to be no increasing expression of any of the five model signals of the enhanced greenhouse effect in the observational record. More recent (post 1995) studies using improved GCMs and including the direct effects of sulphate aerosols does,

Figure 4.9: *Model versus observed pattern correlations. Pattern correlations are calculated from grid point data using 71 overlapping decades (from 1910-19 to 1980-89) relative to a fixed reference decade (1900-09). Results are plotted on the first year of each decade.*

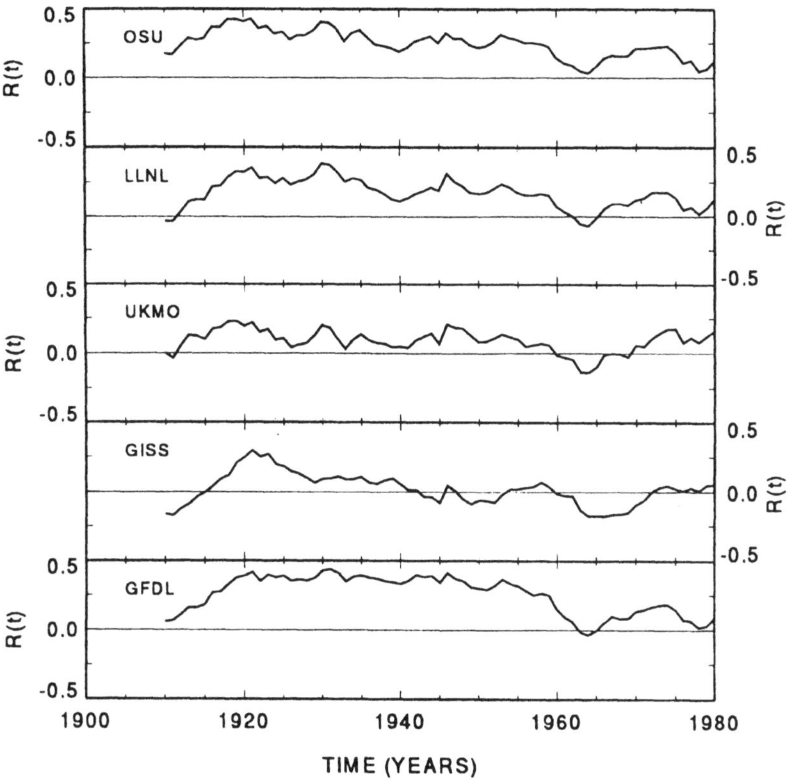

however, lead to an increasing expression of the combined anthropogenic (greenhouse gases plus sulphate aerosols) in the surface temperature record (see Santer et al. 1995).

4.4.3 Detection Studies after 1995

Since the autumn school on which this chapter is based, the climate change detection issue has changed dramatically. Several important studies have been published (Santer et al. 1995, 1996a; Hegerl et al. 1996; Tett et al. 1997) using either surface or upper air data or as is more common now, combinations of many levels of data. Their methods, arguments and conclusions have been widely discussed both at scientific meetings and in many non peer-review reports (e.g. the grey literature and web pages). The issue has

Figure 4.10: *Seasonal and annual results for R(t), the model versus observed pattern correlations for the UKMO model (Wilson and Mitchell, 1987). Pattern correlations calculated as in Figure 4.9.*

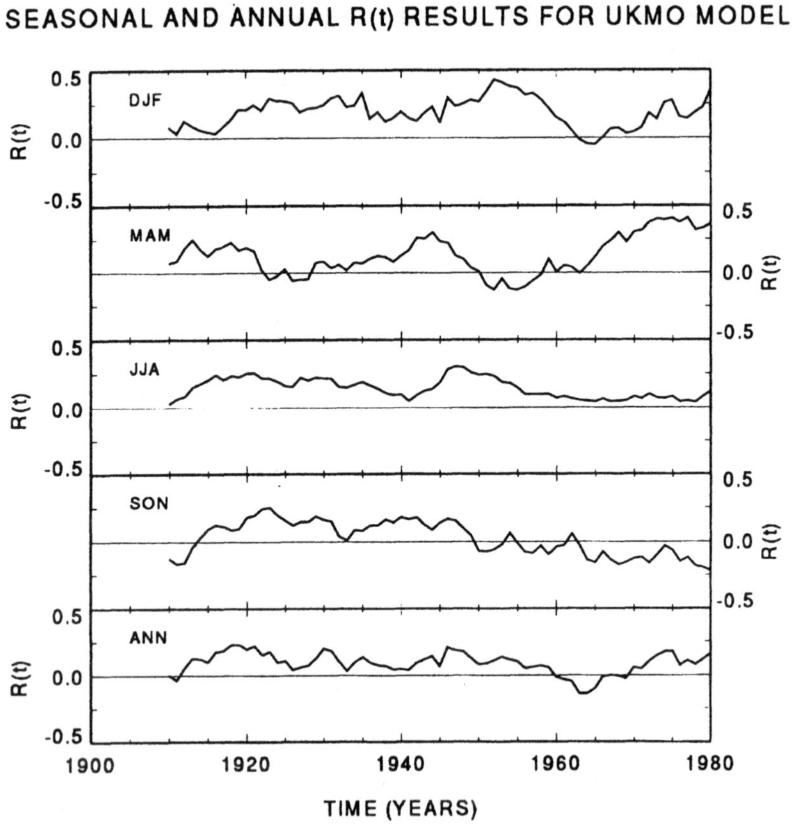

Figure 4.11: *Seasonal and annual results for $R(t)$, the model versus observed pattern correlations for the GFDL model (Wetherald and Manabe, 1990). Pattern correlations calculated as in Figure 5.9.*

SEASONAL AND ANNUAL R(t) RESULTS FOR GFDL MODEL

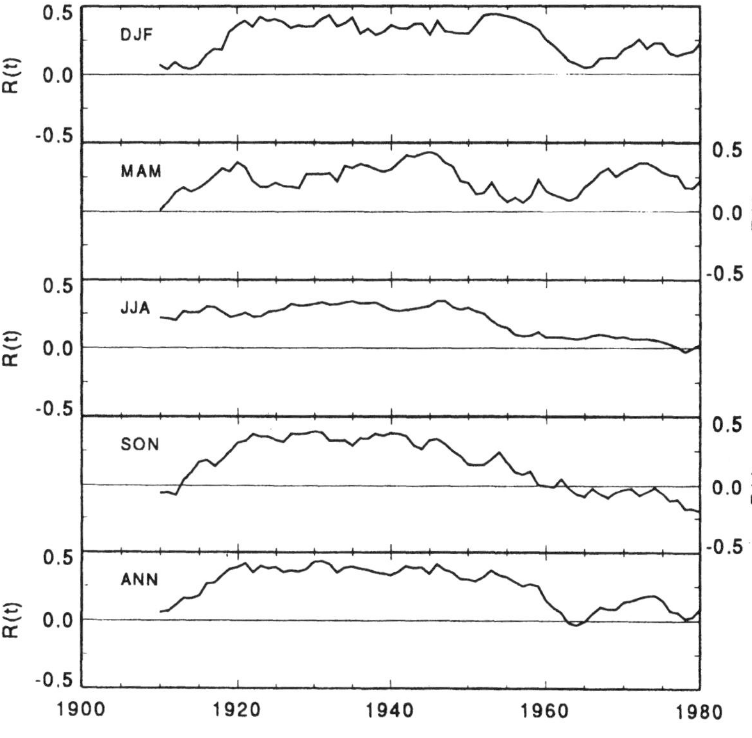

become increasingly complex, not only in the mathematics used, but also in the range of arguments promulgated by both sides (pro- and anti-greenhouse influence in the climate record) in the debate. The most comprehensive review of the subject was produced in the second IPCC report (Santer et al. 1996b), although my use of the word "comprehensive" would be disputed by the anti-greenhouse side (the so-called "greenhouse skeptics"). It has become increasingly difficult to discuss the issue in a reasoned scientific way as the skeptics do not always appear to see the need to publish their original findings in the peer-review literature.

4.5 Conclusions

Considerable care must be taken when dealing with any type of climatic data. Several sources of potential error can easily impair climatic time series, rendering some analyses of doubtful value. The most serious, and often ignored, source of error is due to the change, over the course of time, in the environment around a station, particularly due to urbanization.

Using carfully checked temperature data from over 2000 land-based stations and about 50 million marine observations it has been shown that the average "daily-mean" temperature of the world has risen by about 0.6^oC over the last 150 years. The warming has been uneven in both time and space. Much of the warming occurred in the Northern Hemisphere before 1940 and since 1930 in the Southern Hemisphere, while spatially some regions have warmed by nearly 2^oC while a few areas indicate cooling (Jones and Briffa, 1992; Jones et al. 1999a). Analyses with maximum and minimum temperatures over 1951-93 indicate that three quarters of the rise over this time has occurred during nighttime compared to daytime (Easterling et al. 1997).

The interrelationships of the climate system mean that changes in surface variables will be accompanied by changes in the atmospheric circulation. Additional checks on the homogeneity of any of the surface data sets and associated time series can be made by finding synchronous and connected changes in the other two.

Detection exercises of the kind dicussed here failed before 1995. Since then there have been several successful studies (generally using GCM signals defined from combined greenhouse gas and sulphate aerosol integrations) published (Santer et al. 1995, 1996a; Hegerl et al. 1996 and Tett et al. 1997). These studies led to the most quoted statement of recent decades in climatology - namely "The balance of evidence suggest a discernible human influence on the climate system" (Santer et al. 1996b) and the statement was one of the driving forces behind the international agreement on greenhouse gas emission reductions signed in Kyoto. It is difficult to visualize how the detection and attribution issue will develop over the next five years, however given the efforts of many groups around the world it is likely that even more dramatic changes will take place than between 1994 and 1999.

Chapter 5

Interpreting High-Resolution Proxy Climate Data - The Example of Dendroclimatology

by Keith R. Briffa

5.1 Introduction

Current scientific concern to establish the reality, the nature and the speed of climate changes, believed by many to be the inevitable consequence of human activities, should serve to reinforce our determination to understand similar details of the "natural" (i.e. non-anthropogenic) variability of climate. Reconstructing past climates on all timescales is clearly important if we hope to understand the mechanisms that control climate (Bradley, 1990). However, the scope and the rapidity of the changes foreseen in many scenarios of an "enhanced-greenhouse" world, highlight the particular relevance of palaeoclimate studies that focus on recent centuries and millennia (Eddy, 1992). When considering the question "Can we detect an enhanced greenhouse signal?", natural records of past climate variability (so-called "proxy"

Keith R. Briffa is currently funded by the European Community Environment Programme under grant EV5V-CT94-0500 (DG12 DTEE) climate data) that are annually resolved and that capture decadal-to-century timescale variability, represent an essential basis for comparison with relatively short modern climate records which are rarely longer than a hundred years.

When assessing the value of proxy climate data, we should consider a number of general questions:

- How well can the evidence be dated?
  - Is it continuous or erratic?
  - Is the resolution seasonal, annual, millennial, etc.?
- What is the statistical quality of the data?
  - Is the series replicated?
  - What is the strength of the common signal?
  - What is the inherent error?
- What can be deduced from the data?
- What is the nature of climate control - precipitation, temperature, etc.? Seasonal, annual?
- How strong is the forcing signal and at what timescales?
- Is the full spectrum of the forcing under investigation represented?
- What is the response time?
- What is the spatial representativeness?

These questions are relevant to all forms of proxy data interpretation but this discussion will address some of them in the context of one particular proxy source, that of tree rings.

Tree-ring chronologies can provide high-resolution (annual or even seasonal), continuous, relatively-long (up to millennial), absolutely-dated, well-replicated and sensitive (i.e. accurate) indicators of climate variability from large areas of the globe. This gives tree rings a special status among the various proxy climate data sources. The study of the annual growth of trees and the consequent assembling of long, continuous chronologies for use in dating wood (such as in an archaeological context) is called dendrochronology. The study of the relationships between annual tree growth and climate is called dendroclimatology.

This chapter attempts to give a brief overview of dendroclimatology, focusing on some basic concepts and statistical techniques which have implications for gauging the value of dendroclimatic reconstructions. For more detailed general references which provide much discussion on the biological background, methodology and scope of tree-ring-related work, see Fritts (1976), Hughes et al. (1982), Schweingruber (1988), Cook and Kairiukstis (1990) and Hughes (1990).

5.2 Background

The annual growth of a tree is the net result of many complex and interrelated biochemical processes. Trees interact directly with the micro environment of many leaf and root surfaces. The fact that there is a relationship between these extremely localized conditions and larger scale climate parameters offers the potential for extracting some measure of the overall influence of climate on growth from year to year.

Growth may be affected by many aspects of the micro climate: for example, sunshine amounts, precipitation, temperature, wind speed and humidity. There are other non-climate factors that may exert an influence, such as competition, defoliators and soil nutrient status. Nevertheless, wherever tree growth can be dated absolutely and is limited directly or indirectly by some climate variable, dendroclimatology can be used to reconstruct some information about past environmental conditions.

The general approach taken in dendroclimatic reconstruction work is:

- to collect (sample) data from a set of trees which have been selected on the basis of site and phytosociological characteristics so that climate should be the major growth-limiting factor;
- to assemble the data from individual trees into a composite site chronology by cross-dating and averaging the individual series after the removal of age effects (standardization);
- to build up a network of site chronologies for a region;
- to identify statistical relationships between the chronology time series and instrumental climate data for the recent period (the calibration period);
- to test, or verify, the resulting reconstruction against independent data; and
- finally, to use these relationships to reconstruct climate information from an earlier period (the reconstruction period) covered by the tree-ring data.

5.3 Site Selection and Dating

Not all trees are suitable for dendroclimatology. It is axiomatic that the rings be dated absolutely. In general, dendroclimatology is difficult to apply in the tropics where seasonal growth cycles tend not to occur (or not to occur with any consistency). There is rarely a marked seasonal cycle in climate conditions in the tropics. Plants without radial cambium, such as palms, do not produce annual rings and cannot be used.

Most studies have been concentrated in temperate to mid-latitude regions where there is a wealth of species which exhibit clear annual rings. Valuable information has, however, also been obtained from higher latitudes and from certain subtropical species.

In harsher, more marginal, regions, often near the edges of the ecological range of tree species, the seasonal limitation on growth may become so strong that radial growth may not occur around the entire tree. This will produce partially "missing" rings which can only be detected by comparison with samples from other parts of the tree or other trees. Similarly, severe and abrupt weather conditions can sometimes cause cell growth to slow down or even stop during the growing season, resulting in what appear to be "double" rings in particular years. Again, these are detected by the comparison of the ring series from many trees. This meticulous "cross dating" of replicate samples is an extremely important aspect of dendrochronology and distinguishes it from many other so-called high-resolution proxy studies. It is important to appreciate the importance of this point. Other palaeoclimate data sources are capable of providing annual data e.g. lake varves (Petterson et al., 1993), ice cores (Thompson, 1990) and corals (Dunbar and Cole, 1993), but tree-ring data are dated, absolutely, to the year, by reference to a continuous, replicated, chronology, stretching unbroken to modern times. The comparison of temporally overlapping ring-width series enables mean chronologies, thousands of years long, to be assembled from many individual-tree series that may typically have only 200-300 rings (e.g. Baillie 1982). Merely counting back rings (or layers in ice cores or corals) from the present is not sufficient to ensure that rings have not been overlooked or are not missing from the sample series. Only crossdating ensures the absolute timescale.

5.4 Chronology Confidence

5.4.1 Chronology Signal

The replication of data inherent in constructing an average chronology for a site or region is another axiom of dendrochronology. It is only by comparing replicate series that one can judge the strength of common variability (and hence forcing) within any proxy series, and it is only by averaging data series that non-common variability, "noise", can be reduced or eliminated. The variability in common between contemporaneous data series represents an empirical signal - a purely statistical measure of common growth forcing among a group of trees. This can be measured using Analysis of Variance techniques or simply as the mean correlation coefficient ($RBAR$) of all replicate comparisons (in fact it is possible to calculate separate quantities representing the strength of common growth forcing within and between trees at a site, but, for simplicity, we will consider only the one between-tree value here. For further details see Fritts, 1976; Wigley et al., 1984; Briffa and

Jones, 1990).

For relatively high-frequency data, $RBAR$ is unbiased and provides an accurate measure of chronology signal, but for low-frequency (i.e. long timescale) variability it is a poor measure (i.e. it has wide confidence limits) dependent on the series autocorrelation. For low-frequency data, multiple sub-sample replication (i.e. bootstrapping; Efron, 1979) can provide an alternative measure of chronology confidence (e.g. Guiot, 1990; Cook, 1990).

$RBAR$ for a group of trees could, in theory, range from -1.0 to 1.0, though, in practise, only positive values are meaningful (negative values indicating a complete lack of common growth forcing). The higher the value, the stronger is the underlying common signal; then the less variance within each series represents noise and fewer series need to be averaged to reduce the noise remaining in the final mean chronology to an "acceptable" level (see later).

5.4.2 Expressed Population Signal

The statistical quality of a mean chronology may be gauged by calculating the degree to which it represents the hypothetical perfect (noise-free) chronology (i.e. one that is infinitely replicated). This is given by the Expressed Population Signal (EPS) as

$$EPS(t) = \frac{t \cdot RBAR}{t \cdot RBAR + (1 - RBAR)} \tag{5.1}$$

where t is the number of tree series averaged and $RBAR$ is the mean inter-tree correlation coefficient (Figure 5.1a). For further discussion of the relationship between EPS and earlier defined chronology signals, see Briffa and Jones (1990). EPS ranges from zero to 1.0 (ignoring negative $RBAR$ values - see above). There is no unequivocal answer to what value of EPS constitutes "acceptable" confidence, but a value of 0.85 has been tentatively suggested as desirable (Wigley et al., 1984). It is clear (cf. Figure 5.1a) that above about 0.85-0.90 the increase in EPS with increasing replication slows markedly, especially for high $RBAR$ values.

5.4.3 Subsample Signal Strength

It is often the case that tree-ring chronologies are progressively less-well replicated further back in time when older material becomes more difficult to locate. The additional uncertainty, above that represented in the better-replicated sections, can be measured using the Subsample Signal Strength (SSS) calculated as

$$SSS = \frac{t'[1 + (t - 1)RBAR]}{t[1 + (t' - 1)RBAR]} \tag{5.2}$$

Figure 5.1: a) *Expressed Population Signal (EPS) plotted as a function of chronology replication for several example RBAR values. High empirical signal, i.e. high RBAR, means that fewer samples are required to maintain a high quality chronology.*
b) *The replication (t') necessary in a part of a chronology to maintain the Subsample Signal Strength above a value of 0.85. Examples are shown for a range of RBAR values.*

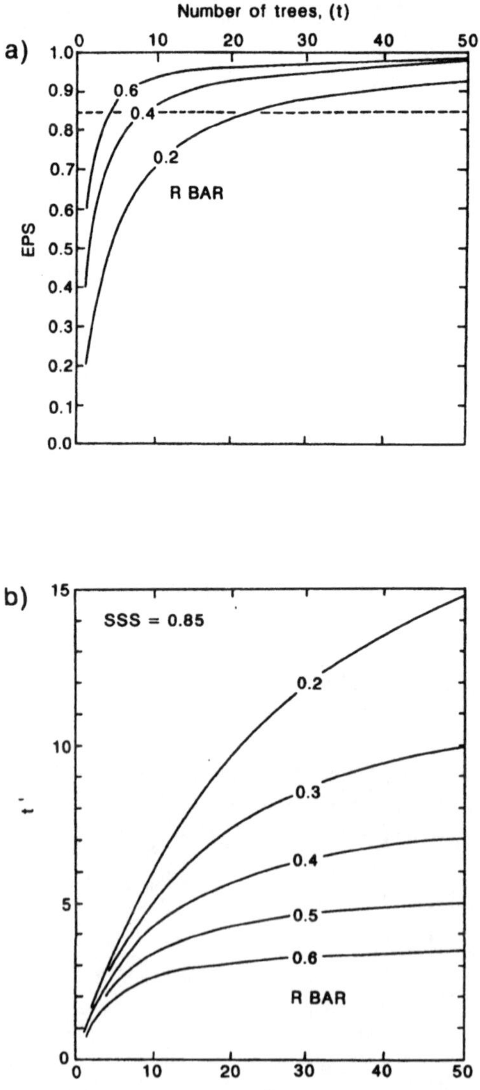

where t and t' are the number of sample series in the optimum and sub-sample parts of the chronology respectively (Figure 5.1b). This can be expressed in terms of the EPS values calculated for both sections of chronology, i.e.

$$SSS = \frac{EPS(t')}{EPS(t)} \tag{5.3}$$

SSS (like EPS) ranges from zero to 1.0 and, again, the question of an acceptable level is somewhat subjective. In the context of climate reconstruction, it is desirable to maintain SSS at a level well above the level of variance explained by climate calibration (see later). This question is discussed in Wigley et al. (1984) and Briffa and Jones (1990).

It should be noted that $RBAR$ (and consequently EPS and SSS) are generally considered as single parameters representative of a chronology as a whole. $RBAR$ is an unbiased estimator of the population parameter and will give an accurate estimate provided it is calculated using data for at least 5-10 series (Briffa, 1984). However, when calculated for a moving time window, noticeable variations can occur. Provided these are based on reasonable sample sizes (say a minimum of 30-50 years), they may represent genuine temporal variability in the strength of common forcing and this should be recognised when considering a single, overall chronology value (Briffa et al., 1987; Briffa and Jones, 1990).

5.4.4 Wider Relevance of Chronology Signal

The concept of a replicated signal, and the requirement to quantify the degree to which such signals are contaminated with noise in constructed time series obviously has relevance for all proxy data. This is dramatically illustrated by the recent discussion of oxygen isotope data from closely adjacent (within 30 km) cores on the summit of the Greenland ice sheet (e.g. Dansgaard et al., 1993; Taylor et al., 1992). For many years, similar "high-resolution" data from another Greenland core (e.g. Dansgaard et al., 1975) were interpreted as evidence of decadal-to-century variability in annual temperature and strongly influenced our perception of the climate history of the northern North Atlantic during recent millennia (e.g. Lamb, 1977). It has now been shown (Grootes et al., 1993) that the correlation in the two parallel $\delta^{18}O$ series from Summit, despite good agreement on the millennial timescale, is very poor over the last 10,000 years. Whether this is the result of a lack of absolute (i.e. demonstrably annual) dating control in the two cores, or an indication of a very poor level of common forcing (which must imply a high degree of noise in one, or both, series), is not yet clear. However, such information is extremely valuable in warning that previous interpretation of single core data, in terms of detailed temperature information over recent centuries, must now be considered to be at least potentially suspect.

Despite the logistic and financial constraints, it is to be hoped that investigations of other proxy sources, such as lake varves and corals, will also address the need to produce replicate data series.

5.5 "Standardization" and Its Implications for Judging Theoretical Signal

5.5.1 Theoretical Chronology Signal

In our discussion until now, we have recognised only a statistically-manifest signal contained in each of a number of tree-ring series and expressed in a mean chronology, but it is important also to appreciate a different concept: that of theoretical signal in the mean chronology. This is an expression of some forcing which is of particular interest to the investigator and will differ according to the aims of the study. This theoretical signal may be very different from the empirical signal represented in the data. Top of the dendroclimatologist's list of important theoretical tree-growth signals is obviously that of (some form of) climate, but other possible signals include the effects of growth fertilization (e.g. through regional nitrogen deposition or increasing atmospheric CO_2), pollution, insect defoliation, fire and management practices. Each of these may simultaneously affect the growth of all trees within a given site or region. These processes operate on different timescales and one may mask or amplify the effects of others. Careful consideration of the nature of the signal under investigation, particularly its time-series characteristics, can suggest ways of modifying original tree-growth measurement series so as to enhance the signal by removing variance known not to be associated with it.

5.5.2 Standardization of "Raw" Data Measurements

The process of selective removal of unwanted variance in raw measurement series, prior to their being averaged to form mean chronologies, is known in dendroclimatology as "standardization". Expansive discussions of the rationale and techniques used to standardize tree-ring series may be found in Fritts (1976); Fritts and Swetnam (1986); Briffa et al. (1987); Cook et al. (1990a) and many references therein.

Standardization is most commonly employed to remove the effects of tree aging that are almost universally apparent in tree-ring (ring-width or ring densitometric) measurement series. Simply stated, where they are not suppressed by competition, older trees generally put on thinner (and less dense) rings than younger trees as net primary productivity is progressively spread around an increasing ring circumference. The pattern of decreasing ring width often approximates to a negative exponential in ring-width series and to a more linear trend in maximum latewood density series (see Bräker, 1981; Briffa et al., 1992a). Decadal and interannual variability is superimposed on this clearly non-climatic trend.

Were it possible to average data from an unlimited (or at least very large) number of tree-ring series, randomly distributed through time, this age bias would cancel out. This is rarely, if ever, the case, however, and many "raw-data" average chronologies contain significant bias resulting from coincidence in the temporal distribution of equal-age samples, particularly the concentration of young trees forming the early sections.

Many standardization techniques have been developed to "detrend" raw tree-ring series during chronology construction. It is not possible to give even a brief description of the mathematical details of these here, but a good review is given in Cook et al. (1990a). All the methods model the unwanted (generally low-frequency) variance in the raw measurement series (using some deterministic equation or data dependent filter) and transform the data to a series of indices by taking either the difference (or quotients) of actual minus (divided by) modelled estimate for each year. The standardized tree-growth indices are generally stationary and have generally constant variance, both of which are desirable characteristics if data are to be averaged to form a mean chronology. (In practice, other more robust techniques are also used to form the mean value function and to standardize chronology variance; see Cook et al., 1990b; Shiyatov et al., 1990.) The essential implication of all of these techniques, however, is that they modify the spectral characteristics of the tree-ring indices, imposing a low-frequency limit on the variability expressed. The degree to which long-timescale variability is therefore removed from all sample series and from the mean chronology constructed from them (and necessarily from any subsequent climate reconstruction made using this chronology) is dependent on the lengths of the sample series and the particular standardization technique employed (e.g. Briffa et al., 1987; Cook and Briffa, 1990).

Different methods are used for different purposes. Where the only concern is to crossdate different series (by assessing the statistical significance of correlations calculated between series at many overlap positions), it is desirable to remove all relatively low-frequency variance in the series, perhaps that corresponding to variability on timescales longer than a few decades. This generally increases (even optimises) $RBAR$ in the resulting indices and increases the power of the comparison tests by eliminating chance, spuriously high, correlations (i.e. those based on coincident trend) that would suggest an incorrect match between two series (Munro, 1984; Wigley et al., 1987). Having dated the series, the original measurement data might, however, be standardized using a technique designed to maintain much longer timescale variations in the final chronology, so preserving the potential for reconstructing decadal-to-century timescale variations in some other signal, such as temperature. This might then lead to a lower $RBAR$ (and EPS) in the chronology.

In constructing chronologies for dendroclimatic reconstruction work, there is often a trade-off between loss of generally low-frequency variance which might represent the theoretical signal and an increase in chronology confidence. If the required theoretical signal involves long-timescale variability, a very conservative approach must be adopted when standardizing, but the cost might be low statistical signal which would then require high sample replication to maintain chronology confidence. Where it is not possible to achieve this, the uncertainty inherent in conservatively standardized chronologies should be explicitly recognised as contributing to uncertainty in climate reconstructions produced from them.

These uncertainties, and the general implications of tree-ring standardization, are recognised by dendroclimatologists, but they are often not explicit in the literature. It is not widely appreciated, outside of the field, that published climate reconstructions do not necessarily represent the complete spectrum of climate forcing of tree growth. Just because a tree-ring chronology (or a climate reconstruction derived from it) is hundreds or even thousands of years long, it does not follow that century- or millennial-timescale variability is represented in it. The standardization of the primary data may preclude this.

5.5.3 General Relevance of the "Standardization" Problem

In tree-ring work the need to standardize measurement series is explicit. However, in other proxy data fields, it may be necessary to recognize the modification of long timescale signals by processes unrelated to the forcing of interest. Lake varve thickness may be moderated slowly by changes in the surface area of deposition or alteration of the surface run off into the lake (as could occur because of isostacy or surrounding vegetation change). Similarly the thickness of annual layers in ice cores may become thinner at depth or be affected by ice flow within an ice sheet because of compaction or bottom topography. Growth of marine corals might be influenced by long-term land run-off changes that are unrelated to climate change or may be difficult to interpret because of their complex three-dimensional shape. There are only speculations, but as replicated series of data derived from such sources become increasingly available for centuries and millennia, it will be incumbent on researchers in these fields to establish the degree to which they may need to be standardized to take account of these (or other factors) and the degree to which such standardization may limit the timescales of climate variability that can be subsequently reconstructed from them.

5.6 Quantifying Climate Signals in Chronologies

The principal consideration in any tree sampling strategy is to select sites where the climate variable of interest is expected to exert a strong limitation on seasonal growth (La Marche, 1982). Whatever the expectation, however, it is still necessary to demonstrate the actual strength and character of the relationship between the tree-ring chronology and the climate data of interest. This is achieved empirically by the use of regression analyses. For general references that review various techniques used in dendroclimatology see for example: Fritts (1976, 1991); Guiot (1990); Cook et al. (1994).

Most analyses use series of monthly mean climate parameters, principally because of their relevance to climatologists and their general availability. However, this must always be a compromise as trees do not respond directly to such crude variables. Rather they integrate much more subtle changes in many micro-environmental factors such as available soil moisture, radiation, humidity etc., all of which vary over very short timescales (as short as minutes and even seconds). These variations are only indirectly represented by available meteorological data.

Linear equations are used even though they may not be entirely appropriate for modelling the full range of tree-growth/climate relationships. This further compromise is necessary because non-linear expressions could not be inverted to give unique estimates of past climate. Acutely aware of these facts and of the pitfalls that await the unwary user of regression analysis (Draper and Smith, 1981; Rencher and Pun, 1980), dendroclimatologists employ a rigorous approach involving the validation or "verification" of fitted or "calibrated" regressions.

5.6.1 Calibration of Theoretical Signal

Calibration in this context may involve deriving an equation which expresses variability in some tree-growth parameter as function of an ensemble of climate variables (e.g., monthly mean temperatures and precipitation totals). This is the so-called response function: a set of regression coefficients whose signs and magnitudes suggest the pattern of climate influence on growth (Figure 5.2). Discussions of the use of response functions may be found in Fritts et al. (1971); Fritts (1976); Guiot et al. (1982); Blasing et al. (1984); Briffa and Cook (1990) and Serre-Bachet and Tessier (1990). Response function analyses are often used to identify the climate variable or season over which this variable may be averaged to produce an optimum "season" for reconstruction. The equation which expresses this variable as a function of tree growth (at one or more sites) is the "transfer function": a set of regression weights which, when multiplied by past tree growth data, provide estimates of past climate (Fritts et al., 1971; Blasing, 1978; Lofgren and Hunt, 1982).

A range of regression techniques is employed: from simple linear regression, through multiple regression with a single predictand up to the most complex spatial problem where multiple predictands are estimated using multiple predictors.

The use of multiple regression is most often the norm in dendroclimatology and the danger of artificial predictability in calibrated regressions is very

Figure 5.2: *Examples of different tree-ring response functions: one based on simple correlation coefficients of tree growth with monthly mean temperatures, the other on principal components regression using the same temperature data as predictors. The growth variables in each of the two years immediately preceding the growing season are included among the predictors to quantify the persistence in the tree-ring timeseries: The autocorrelations with lags one and two years are labelled A1 and A2.*
The sign and magnitude of the coefficients provide a simple picture of the tree-growth/climate relationships. Calculating response functions for different periods gives an indication of the temporal stability of the response pattern.
The examples shown are for a site in the northern Ural Mountains, Siberia. Significant ($p = 0.05$) coefficients are indicated by the dashed lines (simple correlations) and dots (PC regression). The summer response in ring width is predominantly in June and July. For density it is over a longer May-September season.

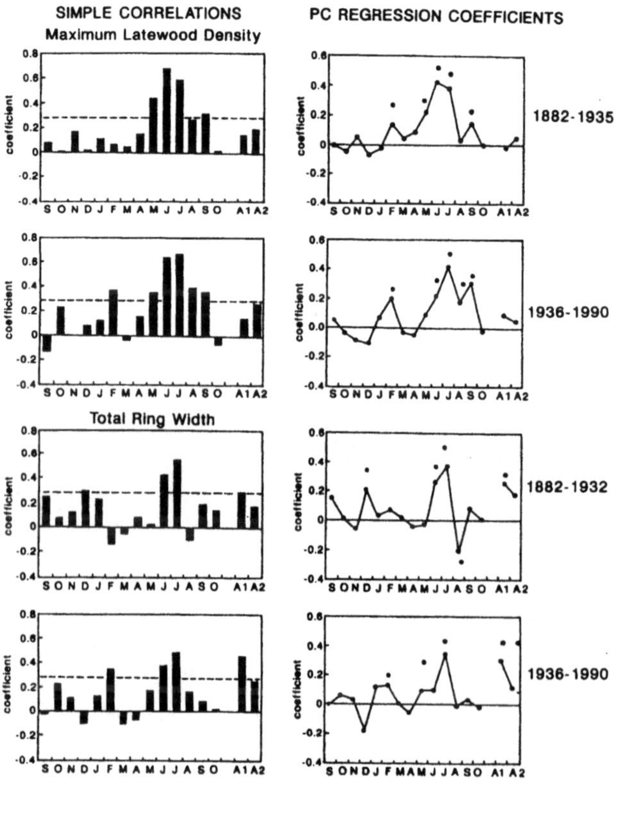

real. Even when using tree-ring data from a single site to estimate a single climate variable, simple regression may not be appropriate. This is because tree growth in one year is often influenced by climate variability over one or more previous years (seasons). Prior "growing-season" climate can predispose the tree's response to current climate in many ways, such as through its effect on stored resources, leaf development (and hence photosynthetic potential), frost hardiness, drought resistance etc. Attempts to model such effects, often manifest as statistical persistence in tree-ring time series, can involve the use of multiple tree-ring predictors so that climate in one year is estimated as a function of tree growth in some combination of preceding, current and following years. Where the dependent data series displays significant persistence, this lagged-predictor approach has the potential to reduce the regression degrees of freedom severely.

In addition to lagged predictors, reconstructions of single area-average climate series might use a number of chronology predictors from different chronology sites. The typically restricted length of many observational climate series (probably no more than 100 years and often considerably less) then exacerbates the problem of low degrees of freedom.

To reduce the effective number of predictors in such regressions, principal components analysis, PCA (see Section 13.3) is often used on the predictor data set. Only the amplitudes[1] of the "significant" predictor components are then offered as candidate predictors. As is frequently the case in PC regression (e.g. Preisendorfer et al., 1981) various criteria have been used to judge "significance" (e.g. see Cook et al., 1994; see also the discussion in Section 13.3.3). High intercorrelation among the original predictors ensures that a notable reduction in the number of remaining "candidate" PC predictors and a corresponding improvement in the regression degrees of freedom is achieved. However, in many applications this is often followed by further elimination of candidate predictors, with only those PC amplitudes correlating with the predictand better than some predetermined significance threshold being retained. Because this further reduction in predictors is clearly based on an a posteriori decision, however, it would not be appropriate to assume a further corresponding improvement in the regression degrees of freedom (Barnett and Hasselman, 1979).

The most complex reconstruction scenario is that in which a network or grid of climate data series is directly expressed as a function of tree-ring variability over a number of sites (Figure 5.3). The final transfer function matrix (comprising a set of individual equations, one for each predictand point expressed as a function of variability at each chronology site) is derived through the initial orthogonalization of the dependent network; selection of "significant" predictand amplitudes; calibration of only the significant predictand components - either individually (i.e., Orthogonal Spatial Regression; Briffa

[1]In the terminology of Chapter 13 the predictors are the EOFs $\vec{p}^{\,i}$ and the amplitudes are the EOF coefficients α_i.

et al., 1983, 1986) or as canonical variates (Fritts et al., 1971) using only the significant predictor PCs as candidate predictors; and ultimate backtransformation of the selected PC regression equations (e.g. Blasing, 1978). For a comprehensive review of the spatial reconstruction problem, see Cook et al. (1994).

In common with one-dimensional variable reconstruction, there is the potential for severe over-calibration of the two-dimensional transfer function. Diagnostic statistics calculated using the calibration data, when based on multiple predictors, invariably overestimate the true predictive power of the calibrated equations. This is true also for measures of the goodness of fit between actual and estimated data that take account of the effective degrees of freedom (Cramer, 1987; Helland, 1987) even assuming that these can be defined properly (as discussed above). In dendroclimatology, experience has shown that a much more realistic idea of the likely predictive power of the calibrated regression is provided by model verification (Figure 5.4).

5.6.2 Verification of Calibrated Relationships

Verification is the testing of the validity or effectiveness of the calibrated equation by comparing the regression estimates of the predictand data outside of the period used to fit the equation(s). Observational data (tree-ring data where the response function is being verified, and climate data in the case of the transfer function) are withheld from the calibration especially for this purpose. A number of statistics are commonly employed to quantify verification performance. General reviews can be found in Fritts (1976); Gordon (1982); Fritts et al. (1990) and Cook et al. (1994). They range from a simple non-parametric test of very high-frequency association, through to a true measure of the variance in common between the estimated and observed data. The power of these tests, however, is proportional to the length of verification period. Where observed (instrumental) climate series are relatively long (say 100 years), cross calibration/verification can be used on separate halves of the data alternately and a reasonable impression is gained of the fidelity with which interannual and decadal timescale variability is captured.

In situations where climate data for only 40-60 years are available, either very short verification periods are used or sub-sample techniques (e.g. the jackknife - see Gordon, 1982) are employed. In both cases, the verification is strictly limited to testing the highest frequency (i.e. interannual) performance of the calibration model(s) and the stability of the form of the model.

A powerful procedure for testing model stability is to compare the alternative reconstructions themselves (i.e. early estimates derived separately from equations fitted on each half of the observational data). Any serious differences in the alternative reconstructions would cast doubt on the veracity of one or both models (e.g. Briffa et al., 1992b).

It is important to remember that even where verification is performed using independent observational data, parametric statistics assume normality

Figure 5.3: *A schematic representation of the Orthogonal Spatial Regression (OSR) technique for direct dendroclimatic reconstruction of geographical patterns of climate variability. Further details are given in the text.*

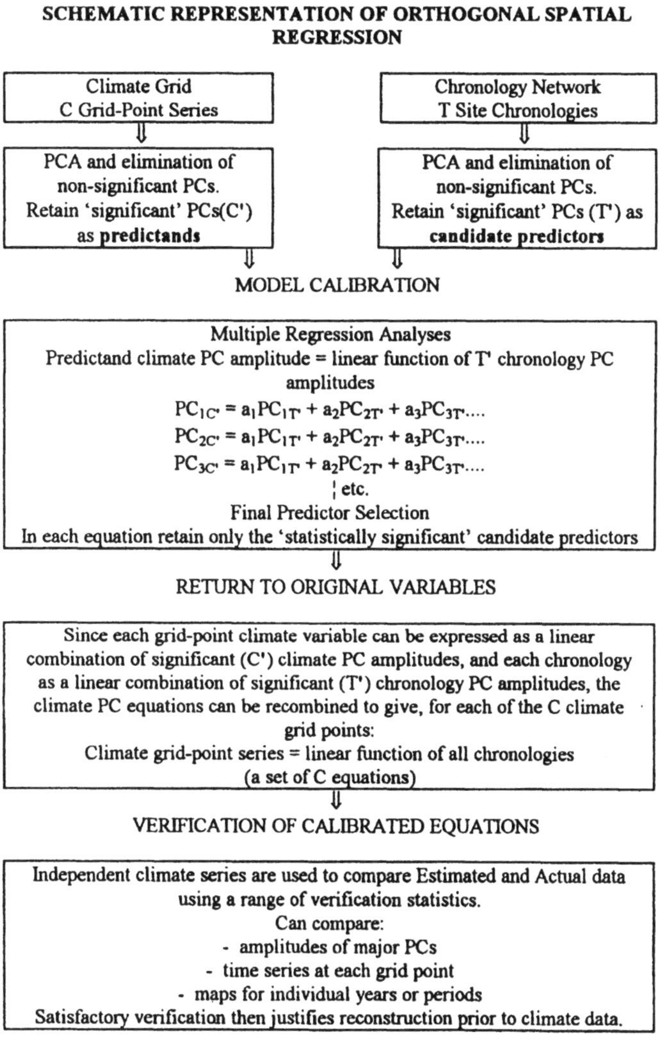

Figure 5.4: *An illustration of "over calibration" in a transfer function for the reconstruction of western N. American summer temperatures based on a network of 53 chronologies. The calibration* R^2 *map indicates higher levels of explained variance than are achieved in verification against independent temperature data. For further details see Briffa et al. (1992b).*

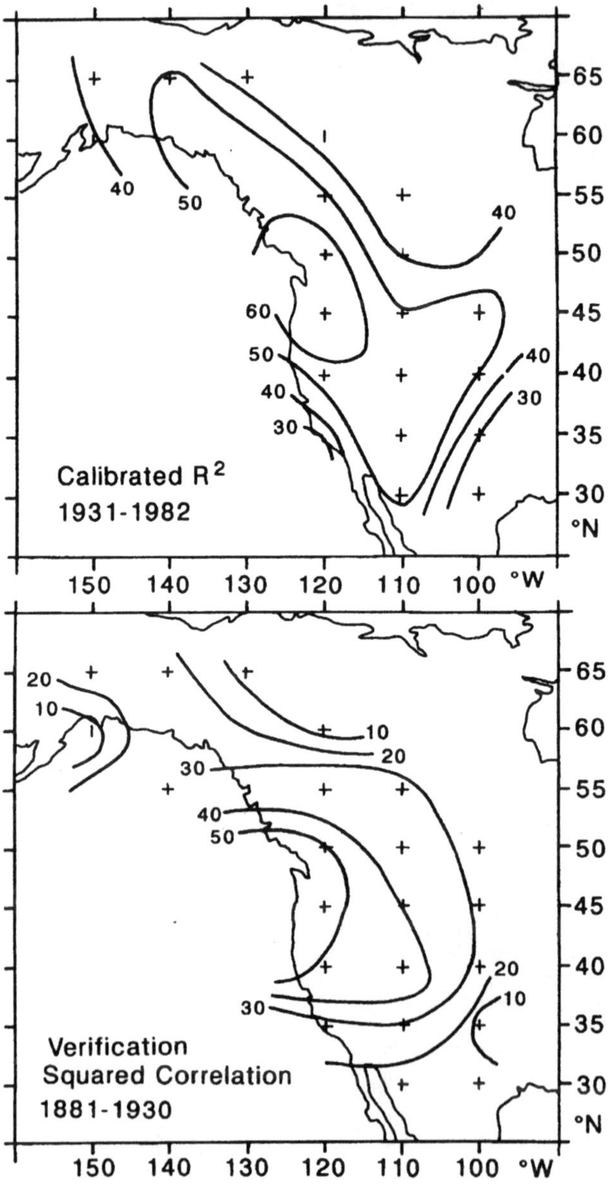

in the predictand and estimated data and subsequent assumptions on the likely predictive performance of the regression model are still strictly valid only for regression estimates that fall within the range of the observational (predictand) data. Estimates that fall outside of this range are extrapolations and their accuracy is dependent on the assumption that such extrapolation is valid. The more extreme the extrapolation the greater is the likelihood that these estimates are in error (Graumlich and Brubaker, 1986).

Verification of the theoretical (or assumed) temporal signal is almost non-existent outside of dendroclimatology (though it could be argued, with some justification, that no other proxy science is as vulnerable to the dangers of overcalibration of regression models). Nevertheless, other forms of proxy data have been interpreted on the basis of calibration of multiple regression statistics with little or no formal testing of the regression veracity.

5.7 Discussion

The intention in writing this chapter was not to catalogue all of the numerous mathematical techniques which are used to crossdate and standardize tree-ring chronologies, or to calibrate and verify dendroclimatic reconstructions. To do so would have resulted in many pages of abstract equations with little apparent relevance for students in other fields. Those who seek the mathematical corpus may delve into the numerous references cited. Neither is this chapter an attempt to list even a cross section of the hundreds of reconstructions of climate-related variables that form the product of the science to date. A recent selection is provided in Bradley and Jones (1992). Instead, this chapter has discussed several concepts that underpin dendroclimatology and which have important implications for viewing published work in the context of wider palaeoclimatic research. It is hoped that the relevance of these concepts to other fields of palaeoclimatic research is also apparent. Lack of space has prevented discussion of some questions listed at the start of the chapter: specifically of seasonality and spatial representativeness of data. For some discussion on these points see Briffa and Jones (1993) and Hughes and Diaz (1994).

Before closing the discussion, however, it is necessary to highlight a fundamental tenet of dendroclimatic research which, while always assumed, is frequently difficult, perhaps impossible to demonstrate in practice: that of uniformitarianism. In a dendroclimatic context, this states that "the processes which control tree growth at present have operated unchanged in the past".

Even when sampling living trees one can envisage the possibility that the characteristics of the site (evolving soil fertility; changes in water holding capacity of the soil etc.) may have changed during the life of the trees and modified their response to climate. The actual process of growth itself (root and foliage development, changing competitional status etc.) can impart age-related bias in a tree's climate response. Fortunately, this bias is small and replicate samples of different age will overcome it. However, the age of some trees can be extremely great (e.g. Stahle et al., 1988; Villalba, 1990; Lara and Villalba, 1993) and effort is increasingly being directed towards the development and climatic interpretation of multi-millennial chronologies, perhaps constructed from a combination of living, historical and even sub-fossil material (e.g. Schweingruber et al., 1988; Briffa et al., 1992a; Graybill and Shiyatov, 1992; Cook et al., 1992). In such chronologies there must always be uncertainty regarding the long-term stability of (non-climate) environmental influences or differing climate sensitivity due to inhomogeneity in the site characteristics of the samples. Attempts to reconstruct long timescale climate change (centuries to millennia) using tree-ring data should focus attention on these problems, as well as on the importance of standardization and the recognition that the reconstructions may represent extrapolations beyond the range of the calibration (and verification) data.

Notwithstanding these difficulties, the construction of millennial tree-ring chronologies offers the exciting prospect of reconstructing inter-annual, decadal, and century timescale variability, and the potential for comparing century-to-millennial climate inferences with other less well resolved proxy evidence such as that provided by palynology and glaciology.

5.8 Conclusions

- When interpreting proxy data it is necessary to appreciate the difference between empirical and theoretical signal.
- Empirical signal is a statistical property of the data and varies according to how they are processed. Its expression is dependent on replication.
- Theoretical signal needs to be established with care and can sometimes be enhanced or masked by chronology processing methods and by other non-climate forcing.
- Theoretical signal should be demonstrated and quantified, e.g. through regression. It may be limited to a particular "season".
- Regression skill should be assessed realistically, ideally through the use of independent validation, the assumptions and limitations of which should be recognised.
- Regardless of regression skill, remember that "standardization" of the primary data may restrict the timescales represented in the reconstruction.
- Always remember the underlying assumption of uniformitarianism in the response of data to climate forcing. This may be hard or impossible to prove.

Chapter 6

Analysing the Boreal Summer Relationship Between Worldwide Sea-Surface Temperature and Atmospheric Variability

by M. Neil Ward

6.1 Introduction

The ocean and the atmosphere can be viewed as two sub-systems of the earth's environmental system. The ocean and atmosphere interact at the air-sea boundary. This chapter is oriented towards analysing the impact of the ocean on the atmosphere. The timescales analysed are from the seasonal to the multidecadal. Statistical analysis is playing an important role in advancing our knowledge of air-sea interaction and its influence on worldwide

Acknowledgments: Mike Hulme kindly provided rainfall data under UK DoE Contract PECD 7/12/78. Discussions with Chris Folland, Brian Hoskins and Hadley Centre colleagues have enhanced this Chapter and Bob Livezey made helpful comments on an earlier version of the text.

climate variability. So this chapter is particularly well-suited for illustrating an application of statistics in climatology.

Section 6.2 discusses the physical basis for the impact of the ocean on the atmosphere. Section 6.3 discusses some of the observed sea-surface temperature (SST) variability, and identifies three modes of SST variations. The statistical relationship between each of these modes and the near-surface marine atmosphere is described in Section 6.4. It is believed that SST variability affects atmospheric variability over continental land masses as well as the local marine atmosphere. Statistical evidence for this is provided in Section 6.5, which describes the statistical relationship of rainfall in the Sahel region of Africa with the modes of variability discussed in Sections 6.3 and 6.4.

6.2 Physical Basis for Sea-Surface Temperature Forcing of the Atmosphere

6.2.1 Tropics

Lindzen and Nigam (1987) developed a simple model (hereafter referred to as the LN model) assuming that the tropical marine boundary layer was well mixed. They therefore argued that temperature gradients in the atmospheric boundary layer would reflect SST gradients. To look solely at the component of forcing of the boundary layer circulation from this effect of SST gradients, as opposed to forcing from upper level circulations, they set the top of the boundary layer fixed. It then follows from the hydrostatic relation that surface pressure gradients simply reflect SST gradients. Neglecting some smaller terms (Neelin, 1989; Philander, 1990), near-surface wind in the LN model is related to pressure gradients:

$$ku - fv + \frac{\partial \Phi}{\partial x} = 0 \qquad (6.1)$$

$$kv + fu + \frac{\partial \Phi}{\partial y} = 0 \qquad (6.2)$$

where Φ is geopotential, k is a coefficient of surface resistance representing friction and f is the Coriolis parameter. These equations therefore predict that near-surface divergence of wind will be a function of the Laplacian of the SST (this also implies that anomalous divergence will be a function of the Laplacian of the anomalous SST pattern), plus a contribution due to the variation of f with latitude. The Laplacian SST relationship will tend to give maximum anomalous convergence over maxima in the SST anomaly field. Two potentially important processes, both of which will tie near-surface convergence anomalies more closely to maxima in SST actuals, are not explicitly included in the LN model. Firstly, the lower and upper levels of the atmosphere are coupled, and diabatic heating at upper levels as air rises and condenses moisture can lead to a positive feed-back intensifying the circulation. This clearly will tend to occur only in regions of significant deep penetrative convection with ample moisture supply, and these regions tend to be in regions of SST maxima. Secondly, and related, convergence of moisture is perhaps most crucial for the large scale atmospheric response. Convergence of moisture will be influenced by the SLP pattern, but also by such factors as the magnitude of the SST (higher SST leads to a warmer near-surface atmosphere that can hold more water in gaseous form), the air-mass characteristics such as stability and the wind strength.

The release of latent heat by condensation of moisture into cloud droplets in tropical regions of strong convection can lead to remote atmospheric responses. This has been shown in simple models of the tropical atmosphere (e.g. Gill, 1982). So an anomalous SST maximum in the western Pacific may lead to enhanced convection and latent heat release locally, which in turn can influence atmospheric variability in many regions of the globe. Connections of tropical heating to extratropical circulation have also been proposed (e.g. Hoskins and Karoly, 1981).

6.2.2 Extratropics

The impact of extratropical SST variations on atmospheric variability is more controversial. Compared to the tropics, atmospheric variability in mid-latitudes seems to have a stronger component generated by internal atmospheric dynamics, with the SST often in the main responding to, rather than significantly altering, the low frequency atmospheric variability. Nonetheless, evidence has been presented for the influence of variations in SST gradients on atmospheric baroclinicity in important cyclogenesis regions such as near Newfoundland (Palmer and Sun, 1985). These ideas are not developed in this chapter, but evidence is presented of an SST fluctuation that involves changes in the relative temperature of the Northern Hemisphere compared to the Southern Hemisphere. It has been suggested that such swings in this interhemispheric temperature gradient lead to changes in the location and intensity of the inter-tropical convergence zone (ITCZ) (Folland et al., 1986a; Street-Perrott and Perrott, 1990). A warmer Northern Hemisphere draws the ITCZ further north and/or makes it more active, bringing more rain to Northern Hemisphere semi-arid regions such as the Sahel region of Africa.

6.3 Characteristic Patterns of Global Sea Surface Temperature EOFs and Rotated EOFs

6.3.1 Introduction

Various statistical methods have been used to identify the nature of temporal and spatial SST variability. EOF analysis of SST anomaly data identifies patterns of co-variability in the data, and provides time-series (time coefficients) that describe each pattern's temporal variability. This Section describes an EOF analysis of global SST data (Folland et al., 1991), and presents three of the EOF patterns that turn out to have strong atmospheric variability associated with them (discussed in Sections 6.4-6.5).

6.3.2 SST Data

Ship observations of SST are now compiled into large computerised datasets. The data used here are taken from the Meteorological Office Historical SST dataset (MOHSST) (Bottomley et al., 1990). Prior to about 1942, observations were taken using uninsulated canvas buckets, and this led to measured temperatures being too low. Corrections, which are of the order of 0.3^oC, have been calculated and applied to the data used here (Folland, 1991). The corrected data are averaged into seasonal anomalies from a 1951-80 climatology for each 10° *lat* $\times$ 10° *long* ocean grid-box. Ship observations do not cover the whole ocean, so the data set has no data for some grid-boxes in some seasons. Data coverage is particularly poor in the Southern Ocean. The EOF analysis summarised in this Section (Folland et al., 1991) used data for 1901-80. All grid-boxes with less than 60% of data present 1901-80 were excluded from the analysis. There were 297 grid-boxes that qualified for the analysis. In EOF analysis, the time-series for each grid-box must have no missing data. So missing data in the 297 time-series were interpolated using Chebychev Polynomials.

6.3.3 EOF Method

The concept of Empirical Orthogonal Functions (EOFs) is spelled out in some detail in Section 13.3, here only a brief summary is given to clarify the notations. The basic EOF model can be written

$$\alpha_k(t) = \vec{\mathbf{X}}_t^T \vec{p}^k \tag{6.3}$$

Here,

- $\vec{\mathbf{X}}_t$ is the m-dimensional vector of the sea-surface temperature anomalies (i.e., each time-series has its mean subtracted prior to analysis) containing values for season t at the $m = 297$ 10° *lat* $\times$ 10° *long* grid-boxes. For each year, the four seasons DJF, MAM, JJA and SON are formed.

- $\vec{p}^k$ is the kth EOF of $\vec{X}$, i.e., it is the kth eigenvector of the estimated covariance matrix of $\vec{X}$. The EOFs are normalized so that $(\vec{p}^k)^T\vec{p}^k = 1$. The EOFs have been derived from seasonal time series (4 values per year) for 1901-1980.

- α_k contains the time coefficients (which form a time-series) for the kth EOF pattern. The coefficient $\alpha_k(t)$ can be viewed as measuring the strength of EOF pattern k in the observed SST anomaly field for season t.

- After the EOFs $\vec{p}^k$ have been determined, time coefficients α_k can be calculated, by means of (6.3), for any dataset of $10^o \times 10^o$ SST anomalies $\vec{X}$. So for study of the boreal summer season in this chapter, a dataset of 1949-90 JAS SST anomalies has been used to yield α_k time coefficients for JAS 1949-90.

6.3.4 EOFs $\vec{p}^1$-$\vec{p}^3$

The weights in $\vec{p}^1$ are all the same sign, so the EOF time-coefficient represents a weighted average of SST anomalies across the analysis domain. The time-coefficients contain an upward trend, describing the pattern of global warming through the historical period. This is not the subject of this chapter, so $\vec{p}^1$ is not discussed further.

The weights of $\vec{p}^2$ describe the SST covariance associated with ENSOThe weights in the EOF have been plotted on a map and contoured to show the pattern of $\vec{p}^2$ (Figure 6.1a). The pattern suggests that warm events in the central/eastern Pacific usually accompany cold SST anomalies in the extratropical Pacific and warm anomalies in the Indian Ocean. The EOF suggests that the reverse pattern of anomalies occurs in cold events. Note that the two opposite phases of a mode described by an EOF may have some asymmetry, but this cannot be deduced from an EOF analysis like that reported here. The JAS $\vec{p}^2$ time coefficients $\alpha_2(t)$ (Figure 6.1a) clearly describe the known major warm and cold events.

EOF $\vec{p}^3$ (Figure 6.1b) has negative weights in the North Atlantic and North Pacific and positive weights in the South Atlantic, Indian Ocean and southwestern Pacific. The time coefficients $\alpha_3(t)$ for $\vec{p}^3$ (Figure 6.1b) have strong variability on decadal to multidecadal timescales. In particular, they describe a large climate fluctuation that has occurred over the last 45 years, with the recent 25 years marked by a warmer Southern Hemisphere (including all the Indian Ocean) and cooler Northern Hemisphere.

Figure 6.1: *Top four diagrams ((a) and (b)): EOFs $\vec{p}^{\,k}$ (derived from all seasons, 1901-80 data) and time coefficients $\alpha_k(t)$ for $k = 2, 3$ and $t =$ all JAS-seasons 1949 - 1990. Smooth line is low pass filter (see text). These two EOFs are statistically distinct according to the test of North et al. (1982), but see discussion of effective sample size for the test in Folland et al. (1991). $\vec{p}^{\,2}$ represents 7.8% of variance and $\vec{p}^{\,3}$ 3.8%.*
Bottom two diagrams (c): 2nd Varimax-rotated EOF $\vec{p}_R^{\,2}$, derived from the raw EOFs $\vec{p}^{\,4}$ to $\vec{p}^{\,13}$, and its time series $\alpha_2^R(t)$. $\vec{p}_R^{\,2}$ represents 2.5% of the original variance. Contour map units are in thousands.

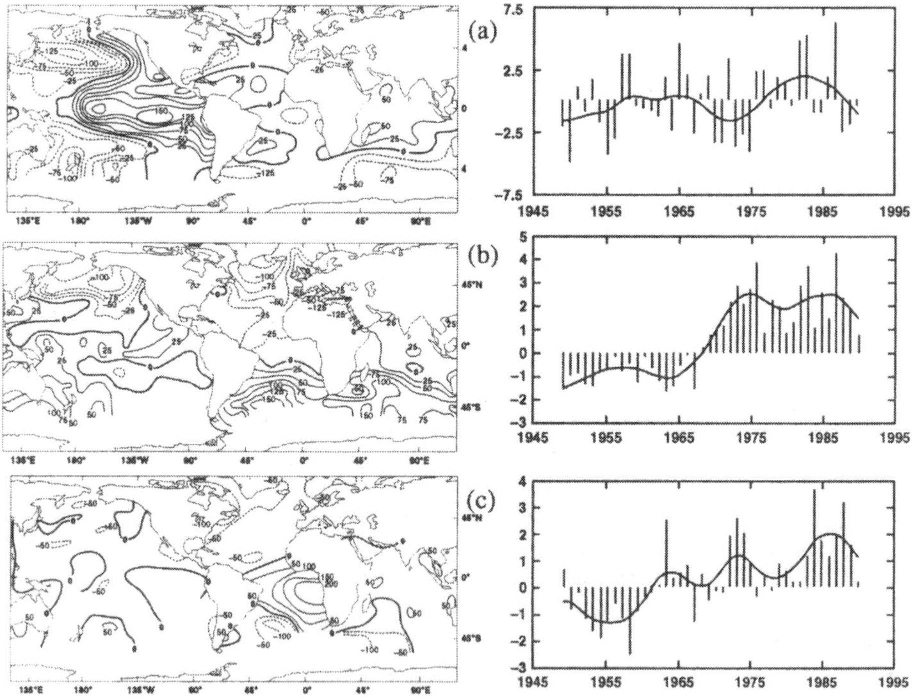

6.3.5 Rotation of EOFs

The patterns of EOFs $\vec{p}^{(k)}$ with indices $k > 3$ were more noisy and less global in scale. Rotation (cf. 13.2) can often yield clearer and more stable patterns, especially when the patterns of interest are on a scale that is smaller than the analysis domain. There are various criteria used for rotation, the most common being VARIMAX (13.12). Here, EOFs $\vec{p}^{1}$-$\vec{p}^{3}$ were left unrotated because they are readily interpretable as global scale patterns with physical significance. So EOFs $\vec{p}^{4}$-$\vec{p}^{13}$ were VARIMAX rotated. Inspection of the eigenvalues revealed a sharp drop in variance explained between EOFs $\vec{p}^{13}$ and $\vec{p}^{14}$, and this guided the choice of excluding EOFs $\vec{p}^{k}$ with $k \geq 14$ (see O'Lenic and Livezey, 1988, for discussion of which EOFs to retain in rotation).

The first rotated EOF ($\vec{p}_R^1$) was a pattern mainly in the extratropical North Atlantic. It seems to mainly reflect a response to the North Atlantic Oscillation atmospheric variation (not shown here). It is not discussed further in this chapter.

$\vec{p}_R^2$ (Figure 6.1c) has strong positive weights in the southern and eastern tropical Atlantic. There is a slight opposition of weight in the northern tropical Atlantic. The time coefficients $\alpha_2^R(t)$ (Figure 6.1c) mainly measure the strength of warm events and cold events associated with the Benguela current in the South Atlantic.

6.4 Characteristic Features in the Marine Atmosphere Associated with the SST Patterns $\vec{p}^{2}$, $\vec{p}^{3}$ and $\vec{p}_R^2$ in JAS

6.4.1 Data and Methods

Ship observations of sea-level pressure (SLP) and near-surface zonal (u) and meridional (v) wind are used to identify the atmospheric variability associated with each of the three characteristic SST patterns $\vec{p}^{2}$, $\vec{p}^{3}$ and $\vec{p}_R^2$ shown in Figure 6.1. The ship observations of the atmosphere, like the SST, are now held in computerised datasets. For the analysis here, data have been taken from the Comprehensive Ocean-Atmosphere Dataset (COADS, Woodruff et al., 1987) and processed as described in Ward (1992, 1994). The data are formed into seasonal anomalies on the 10° *lat* × 10° *long* scale for each JAS season 1949-88. Like the SST data, it was necessary to apply a correction to the wind data to remove an upward trend in wind speed that was caused by changes in the way winds have been measured aboard ships (Cardone et al., 1990; correction method described in Ward, 1992, 1994). A dataset of 10° *lat* × 10° *long* near-surface wind divergence has been calculated using the corrected wind dataset, and these divergence time-series are also analysed below.

The $\vec{p}^{\,3}$ time-series $\alpha_3(t)$ in Figure 6.1 was clearly associated with decadal and multidecadal variability, while SST $\alpha_2(t)$ and $\alpha_2^R(t)$ contain much variability on timescales less than 11 years (Table 6.1). To help identify the atmospheric variability associated with each of the EOFs, all data have been filtered into approximate timescales $<$ 11 years (high frequency, HF) and $>$ 11 years (low frequency, LF) using an integrated random walk smoothing algorithm (Ng and Young, 1990; Young et al., 1991). The smooth lines in Figure 6.1 show the LF filtered data for the EOF coefficients - the HF data are the difference between the raw series and the LF component. This filtering process has also been applied to every 10° *lat* $\times$ 10° *long* SLP, u, v, and divergence time-series.

The relationship of SST $\vec{p}^{\,2}$ and SST $\vec{p}_R^{\,2}$ with the near-surface atmosphere is summarised by correlating the HF EOF time coefficients $\alpha_2(t)$ and $\alpha_2^R(t)$ with every HF 10° *lat* $\times$ 10° *long* atmospheric time-series. The correlations for each atmospheric variable are plotted in Figure 6.2 for SST $\vec{p}^{\,2}$ and Figure 6.4 for SST $\vec{p}_R^{\,2}$. For SST $\vec{p}^{\,3}$ the LF components of the series are used to calculate the correlations (Figure 6.3). Although not studied in this chapter, the LF components of $\alpha_2(t)$ and $\alpha_2^R(t)$ may also describe important multidecadal climate variability, and indeed $\alpha_2^R(t)$ actually has more variance in the LF series than in the HF series (Table 6.1)

Table 6.1: *Some statistical characteristics of the SST EOF time coefficients $\alpha(t)$. The time coefficients were filtered into High Frequency (HF) and Low Frequency (LF) series, with the 50% cut-off frequency set to about 11 years. The table shows the percentage of variance in $\alpha(t)$ that is in the HF series and the lag one serial correlation for the HF series and the LF series.*

	$\alpha_2(t)$	$\alpha_3(t)$	$\alpha_2^R(t)$
Percentage of variance in High Frequency	80.8%	17.0%	39.1%
Lag one serial correlation for High Frequency	-0.16	-0.28	-0.03
Lag one serial correlation for Low Frequency	0.92	0.96	0.96

Correlation maps with vector wind are presented by forming a vector from the individual correlations with the u-component and v-component, thereby enabling the maps to be presented more concisely and facilitating visual interpretation.

When two time-series have serial correlation, then the effective sample size

Figure 6.2: *JAS Correlation (1949-88) between high-frequency (< 11 years; HF) component of SST* $\vec{p}^{\,2}$ and
(a) HF sea-level pressure,
(b) HF near-surface wind,
(c) HF near-surface divergence.
Correlations significant at the 5% level are shaded. Contour map units are in hundredths.

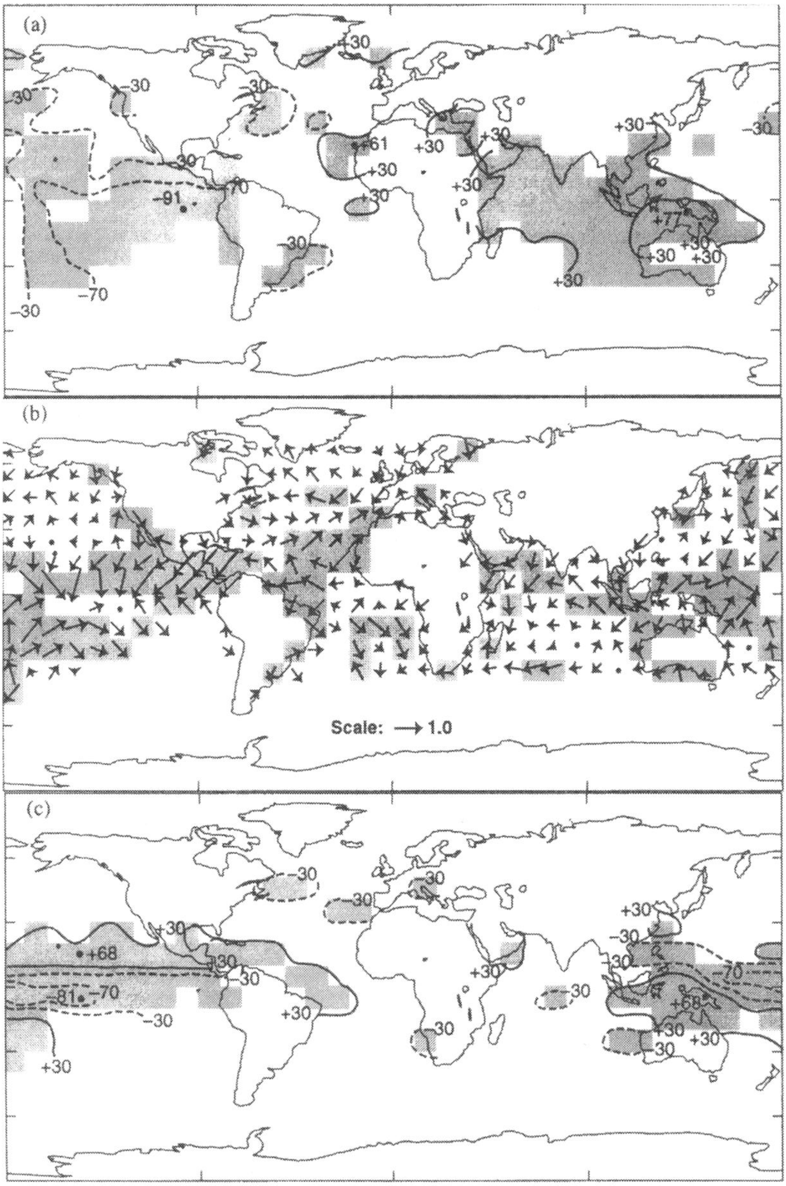

Figure 6.3: *JAS Correlation (1949-88) between the low-frequency (> 11 years; LF) component of SST $\vec{p}^{\,3}$ and*
(a) LF sea-level pressure,
(b) LF near-surface wind and
(c) LF near-surface divergence.
Contour map units are in hundredths.

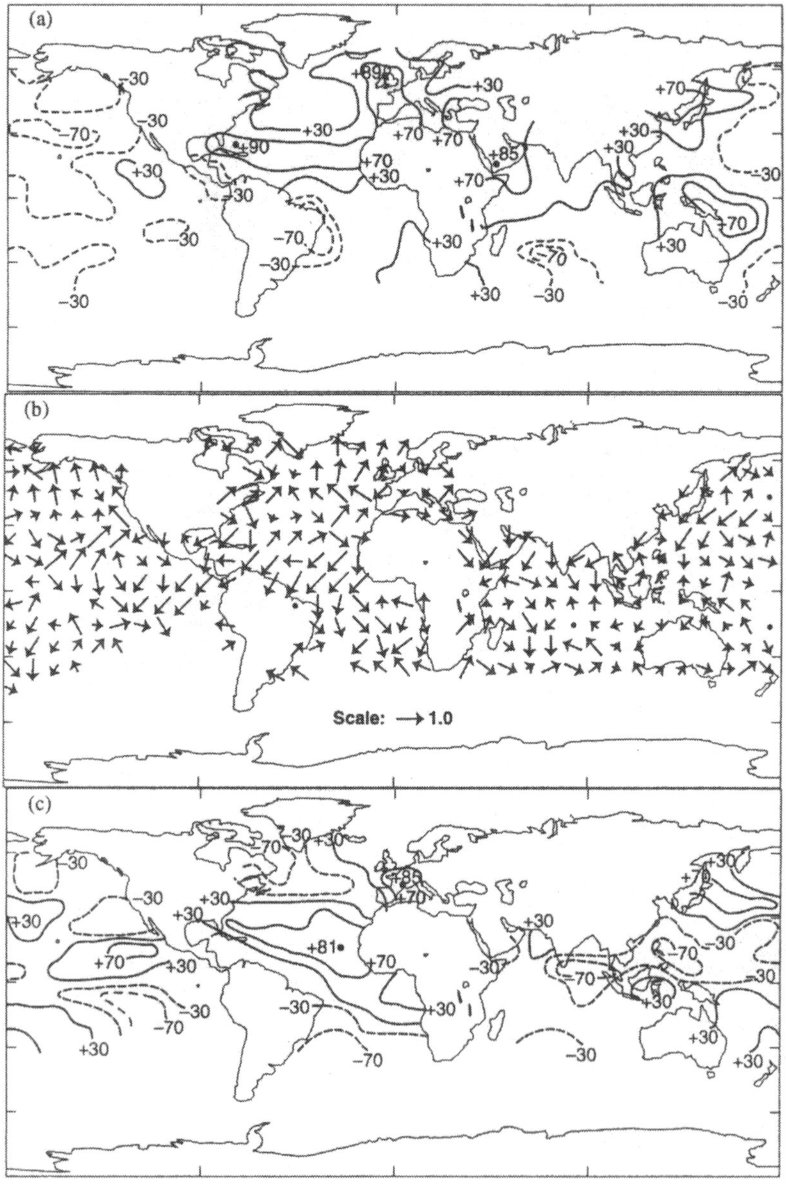

Figure 6.4: *JAS Correlation (1949-88) between the high-frequency (< 11 years; HF) component of SST* $\vec{p}_R^2$ and
(a) HF sea-level pressure,
(b) HF near-surface wind and
(c) HF near-surface divergence.
Correlations significant at the 5% level are shaded. Contour map units are in hundredths.

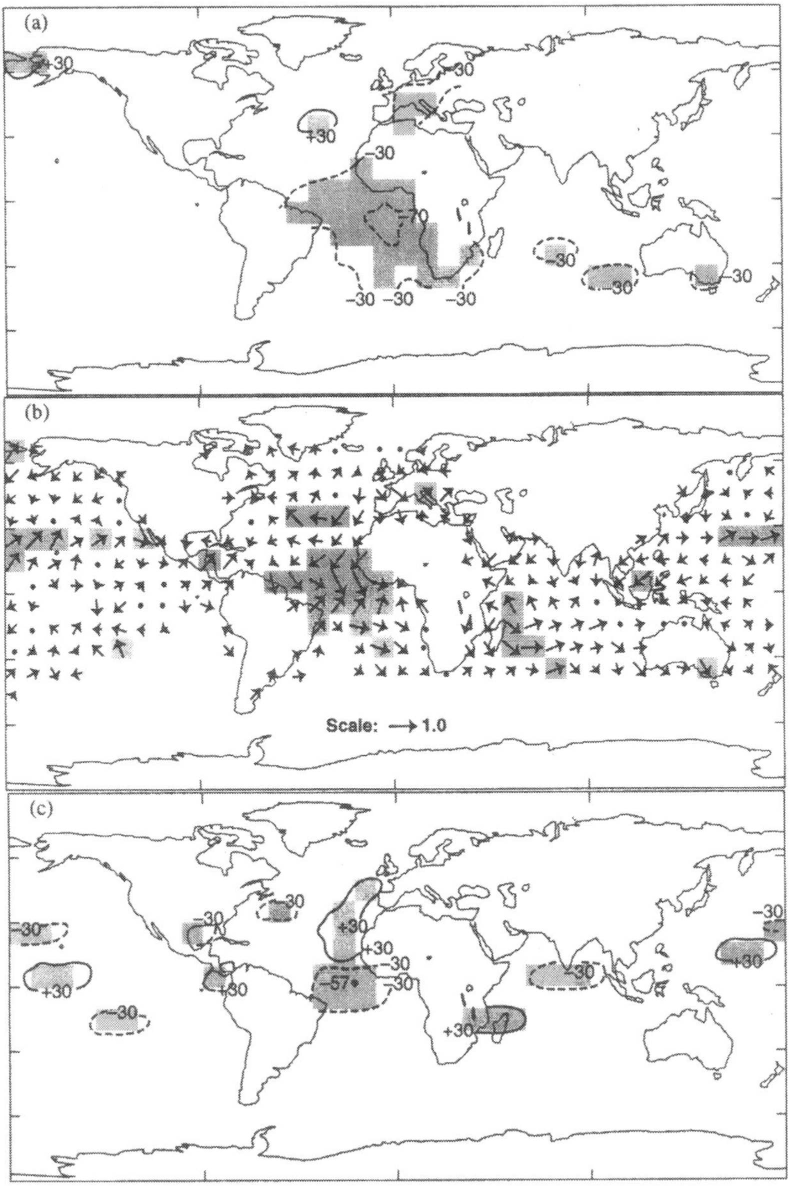

is reduced from the number of time points, and should be considered when assessing statistical significance (see Section 9.4). In this chapter, cross-correlations are calculated between series that are filtered into HF and LF. The HF components have near zero serial correlation (Table 6.1), so reductions in degrees of freedom are trivial and hence ignored, assuming $n - 2$ degrees of freedom throughout, where n is the number of time points in the correlation. This is a particularly useful by-product of filtering the data. If two time-series have lag-one serial correlation even as low as about 0.2, the degrees of freedom can be substantially reduced. So serial correlation should always be checked for when assessing the statistical significance of correlations. The LF data have very high serial correlation (Table 6.1), so cross-correlations between LF series have very few degrees of freedom indeed. No attempt has been made to assess significance; the maps should be viewed as purely descriptive.

6.4.2 Patterns in the Marine Atmosphere Associated with EOF $\vec{p}^{\,2}$

In the tropical central and eastern Pacific, the correlations with SST $\alpha_2(t)$ (Figure 6.2) suggest that the atmosphere is responding to warming and cooling of the SST in a way similar to that predicted by the LN model. Warm events accompany lower sea-level pressure, anomalous atmospheric flow down the pressure gradient, deflected by the Coriolis force, and anomalous convergence over the maximum in SST anomalies straddling the equator at $10^oS - 10^oN$. At $10 - 20^oN$ there are strong correlations with anomalous divergence.

Near and to the south of the equator in the western Pacific, SST $\alpha_2(t)$ correlates with high sea-level pressure and near-surface divergence. Some El Niños and La Niñas strongly modulate the gradient of SST into this region of the western Pacific (Fu et al., 1986; Ward, 1994; Ward et al., 1994), and it is these events that give rise to the strongest variability in the western Pacific. Correlations with near-surface convergence in the China Sea may reflect a direct anomalous circulation driven from the Indonesian region to the south (see Gill, 1982, p. 466). Anomalous meridional overturning over these longitudes has been found to be an important component of variability in a version of the UK Met. Office's GCM (Rowell, 1991; Ward et al., 1994).

Connections between SST $\vec{p}^{\,2}$ and the marine atmosphere are found in locations remote from the direct SST forcing in the Pacific. For example, there are correlations with high sea-level pressure in the North Indian Ocean and reduced southwesterly monsoon flow. In the tropical Atlantic, there are correlations with higher sea-level pressure near West Africa and reduced southwesterly monsoon flow near West Africa at $10^oN - 20^oN$.

SST $\vec{p}^{\,2}$ is also related to some atmospheric variability in the extratropics though to prove that the associations are significantly more numerous and stronger than expected by chance would require a field significance test on the extratropical correlations (see Chapter 9; field significance examples are given in Section 6.5.3). In the North Pacific, there are negative correlations with sea-level pressure, and on the western flank of this region there are significant correlations with anomalous northerly wind, consistent with the geostrophic wind relation. In the Atlantic, positive sea-level pressure correlations at $65^o N$ and negative correlations near $35^o N$ are geostrophically consistent with the correlations with easterly wind anomalies over $40 - 60^o N$. The remote tropical and extratropical teleconnections with SST $\vec{p}^{\,2}$ may in the main be conditional on El Niño and La Niña events modulating convection in the western Pacific, and triggering the anomalous meridional overturning from Indonesia to the China Sea. The associated strong diabatic heating anomalies in this region may modulate the tropic-wide circulation, and lead to some of the extra-tropical connections in the process. It is also possible that the connection between warm events (cold events) in the Pacific and near-surface divergence (convergence) in the tropical Atlantic results from a direct divergent zonal overturning of the atmosphere from the east Pacific to the Atlantic.

6.4.3 Patterns in the Marine Atmosphere Associated with EOF $\vec{p}^{\,3}$

The time coefficients of SST $\vec{p}^{\,3}$ in recent decades mainly describe a long period climate fluctuation (Figure 6.1b). The correlations with SST $\vec{p}^{\,3}$ (Figure 6.3) should be viewed as descriptive of the atmospheric changes that have accompanied the low frequency SST change. The strongest tropical atmospheric change is in the Atlantic, though changes in the Indian and Pacific Ocean are evident too. In the Atlantic, the correlations suggest that a cooler Northern Hemisphere has accompanied higher sea-level pressure and anomalous near-surface divergence in the North Atlantic, lower sea-level pressure and anomalous near-surface convergence close to Brazil, and connecting near-surface wind anomalies all with a northerly component from $20^o S$ to $20^o N$, west of $10^o W$. This wind pattern may in part be a response to the SST changes, and may in part be maintaining the SST changes, since the wind anomaly is likely to have modified the cross-equatorial ocean current and the associated ocean heat fluxes that are important for the relative temperature of the North Atlantic and South Atlantic (some early ideas are found in Bjerknes, 1964).

In the northeastern Pacific, the correlations suggest that the cooler Northern Hemisphere accompanied anomalous northerlies near $10^o N$ with anomalous near-surface divergence. In the Indian Ocean, there is evidence for a reduced strength of the Indian monsoon, with higher sea-level pressure in the northwestern Indian Ocean, and reduced southwesterly flow (i.e. anomalous northeasterly) near $15^o N$ in the Arabian Sea. It is also known that rainfall in India showed some reduction in the later period, and that the African continental monsoon brought less rain to the semi-arid region south of the Sahara desert, the Sahel (see Section 6.5). So it appears that convergence into the ITCZ is reduced and/or its latitude is further south over all longitudes except for the western Pacific. Here there is enhanced near-surface convergence near $0 - 20^oN$, with anomalous divergence to the south, extending into the SPCZ. It is interesting that this tropic-wide atmospheric teleconnection pattern has some similarities with the one associated with SST $\vec{p}^{\,2}$. Indeed, it may be that there is a tropic-wide pattern of boreal summer atmospheric variability that can be triggered by different forcings (Ward et al., 1994).

Extratropical circulation changes have also occurred in association with SST $\vec{p}^{\,3}$. In the North Atlantic, the cooler waters have accompanied higher SLP and more anticyclonic circulation over the western Atlantic and Western Europe. Over these longitudes there has been anomalous near-surface divergence, while on the western side of the basin, anomalous near-surface convergence has occurred south and southwest of Greenland.

6.4.4 Patterns in the Marine Atmosphere Associated with Rotated EOF $\vec{p}_R^{\,2}$

SST $\vec{p}_R^{\,2}$ is a more localised pattern of SST anomalies than EOFs $\vec{p}^{\,2}$ and $\vec{p}^{\,3}$. The associated atmospheric anomalies (Figure 6.4) are also more localised. The negative correlations with SLP in the tropical South Atlantic mirror the SST pattern, and are consistent with the LN model. However, atmospheric wind and divergence anomalies are strongest further north, probably reflecting strong interaction with the modified atmospheric circulation and rainfall patterns that SST $\vec{p}_R^{\,2}$ seems to force over West Africa (see Section 6.5). The region of largest weights in SST $\vec{p}_R^{\,2}$ is a region of very stable southeasterly trades. Higher than normal SST in this region is likely to enhance the moisture content of the trades so that shallow convection will release more latent heat and begin to erode the stability of the trades. This enhanced potential instability of the air mass may lead to the enhanced rainfall on the Guinea Coast that accompanies the positive phase of SST $\vec{p}_R^{\,2}$. Enhanced rainfall on the Guinea Coast is associated with reduced rainfall in the Sahel. The wind and divergence anomalies in the northern and equatorial tropical Atlantic are probably to a large part directly responding to the circulation and heating anomalies over West Africa associated with these rainfall anomalies. Furthermore, the near-surface wind anomalies over the ocean may then feed-back on the ocean to modify the SST anomalies. This interpretation requires verification with coupled ocean-atmosphere models. Nonetheless, the hypothesis illustrates how statistical relationships between ocean and atmosphere can result from a combination of local air-sea forcing that generates atmospheric anomalies, remote teleconnections internal to the physics and dynamics of the atmosphere, and remote teleconnections due to further air-sea interaction.

There is one important extratropical teleconnection with SST $\vec{p}_R^{\,2}$. Significant negative correlations with SLP are found in the central Mediterranean. This may result via SST $\vec{p}_R^{\,2}$ modifying West African rainfall, which creates circulation and diabatic heating anomalies that influence atmospheric circulation in the Mediterranean.

6.5 JAS Sahel Rainfall Links with Sea-Surface Temperature and Marine Atmosphere

6.5.1 Introduction

There is now strong evidence for sea-surface temperature (SST) anomaly patterns forcing a substantial fraction of the seasonal to multidecadal atmospheric variability over at least some continental regions in the tropics. The body of evidence has grown iteratively through the symbiosis of observational studies and numerical model experiments since the 1970s (Rowntree, 1972; Hastenrath, 1984; Lamb, 1978a,b; Folland et al. 1986a; Lough, 1986; Palmer 1986; Ropelewski and Halpert, 1987, 1989; Palmer et al., 1992). This Section summarises the observed SST and near-surface marine atmospheric relationships with seasonal rainfall in the Sahel region of Africa. The empirical relationships presented should be interpreted alongside the numerical modelling evidence for the influence of SST on Sahel rainfall. For ten past years, Rowell et al. (1992) show that a GCM forced with the observed worldwide SST pattern can simulate the observed JAS rainfall total in the Sahel with great accuracy. This suggests that many of the empirical relationships described in this Section are likely to represent stages in the chain of physical forcing of the Sahelian atmosphere by the remote SST.

For this study, monthly station rainfall data for Africa have been formed into 2.5^o *lat* $\times$ 3.75^o *long* grid-box averages (Hulme, 1992 and personal communication).

6.5.2 Rainfall in the Sahel of Africa

The Sahel region of Africa (Figure 6.5a) lies just south of the Sahara desert and receives 80-90% of its annual rainfall in the JAS season, when the Intertropical Convergence Zone (ITCZ) is at its most northerly location. During these months, the coastal region to the south, which will be called the Guinea Coast (Figure 6.5a), experiences a relatively dry season, though the JAS rainfall total varies substantially from year to year. There is some tendency for the rainfall anomalies on the Guinea Coast and the Sahel to vary out of phase (Nicholson, 1980; Janowiak, 1988; Janicot, 1992a; Nicholson and Palao, 1993; Ward, 1994), and this has guided the definition of the Sahel and Guinea Coast regions marked on Figure 6.5a. For the two regions, a mean standardised anomaly rainfall time-series has been calculated (for basic method see Nicholson, 1985; adapted for grid-box time-series in Ward, 1994). The time-series for the two regions (Figure 6.5b,c) have been filtered and are used in the analysis below.

Figure 6.5: *(a) JAS mean rainfall 1951-80 (units are mm). The solid lines delineate regions used for standardised rainfall indices: northern region is Sahel, southern region is Guinea Coast.*
(b) Mean standardised rainfall anomaly time-series for the Sahel region, (c) Same as (b) but for Guinea Coast region.

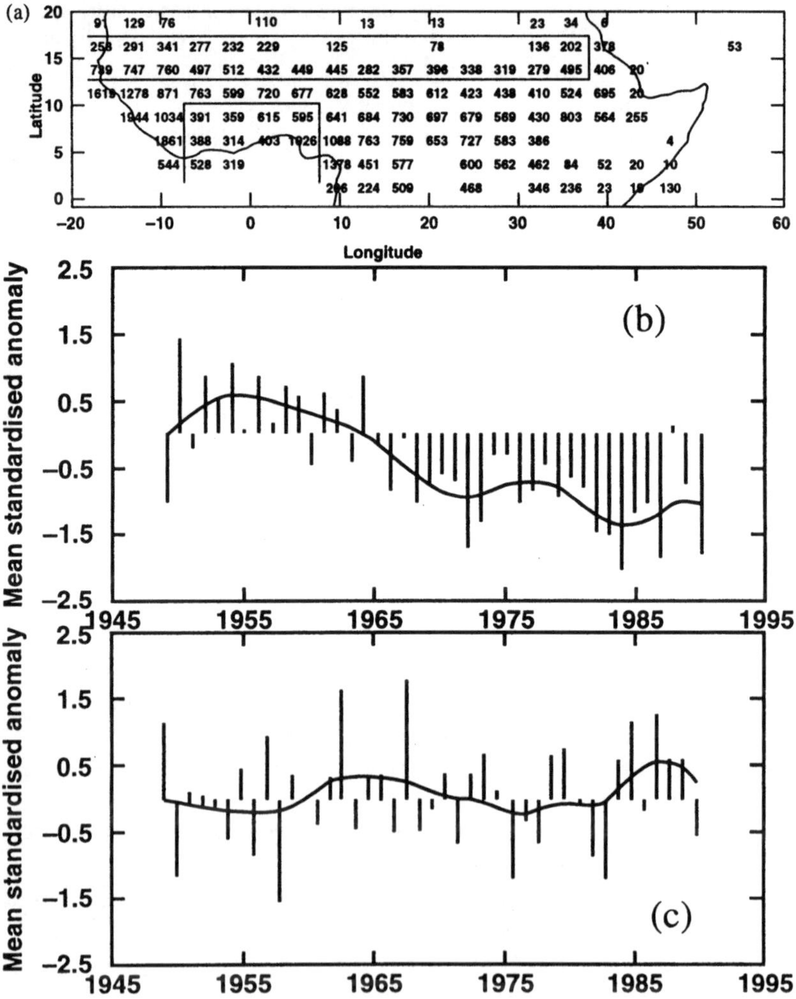

6.5.3 High Frequency Sahel Rainfall Variations

a) Field Significance

Plots have been made of the correlation between the high-frequency (< 11 *years*; HF) Sahel rainfall series and each grid-box series of HF SST (Figure 6.6a) and HF SLP (Figure 6.6b). There are many correlations that are locally statistically significant. However, one would expect on average 5% of the correlations in each figure to be significant at the 5% level by chance (cf. Section 9.2.1). So to assess the null hypothesis of no association between Sahel rainfall and the marine fields, it is necessary to assess the probability of achieving the number of statistically significant boxes in Figures 6.6a,b by chance. This test is called testing "Field Significance" (Section 9.2.1). The SST and SLP time-series contain strong spatial correlations. This raises the likelihood of achieving by chance a large fraction of statistically significant correlations. To assess field significance in this situation, a Monte Carlo method has been used, as recommended by Livezey and Chen (1983). Alternative approaches are permutation procedures discussed in Chapter 9.

- **Sea-Surface Temperature**
  Following Livezey and Chen (1983), 500 random normal rainfall series, each of length 40, have been simulated using a random number generator. These 500 simulated rainfall series were used to create 500 correlation maps like the one in Figure 6.6a using the observed SST data. Table 6.2 shows the fraction of occasions in the Monte Carlo simulation on which the area covered by locally significant 10° *lat* $\times$ 10° *long* SST boxes exceeded the significant area in the Sahel-SST correlation map (Figure 6.6a).

  For all available regions, 32.2% of the sampled ocean area is covered by significant correlations between Sahel rainfall and SST (Figure 6.6a). Such a coverage was achieved in less than 0.2% of the Monte Carlo simulations i.e. none of the 500 correlation maps contained more area covered with significant correlations. So field significance is very high indeed, and the null hypothesis of no association between Sahel rainfall and SST can be rejected with great confidence. Sub-domains of the global ocean were also tested and found to be highly field significant, including the extratropics.

- **Sea-Level Pressure**
  The field significance analysis was repeated for the SLP correlations in Figure 6.6b. Again, field significance is very high for the whole analysis domain. The correlations north of 30^oN are field significant at the 3.0% level, suggesting Northern Hemisphere extratropical circulation is probably related to Sahel rainfall variations. The field significance of the tropical Pacific and tropical Indian oceans is very high. The relatively lower significance for the tropical Atlantic is probably due to the

Figure 6.6: *JAS Correlation (in hundredths) over 1949 - 88 between HF Sahel rainfall and HF*
(a) sea-surface temperature and
(b) sea-level pressure.
Correlations significant at the 5% level are shaded.

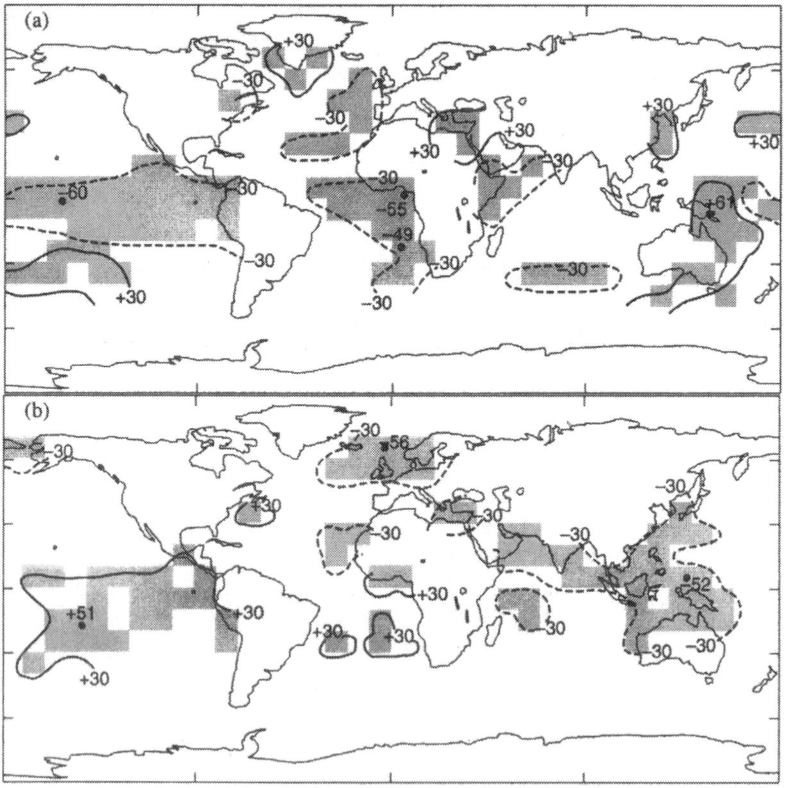

Table 6.2: *Field significance estimates for the maps of correlations between Sahel rainfall and sea-surface temperature (SST; Figure 6.6a) and sea-level pressure (SLP; Figure 6.6b). "% area" is the percentage of ocean region having significant correlations (5% level) with Sahel rainfall. "Prob" is the probability of achieving by chance a larger percentage (estimated using a Monte Carlo test). For the SLP map, field significance is calculated just for the Northern Hemisphere extratropics due to the lack of data in the Southern Hemisphere extratropics.*

Ocean	SST		SLP	
Region	% area	Prob	% area	Prob
All available areas	32.2	<.002	29.2	.002
Tropics (30^oN – 30^oS)	38.0	<.002	37.0	<.002
Extratropics	21.7	<.002		
North of 30^oN			16.0	.030
Tropical Atlantic	35.2	.002	19.1	.068
Tropical Pacific	47.9	<.002	44.7	<.002
All Indian	18.5	.020	30.9	.016

Table 6.3: *Correlation between tropical North African rainfall series and SST EOFs over 1949-90. HF (high frequency; < 11 years) components of the series are used for correlations with SST $\vec{p}^{\,2}$ and SST $\vec{p}_R^{\,2}$, LF (low frequency, > 11 years) components for SST $\vec{p}^{\,3}$. Statistical significance estimated for HF correlations: ** = 1%, * = 5%. Significance for LF correlations not estimated due to too few degrees of freedom.*

	SST $\vec{p}^{\,2}$ HF component	SST $\vec{p}^{\,3}$ LF component	SST $\vec{p}_R^{\,2}$ HF component
Sahel	-0.48**	-0.90	-0.31*
Guinea Coast	-0.06	-0.33	+0.58**

different Atlantic atmospheric patterns that occur in different types of droughts and wet years in the Sahel (Janicot, 1992b; and see Figure 6.7 later). Such a situation reduces the linear correlation over all years between Sahel rainfall and atmospheric circulation in the tropical Atlantic sector.

b) Nature of the Relationships

Through the tropics, the correlation patterns (Figure 6.6) suggest Sahel rainfall is linked to SST and marine atmospheric variability associated with SST $\vec{p}^{\,2}$ and SST $\vec{p}_R^{\,2}$. Indeed, the correlation between each of the EOFs and Sahel rainfall is statistically significant (Table 6.3). In the tropical Pacific, the SST correlations (Figure 6.6a) are very positive in the western equatorial region. This suggests a departure from the SST $\vec{p}^{\,2}$ pattern (Figure 6.1a), which has weak weights in the western Pacific. It was noted earlier that the western Pacific was a strong candidate for triggering tropic-wide atmospheric teleconnections. The correlation patterns support this idea. However, the strong correlations in the eastern Pacific with SST and SLP (Figure 6.6a,b) are also consistent with an influence of a direct circulation from the eastern Pacific to the Atlantic and West Africa.

Outside the tropics, there are two potentially important teleconnections for Europe. Firstly, Sahel rainfall correlates positively with SST and negatively with SLP in the eastern Mediterranean. Secondly, Sahel rainfall correlates with low SLP near and to the north of the UK. Both these connections require theoretical and numerical model investigation.

Janicot (1992b) argued that the tropical Atlantic marine relationships with Sahel rainfall were conditional upon whether rainfall anomalies on the Guinea Coast were of opposite sign to those in the Sahel (DIPOLE years) or were the same sign (NO DIPOLE years). To show that not just the Atlantic but the global relationships with marine variability are different in DIPOLE compared to NO DIPOLE years, Sahel rainfall has been correlated with SLP in DIPOLE years (Figure 6.7a) and NO DIPOLE years (Figure 6.7b). In DIPOLE years, correlations are strongest in the tropical South Atlantic where the pattern is similar (but sign inverted) to the SLP - SST $\vec{p}_R^{\,2}$ map (Figure 6.4a), supporting the idea that positive (negative) values of SST $\vec{p}_R^{\,2}$ force more (less) rain on the Guinea Coast and less (more) rain in the Sahel (see Table 6.3). In NO DIPOLE years, the SLP correlations (Figure 6.7b) are similar (but sign inverted) to those with SST $\vec{p}^{\,2}$ (Figure 6.2a), suggesting that SST $\vec{p}^{\,2}$ tends to be associated with rainfall anomalies of the same sign over West Africa. This also suggests that SST $\vec{p}_R^{\,2}$ and SST $\vec{p}^{\,2}$ explain an independent fraction of Sahel rainfall variance. One way to test this is using multiple regression (a related concept is partial correlation).

Figure 6.7: *JAS Correlation (in hundredths) between HF Sahel rainfall and HF SLP in*
(a) DIPOLE years in the period 1949-88,
(b) NO DIPOLE years in the period 1949-88.
A DIPOLE year is a year in which JAS HF Sahel rainfall anomaly is of opposite sign to HF Guinea Coast rainfall anomaly. All other years are NO DIPOLE.

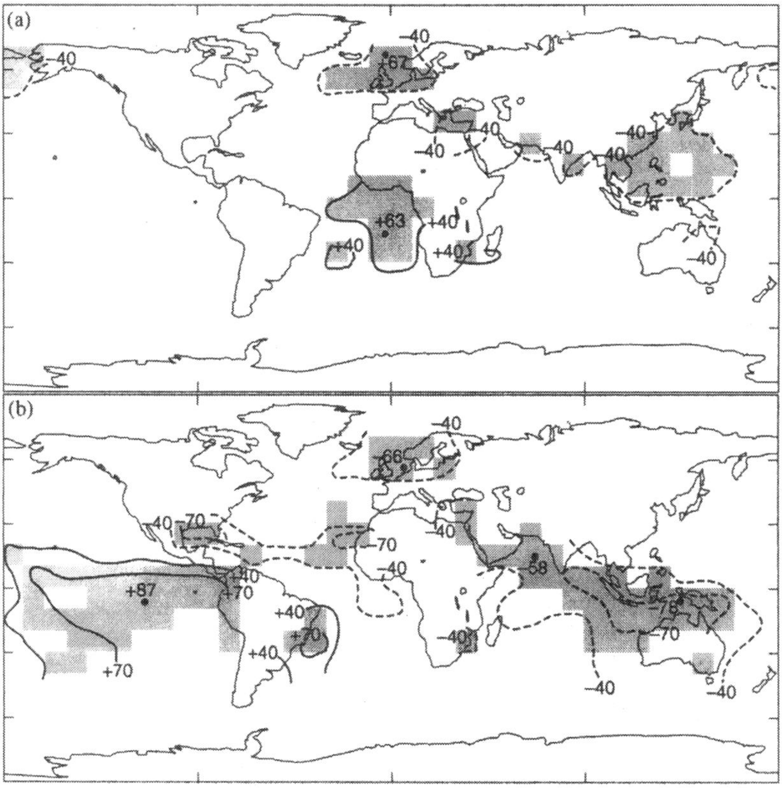

c) SST EOF Multiple Regression

The aim of a multiple regression model is to produce predicted values of a dependent variable from a linearly weighted combination of independent variables. The weights (regression coefficients) are chosen to minimise the mean square error of the predicted values, here for the period 1949-90. The dependent variable here is HF JAS Sahel rainfall and the predictor independent variables are HF JAS values of $\alpha_2^R(t)$ and $\alpha_2(t)$:

JAS HF SAHEL	=	$0.00 - 0.12(\alpha_2)$	-	$0.23(\alpha_2^R)$
significance of each variable:		1%		1%

Multiple r of the model is: 0.60

The null hypothesis that each regression coefficient is zero (separate test for each coefficient) is tested, and rejected at the 1% level for both variables, showing that each explains a significant fraction of Sahel rainfall variability. This in fact was an expected result since the EOF time coefficients are almost uncorrelated (the all seasons coefficients from the original analysis 1901-80 are perfectly uncorrelated). Thus, since both EOFs have significant individual correlations with Sahel rainfall (Table 6.3), we can expect them to combine well in a multiple regression. This is a particularly useful property of EOF time-coefficients for use in multiple regression. However, this whole methodology relies on the key SST patterns being orthogonal. Here, progress has been possible since ENSO and warm/cold events in the tropical Atlantic are largely uncorrelated (at zero lag), and thus the EOF analysis was not significantly hampered by its orthogonality constraint. However, identification of the types of ENSO that most strongly affect Sahel rainfall (the ones with strongest gradients of SST anomaly into the western Pacific) cannot be made using this global EOF analysis of all years, since the two types of ENSO are highly correlated, and only one EOF representing ENSO is possible.

The regression model above is one step from forming a seasonal forecast technique for the Sahel. For seasonal forecasting, pre-rainfall season values of the EOF coefficients are used to form a regression model (see Folland et al., 1991, Ward and Folland, 1991).

6.5.4 Low Frequency Sahel Rainfall Variations

Inspection of the time-coefficients of SST $\vec{p}^{\,3}$ (Figure 6.1b) and Sahel rainfall (Figure 6.5b) is sufficient to identify a clear association between the two on the LF timescale (Table 6.3). Consistent with this observation, the correlations between LF Sahel rainfall and the marine atmosphere (not shown) are very similar to those with SST $\vec{p}^{\,3}$ (Figure 6.3). Thus multidecadal Sahel rainfall fluctuations may be part of the climate variation involving inter-hemispheric SST variations, and tropic-wide circulation anomalies, with teleconnections to the extratropics (see Section 6.4.3).

6.6 Conclusions

Progress in understanding climate variability has arisen through the combined effort of theoretical studies, numerical modelling experiments and statistical analysis of climate observations. This chapter has concentrated on one area of climate variability: the role of air-sea interaction in seasonal to multidecadal climate variability. This area is particularly apt to illustrate the contribution that can be made by statistical analysis of climate observations. At the outset, there is careful analysis and correction of raw observations to build up computerised historical datasets that are as homogeneous as possible. Then statistical analysis can produce information on various aspects of climate variability. This chapter has illustrated the use of EOF analysis to analyse spatial patterns of worldwide SST variability and to describe the temporal variations of those patterns. Diagnostic correlation studies describe the relationship between the SST EOFs, the marine atmosphere, and as a case study, the relationship of marine variability to remote atmospheric variability over a continental region, the Sahel. Some of the analysis provides confirmation of the general applicability of existing theory. Other results provoke the development of more speculative hypotheses for testing in physical and dynamical models of varying complexity. One such hypothesis concerns the existence of a tropic-wide pattern of atmospheric anomalies of which Sahel rainfall is a part. Another hypothesis concerns the effect of tropical South Atlantic SST anomalies on Atlantic and West African circulation anomalies. For the results presented in this study, two developments in the future will be particularly important for further advances: Firstly, much improved data describing the three-dimensional structure of the atmosphere for recent years are about to become available after a re-analysis of the historical data by numerical modelling centres including ECMWF, and secondly, improved understanding of the coupled ocean-atmosphere system will result from the development and application of coupled models of the ocean-atmosphere system to study interactive behaviour and the development of seasonal to multidecadal climate anomalies. To model and understand ocean-atmosphere connections to continental areas, it may also be necessary in some regions to improve the representation of land surface - atmosphere interaction, which may also play a significant role in climate variability (Nicholson, 1988).

Part III

Simulating and Predicting Climate

Chapter 7

The Simulation of Weather Types in GCMs: A Regional Approach to Control-Run Validation

by Keith R. Briffa

7.1 Introduction

General Circulation Models (GCMs) attempt to reproduce the three-dimensional atmospheric processes which result in the surface climate that we experience across different regions of the globe. The time and spatial scales upon which various GCMs operate are much cruder (less-well resolved) than those of the real world. Nevertheless, GCM experiments offer the most promising approach for gaining insight into the physical mechanisms underlying past and potential future climates, particularly at a regional level. It is this regional detail which is needed to develop and test theories about past climates and for making predictions about future climates that might come about as a consequence of increasing greenhouse gases.

It should be self evident that there is a need to establish how well particular GCMs succeed in reproducing modern-day climates (i.e. the accuracy of control-run simulations). This information represents important diagnos-

Acknowledgements: The work described here was funded by the U.K. Dept. Environment under grant PECD/7/12/78.

tic evidence on the performance of the models which should help in their development and continued improvement. However, this information is also important for gauging the degree of confidence that should be attributed to the regional details of climate impact studies based on the output of particular GCMs.

Several recent papers have addressed different aspects of GCM (control-run) validation (for examples, see the papers cited in Hulme et al., 1993). This Chapter describes one particular approach: that of comparing the details of GCM-simulated daily circulation patterns with those observed during a recent run of years. The following description is largely a synthesis of results published in two recent papers (Jones et al., 1993; Hulme et al., 1993). Concentrating on characteristic circulation patterns (or weather types) is a convenient means of assessing the realism of the simulated atmospheric and land surface processes. The examples described focus on the region of the British Isles, so enabling us to make use of an existing body of knowledge about the classification of daily circulation patterns over the region and their relationship with primary climate parameters such as temperature and precipitation: primarily based on the synoptic classification scheme devised by Hubert Lamb (Lamb, 1950; 1972) and the studies which have employed it (e.g. Murray and Lewis, 1966; Jones and Kelly, 1982; Briffa et al., 1990.)

First, the Lamb daily classification (the Lamb weather catalogue) is described and then an "objective" scheme for automatically producing Lamb's weather indices from gridded surface pressure data is outlined and discussed. After a brief description of the GCMs used in this exercise, model-generated temperature and precipitation statistics, weather-type frequencies and the relationships between these are compared with those observed in "real-world" data. A very brief comparison is also made between observed and model-generated average spell-lengths for two major weather types and a similar comparison of modelled versus real-world storm frequencies is illustrated.

7.2 The Lamb Catalogue

The circulation patterns dominating the British Isles (approximately the area within 50° to $60^o N$ and $10^o W$ to $2^o E$) on each day since 1861 (except in the 1860s when no Sunday maps were produced) have been classified according to a scheme devised by Hubert Lamb (Lamb 1950, 1972; and updated regularly by him in *Climate Monitor*). The scheme consists of 26 "weather types" (plus 1 "unclassified") made up from 9 anticyclonic (A) types: A, ANE, AE, ASE, AS, ASW, AW, ANW and AN; 9 equivalent cyclonic (C) types: C, CNE etc.; and the 8 directional types: NE, E, SE, S, SW, W, NW and N. Figure 7.1, reproduced from Lamb (1972), shows typical examples of different weather types defined according to Lamb's manual classification of mean-sea-level pressure and (after World War II) upper air charts. The full digitised classification is available from the Climatic Research Unit. Many studies have used this classification and there is a large literature describing both it and the relationships between the "Lamb type" frequencies and meteorological statistics (e.g., see relevant sections in Barry and Perry, 1973; and Yarnal, 1993).

7.3 An "Objective" Lamb Classification

In order to overcome the inherent subjectivity involved in any manual classification and to expedite the necessarily laborious process of classifying GCM-generated daily pressure fields it is desirable to automate the classification using some objective criteria that can form the basis of a computerized approach. Such a scheme, for automatic classification of daily mean sea level pressure maps in terms of the Lamb catalogue has been suggested by Jenkinson and Collison (1977). These authors devised a set of simple rules appropriate to the classification of pressure charts stored in gridded form on a 5° latitude by 10° longitude grid, the synoptic climatology grid held at U.K. Meteorological Office. Identifying particular gridpoint pressure data with letters, as shown in Figure 7.2, these rules are as follows:

Westerly flow	(W)	=	0.5(L+M) - 0.5(D+E)
Southerly flow	(S)	=	1.74[0.25(E+2I+M)-0.25(D+2H+L)]
Resultant flow	(F)	=	$(S^2+W^2)^{0.5}$
Westerly shear vorticity	(ZW)	=	1.07[0.5(O+P)- 0.5(H+I)] −0.95[0.5(H+I)-0.5(A+B)]
Southerly shear vorticity	(ZS)	=	1.52[0.25(F+2J+N)-0.25(E+2I+M) -0.25(D+2H+L)+0.25(C+2G+K)]
Total shear vorticity	(Z)	=	ZW+ZS

which is Cyclonic when Z is positive and Anticyclonic if negative.

The flow units are geostrophic, expressed as hPa per 10° of latitude at 55^oN ($\approx$1.2 knots). The vorticity units are the same (100 units = 0.55×10^{-4} = $0.46\times$ the Coriolis parameter at 55^oN).

Having derived values for F, Z,W and S, the direction of flow (D) is calculated as:

$$D = tan^{-1}\left(\frac{S}{W}\right) \qquad \text{(in degrees)} \tag{7.1}$$

Where W is positive 180° is added to the calculated value of D. The direction of flow then accords with the 45° sector of the compass within which the final value of D falls. This is illustrated in Figure 7.3 which highlights the sector between 247.5^o and 292.5^o that defines the W type. Synoptic (i.e. C or A types) or pure direction types are identified by the following criteria:

Figure 7.1: *Examples of mean sea level pressure and 500 hPa geopotential height charts for typical Lamb types. (After Lamb, 1972).*

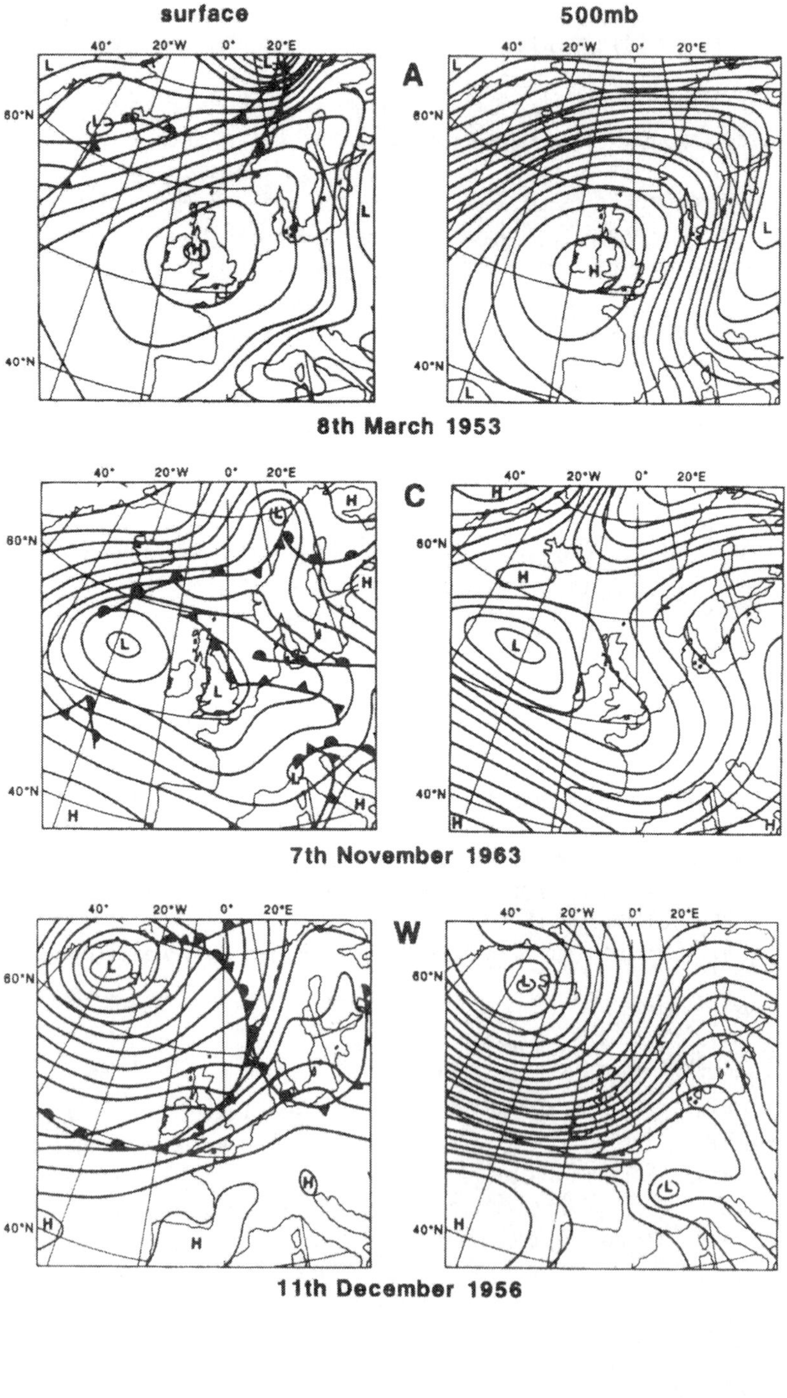

Figure 7.2: *Map of northwest Europe showing the gridpoints (large dots) in the mean sea level fields used to derive the objective Lamb classification. Monthly time series of model-simulated mean temperature and precipitation representative of England and Wales were extracted for the 5° × 5° shaded box for comparison with instrumental data as described in the text.*

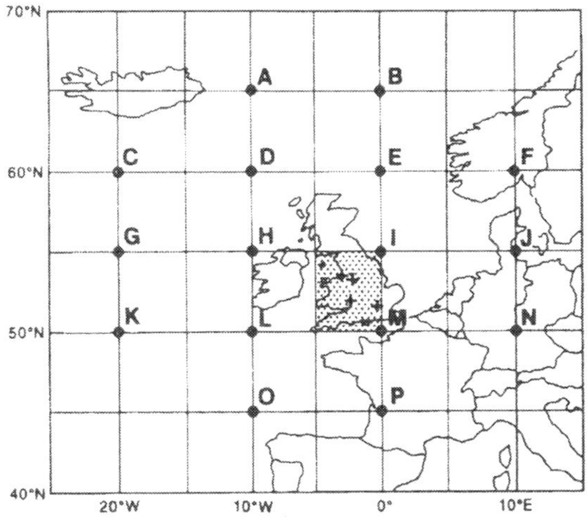

Figure 7.3: *An example identification of the direction of flow as given by equation (7.1). When D falls between 247.5° and 292.5° the flow is Westerly.*

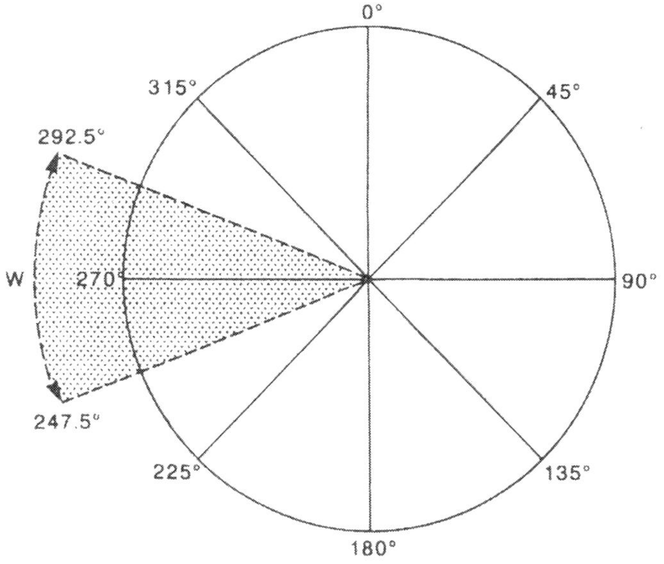

If $\|Z\| < F$	the flow is straight (i.e. a pure directional type).
If $\|Z\| > 2F$	the flow is either strongly cyclonic (where $\|Z\| > 0$) or strongly anticyclonic ($\|Z\| < 0$): the Lamb C or A types.
If $F < \|Z\| < 2F$	the flow is partly (anti) cyclonic and partly directional: a Lamb hybrid type e.g. AS.
If $F < 6$ and $\|Z\| < 6$	the flow is light and indeterminate: a Lamb "unclassified" type, U.

Using this "objective" scheme, each day between the start of December, 1880 to the end of 1989 has been typed (Jones et al., 1993). These daily types have then been totalled to give monthly indices of the 7 basic types defined by Lamb, i.e. anticyclonic (A), cyclonic (C), northerly (N), easterly (E), southerly (S), westerly (W) and northwesterly (NW). The final (NW) type is calculated separately to conform with Lamb's opinion that NW air masses are sufficiently distinct to be considered as a major type. In calculating these major types, hybrid weather types are counted in proportion to the major types of which they are composed, so that CNW would count half to the C total and half to the NW total. Together with the "unclassified" types the resulting totals for each of the 8 types then sum to the number of days in the month.

Seasonal totals of Lamb's manually classified days and the "objective" days have been compared in detail for each of the 8 types over the whole period between 1881 and 1989. They are found to be highly correlated. Individual scatter plots and comparison time series are shown in Jones et al. (1993). The correlations are particularly good for synoptic (i.e. A and C) types but they are not as high for some directional types. This is most likely attributable to Lamb's primary concern with the general steering of weather over a succession of days and his use of upper air charts (only after 1945). Lamb also used two surface charts per day whereas the objective scheme indices are based on only one.

The slight differences in the manual and objective indices are not significant as regards GCM-model validation, however, as the same objective approach is applied to the classification of the observed and generated pressure fields. Also, the correlations between the weather types and observed precipitation and temperature over England and Wales are extremely similar regardless of whether they are manually or objectively defined (Table 7.1). The following description of the comparisons between "observed" and model-based weather type indices are therefore based on the objective classification scheme.

7.4 Details of the Selected GCM Experiments

The comparisons described here are based on the control-run outputs of two GCMs. A summary of the details of these models is given in Table 7.2. UKHI refers to the UK Meteorological Office "high-resolution" model which is an

Table 7.1: *Significant ($p = 0.05$) correlations between seasonal mean weather-type frequencies and mean temperature and total seasonal precipitation over the "British Isles". The correlations are shown for both the manual Lamb classification (L) and the objective scheme (O) described here.*

CENTRAL ENGLAND TEMPERATURE

	Winter		Spring		Summer		Autumn	
Type	L	O	L	O	L	O	L	O
A	-.28	-.27	+.20	+.26	+.66	+.60		
C			-.34	-.36	-.46	-.47		
W	+.70	+.67	+.30	+.24	-.29			+.32
N	-.54	-.46	-.45	-.48	-.40	-.34	-.70	-.61
E	-.64	-.65		-.27				-.21
S	+.29	+.26	+.32	+.31	+.22	+.21	+.46	+.47
NW						-.34		-.24

ENGLAND AND WALES PRECIPITATION

	Winter		Spring		Summer		Autumn	
Type	L	O	L	O	L	O	L	O
A	-.81	-.82	-.69	-.74	-.76	-.76	-.71	-.68
C	+.73	+.69	+.72	+.73	+.80	+.78	+.73	+.76
W							+.23	
N	-.25							
E								
S	+.22	+.31						
NW					-.27			

Table 7.2: *Details of the two GCM control runs. UKHI refers to the "high-resolution" United Kingdom Meteorological Office model and MPI to that of the Max Planck Institut für Meteorologie, Hamburg (see text for further details).*

	UKHI	MPI
Primary Reference	Mitchell at al. (1989)	Cubasch et al. (1992)
Resolution	$2.5^o \times 3.75^o$	$5.6^o \times 5.6^o$
Ocean	Slab	Large-scale geostrophic., coupled
Years analysed	Last 10 years of the control run	Years 61-70 of a 115-year run
Fields analysed	Instantaneous (at 00Z)	Mean of two (00 and 12 Z)

atmospheric model, with fixed cloud properties and simulated wave drag, underlain by a simple "slab ocean" (Mitchell et al., 1989). MPI refers to the Max Planck T-21 coupled atmosphere-ocean model developed in Hamburg which employs the 11-layer, large-scale geostrophic ocean model (i.e. ECHAM-1/LSG) as described by Cubasch et al (1992). The effective resolution of each model and the land ocean configurations over northwest Europe are shown in Figure 7.4. The mean sea level pressure data generated by each model (360 days per year) were interpolated on to the $5^o \times 10^o$ grid shown in Figure 7.2 and monthly mean surface (UKHI) or 2m (MPI) temperatures and total precipitation representing the area $50^o - 55^oN, 0^o - 5^oW$ were also extracted from each model output dataset.

7.5 Comparing Observed and GCM Climates

7.5.1 Lamb Types

Figure 7.5 shows monthly mean frequencies of the 7 basic Lamb types and the additional "unclassified" category calculated for both 10-year GCM control run samples. For comparison, the 110-year observed values are shown together with range bars defined by the highest and lowest concurrent 10-year period means that occurred between 1881 and 1990.

Together, A and C days occur on 162 days of the year (99 and 63 respectively). The models produce an exaggerated annual cycle in the synoptic (A and C) types. This is the result of inaccuracy mainly in the winter, both model outputs being unrepresented by A and dramatically overrepresented by C days in comparison to the real world data (though MPI also overestimates A types in summer). Both models, but especially UKHI, produce too few A types overall and both produce too many C types. Only half of the model-generated means of A and C for each month are within the extreme

Figure 7.4: *The effective grid resolution of the two GCMs. Shading indicates the simplified land configuration employed within each model.*

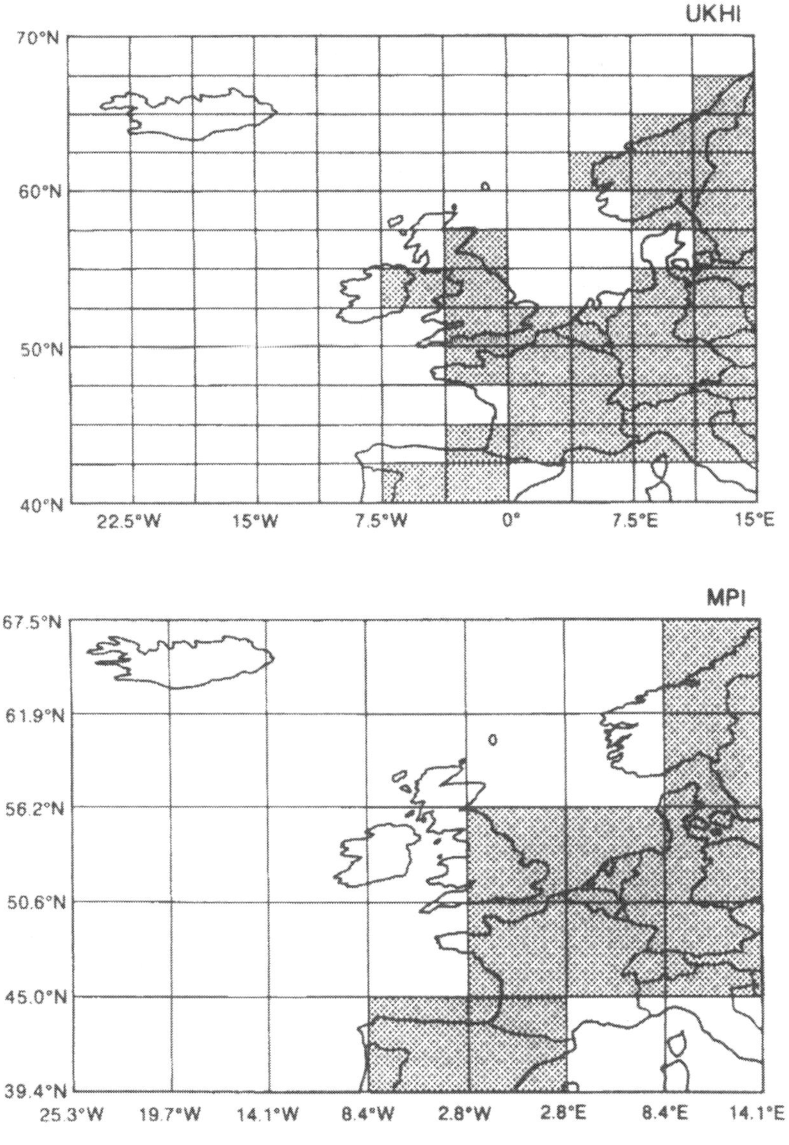

Figure 7.5: *Observed (solid line) and model-simulated (dotted MPI, dashed UKHI) mean monthly frequencies of the major Lamb types. Range bars indicate the highest and lowest 10-year means in the observed record (1881-1990).*

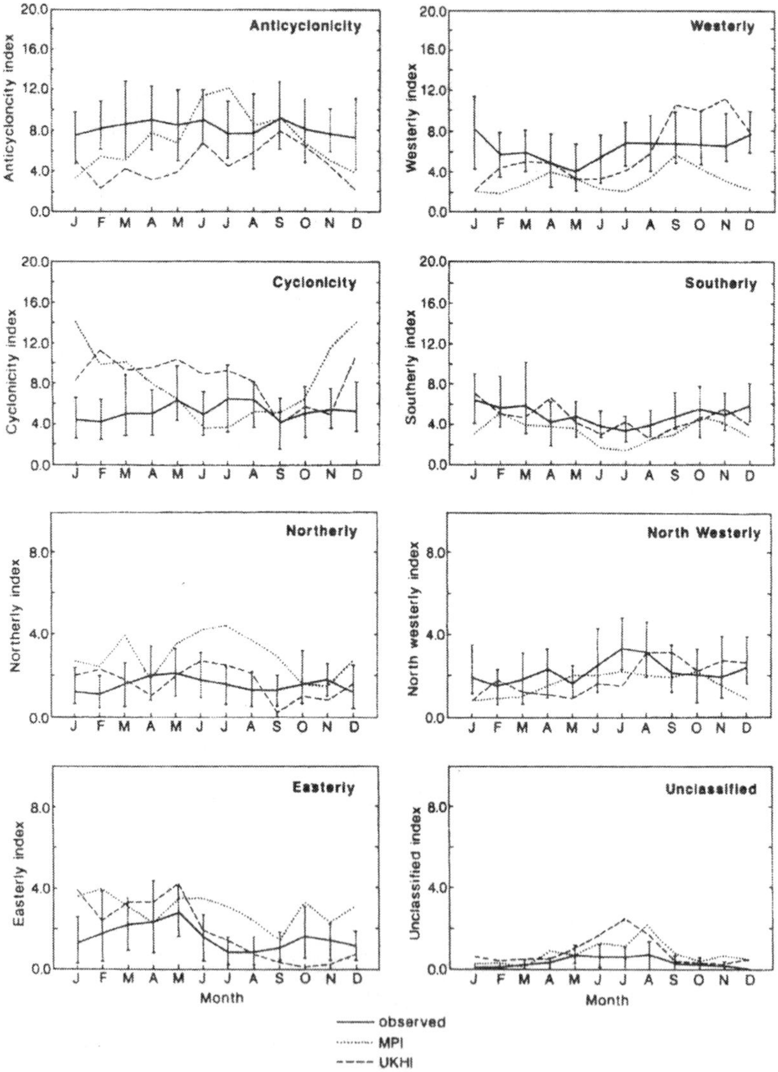

observed range calculated over the last 110 years. The models perform best in late summer and autumn (between August and October), when simulated A and C types for both MPI and UKHI are close to the observed mean values.

As for directional types, W and S are the most important (with annual means of 75.4 and 59.0 respectively). UKHI has an annual W total close to the real-world but it underestimates in January and overestimates in the autumn months, producing a false annual cycle. The MPI annual cycle has the right shape but W days are generally too low in all but the late spring months (April and May). Both models simulate the weak annual cycle in S days very well, and UKHI is remarkably accurate in simulating the correct monthly means. MPI is again consistently low. The simulated monthly means fall outside of the observed range in two winter months (January and December) and all three summer months.

The other directional types (N, E and NW) together occur on average about 64 days of the year (18.8, 18.7 and 26.8 respectively). Annual mean N days are too high in MPI but generally well simulated by UKHI. MPI also overestimates E types whereas UKHI overestimates in late winter and underestimates in autumn (September-November). Both models tend to produce too few NW days (especially MPI) over the year as a whole though they both produce a similar weak annual cycle.

7.5.2 Temperature and Precipitation

Figure 7.6 compares the annual cycles of temperature and precipitation representing observed average conditions over the U.K. with those simulated by each of the GCMs. The observed temperature data are the "Central England" series compiled by Manley (1974; and updated by Parker et al., 1992). The precipitation record is the "England and Wales" series described in Wigley et al. (1984; updated in Gregory et al., 1991). Both of these records are made up of a number of individual site records and are therefore spatially representative and appropriate for comparison with the model-generated data.

Mean annual temperature in the real-world data is 9.3^oC (1881-1990). The equivalent MPI value is very similar at 9.9^oC but the UKHI value is only 6.9^o. The MPI model, in fact, simulates the annual cycle very well, the only discrepancy being an overestimate of temperatures in late autumn, and early winter. Though the shape of the annual cycle is reasonable, the UKHI model is much too cold ($2-3^oC$ lower) in all but the three months October-December, which are well simulated. In contrast, UKHI simulates the monthly totals of precipitation remarkably well. Each is within the observed range. The MPI monthly totals are accurate in autumn and early winter (September-December) but are much too low in the spring and dramatically low in the summer. In fact MPI produces hardly any rain at all during the summer months of June-August whereas the observed values for summer as a whole are over 200 mm (compared with below 40 mm produced by MPI).

Figure 7.6: *Observed monthly mean temperatures and precipitation representative of the area of England and Wales (see references in the text) and equivalent GCM simulated data for the* $5^o \times 5^o$ *gridbox centred on* 52.5^oN, 2.5^oW. *Range bars on the observed data are as described in the legend to Figure 7.5.*

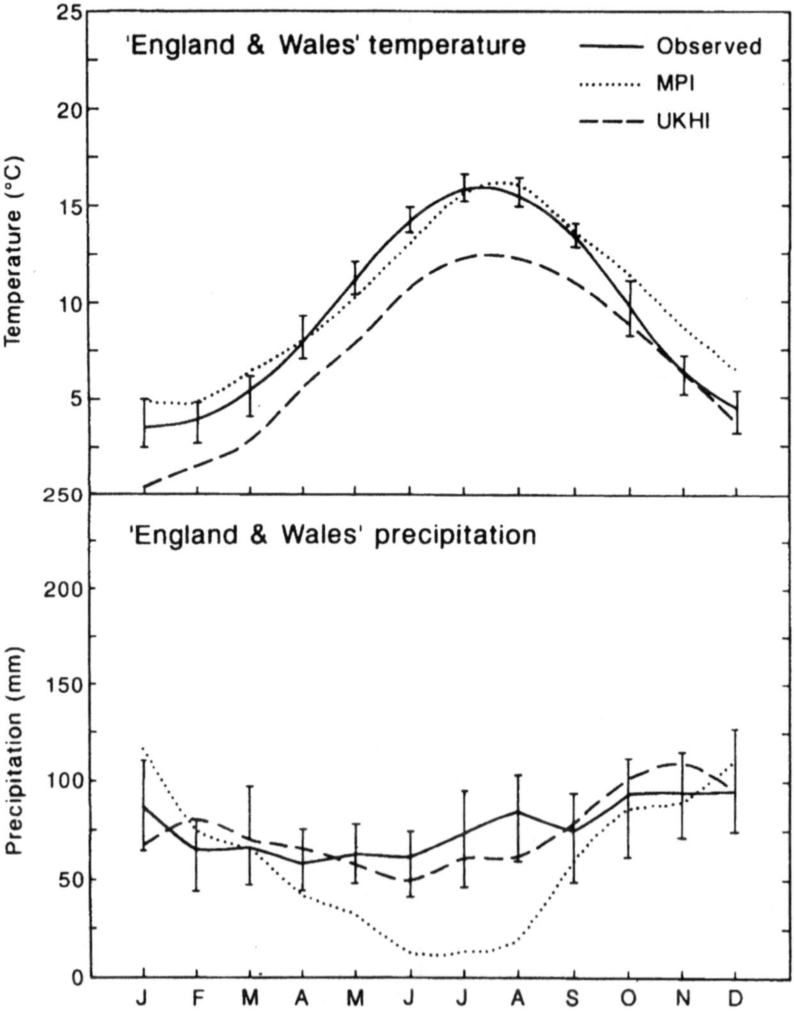

Table 7.3: *Comparison of significant ($p = 0.05$) correlations between seasonal mean weather frequencies and seasonal temperature representing the British Isles. The observed column (O) values are based on the objective typing scheme and instrumental data. The alternate model values (UKHI and MPI) are based on the relevant model-derived temperature data.*

TEMPERATURE

	Winter			Spring		
Type	O	UKHI	MPI	O	UKHI	MPI
A	-.27	-.50	-.46	+.26		
C			+.54	-.36	+.42	
W	+.67	+.53	+.46	+.24	+.41	
N	-.46		-.48	-.48	-.52	
E	-.65	-.45	-.54	-.27	-.51	+.36
S	+.26			+.31		
NW						

	Summer			Autumn		
Type	O	UKHI	MPI	O	UKHI	MPI
A	+.36					-.62
C	-.47				-.49	+.69
W			-.42	+.32	+.49	
N	-.34			-.61	-.40	-.58
E				-.21		
S	+.21			+.47		
NW	-.34			-.24		

7.5.3 Relationships Between Circulation Frequencies and Temperature and Precipitation

Tables 7.3 and 7.4 summarize the relationships between the seasonal frequencies of particular circulation types and mean temperature and total rainfall in the equivalent season. The relationships, quantified in terms of simple correlation coefficients, are shown for the real-world (observational) data and the model-generated circulation types, temperature and precipitation. The relationships are not always consistent in the different seasons (see also Table 7.1). S days are generally warm in all seasons. A-type days are, however, warm in summer but cool in winter. Precipitation variability is dominated by the frequency of synoptic types in all seasons; it being generally wet on C days and dry on A days.

The sign of the temperature relationships are, on the whole, correctly sim-

Table 7.4: *Comparison of significant ($p = 0.05$) correlations between seasonal mean weather type frequencies and seasonal precipitation representing the British Isles. The observed column (O) values are based on the objective typing scheme and instrumental data. The alternate model values (UKHI and MPI) are based on the relevant model-derived precipitation data.*

PRECIPITATION

	Winter			Spring		
Type	O	UKHI	MPI	O	UKHI	MPI
A	-.82	-.44	-.55	-.74	-.46	-.48
C	+.69		+.65	+.73	+.58	+.52
W	+.31	+.77			+.58	
N		-.38			-.43	
E		-.45			-.76	
S						
NW					+.49	

	Summer			Autumn		
Type	O	UKHI	MPI	O	UKHI	MPI
A	-.76	-.44	-.41	-.68	-.73	-.74
C	+.78	+.43		+.76		+.73
W			+.80		+.47	+.36
N		-.42				-.62
E		-.43	-.43			-.41
S		+.48	+.47			
NW						

ulated by the models, though performance in summer is inferior to winter, when both models reproduce the correct sign and strength of relationship between temperature and A, E and W types (and MPI correctly simulates a negative correlation with N). Only UKHI captures the correct spring relationships with W, N and E, though C is incorrectly warm. Both models correctly give cold N in winter but MPI has spurious cold A and warm C relationships. Importantly, neither model generates the correct (but weak) warm S relationship apparent in the observed data in all seasons.

As for the precipitation relationships, both models reproduce realistic correlations with A and C types though the magnitude of the model-related correlations is less than in the observed data and MPI does not reproduce the positive relationship (i.e. more rain) with C in summer while UKHI likewise does not produce wet C days in autumn and winter. The model data demonstrate overly strong relationships between precipitation and directional types, especially in UKHI. Whereas the observed relationships between precipitation and W types are significant (but still quite low at 0.31) only in winter, UKHI simulates a much stronger relationship (0.77) in winter and in spring and autumn. Similarly, MPI shows a very strong (0.80) but unrealistic relationship in summer (remember that MPI also produced too few W days in summer and too little precipitation - see Figures 7.5 and 7.6).

7.5.4 Weather-Type Spell Lengths and Storm Frequencies

Two final brief illustrations of ways in which the model data can be compared with observed data are shown in Figures 7.7 and 7.8. The first examines the lengths of C and A spells in the sample model output. UKHI tends to underestimate short and medium C spells (up to about 6 days) and overestimate longer spells, while MPI appears to overestimate all spells. Both models reproduce A spells quite well, except for the longest spell lengths which they would not be expected to reproduce given the comparatively short output sample (only 10 years).

The observed and modelled pressure fields can be used to generate a Gale index, G. This is calculated using the formula devised by Jenkinson and Collison (1977), i.e.

$$G = \sqrt{F^2 + (0.5Z)^2} \qquad (7.2)$$

where F and Z are calculated as shown in Section 7.3. A gale is defined here as any day when G is greater than 30. Jenkinson and Collison chose this value to accord with the gale frequency observed between 1951 and 1960 (see also Hulme and Jones, 1991 and Hulme et al., 1993).

The number of gales is too low in both models, particularly so in winter.

Figure 7.7: *Mean annual spell lengths of two example circulation types calculated from observed and model-generated data. Observed data range bars are explained in the legend to Figure 7.5.*

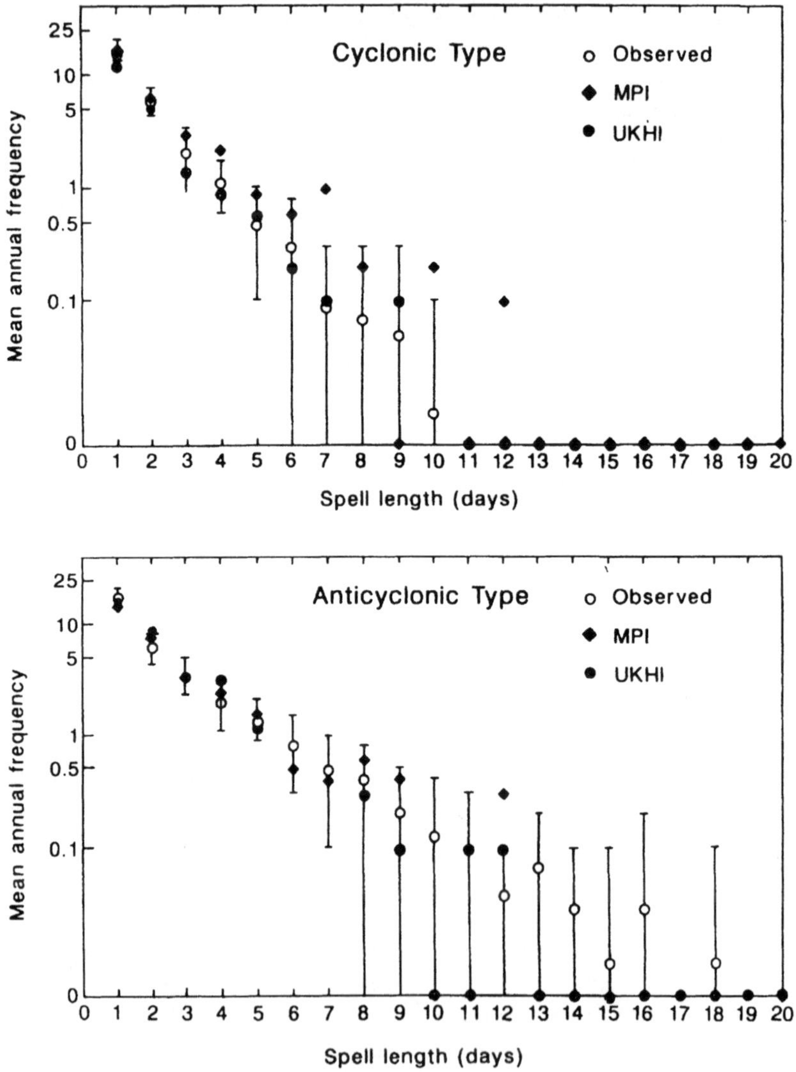

Figure 7.8: *Observed and model-based estimates of mean monthly gale frequencies over the British Isles. Note that the horizontal axis runs from July to June. The observed range bars are described in the legend to Figure 7.5.*

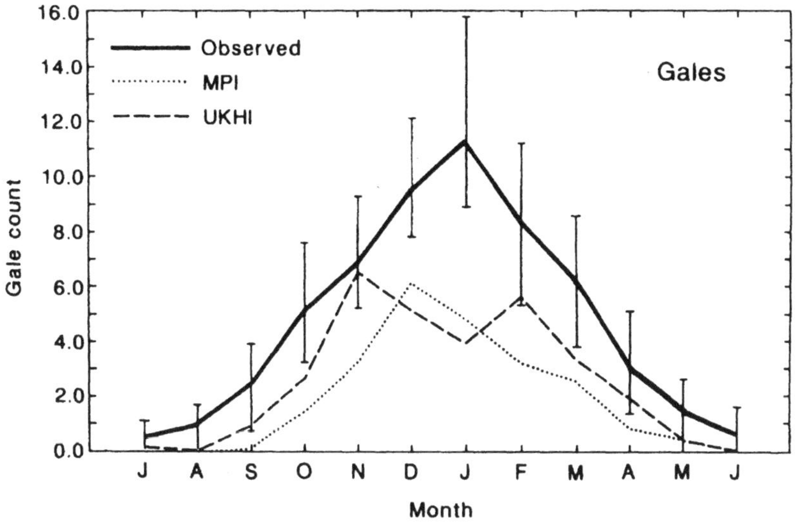

7.6 Conclusions

7.6.1 Specific Conclusions

Specific conclusion relating to the particular model-run comparisons explored here are:

- Both GCMs exaggerate the seasonal cycle of A and C types: MPI reproduces the correct annual total of A types while the UKHI A total is too low.
- UKHI gets the correct number of W days in a year but the seasonal cycle is out of phase. MPI underestimates W in most months, especially in winter.
- For other weather types UKHI is slightly better, although monthly totals fall outside of the observed ranges in some months.
- For precipitation, UKHI performs well and is close to the real mean values in all months. MPI is much too dry in late spring and summer.
- For temperature, MPI performs much better than UKHI, which is consistently too cold in most months.
- Both models correctly indicate the dominance of A and C types in determining rainfall.

- Both models exaggerate the strength of the relationships between directional types and precipitation, particularly for W.
- In both models the circulation relationships with temperature and/or precipitation are most poorly reproduced in summer.
- Both models underrepresent the number of gale days, particularly in winter.

7.6.2 General Conclusions

This chapter has described a simple, objective method of exploring the "internal climate consistency" of GCMs. The models explored here have been used only as illustrative examples and it is clear that this approach could and should be applied to other regions and other models.

Chapter 8

Statistical Analysis of GCM Output

by Claude Frankignoul

8.1 Introduction

In general circulation model (GCM) studies, statistical methods are needed for a number of purposes: to validate a model with observations, to identify its response to anomalous boundary conditions or its sensitivity to changes in model formulation, and to determine its predictive skill. The first two problems are discussed in this Chapter 8 and in the next Chapter 9, while the evaluation of forecasts is discussed in Chapter 10.

In model validation studies, the question is whether a GCM is *consistent* with reality, for example its mean state (climate) and its variability (weather), which are themselves estimated from the observations. One may also ask whether a change in a parameterization improves the model climate, or whether a model is *more realistic* than another one. Although visual comparisons can reveal obvious differences, they become less effective as the model's fidelity increases, and they are inadequate for separating the effect of model inadequacies from that of the observational data uncertainties. Hence, quantitative measures of agreement based on statistical techniques are needed.

In sensitivity or response studies, models are integrated twice in different conditions (e.g., boundary conditions) to document the model response to the prescribed change. The question is whether there is a *significant* climatic impact, and possibly whether the model response is consistent with a response anticipated from observations, theory, or prior experiments. Since the early seventies, many response studies have been made with atmospheric GCMs, and it is in this context that the analysis strategies discussed here were first developed.

The difficulties are that GCMs, like reality, usually have a large natural variability and that the sample is often limited, so that the ratio of the true GCM signal to the sampling variability (the signal-to-noise ratio) is generally small. Also, GCM fields, like observations, have large and complex correlation scales, so individual tests performed at grid points need to be interpreted by taking their multiplicity and their interdependency into account. Finally, the external forcing may be imperfectly known, resulting in GCM response uncertainties that need to be distinguished from the effects of model inadequacies.

If the GCM variables have a known statistical distribution, or approach it sufficiently, standard parametric methods can be used. Many GCM variables are very nearly Gaussian, and hence parametric tests are often used. However, some variables, like precipitation, are not normally distributed, and in this case nonparametric methods (such as permutation techniques) are needed. In most cases, the multivariate statistic methods are more appropriate than the univariate ones, but univariate tests are easier to implement. The univariate and multivariate approaches have advantages and limitations, depending on the problem at hand, and it is often advisable to use both.

In this chapter, the parametric approach is discussed with emphasis on the mean, rather than higher-order moments. In Section 8.2, the t-test on the mean is introduced, and its application to GCM experiments discussed. The multivariate test is introduced in Section 8.3. Because of the large dimension of the GCM fields, the sample size is much smaller than the dimension, hence general strategies had to be devised for the use of the standard multivariate methods: they are discussed in the context of both response studies and model testing. Permutation procedures are discussed in Chapter 9.

8.2 Univariate Analysis

8.2.1 The t-Test on the Mean of a Normal Variable

Let $\{\mathbf{x}_1, \dots \mathbf{x}_n\}$ be a sample of n independent observations on a univariate random normal variable $\mathbf{X}$ with distribution $\mathcal{N}(\mu, \sigma^2)$.[1] Then, the *null hypothesis* H_0 that $\mu = \mu_0$ can be tested against an alternative hypothesis H_A, say $\mu \neq \mu_0$ (two-sided test), by considering the test statistic

$$t = \frac{\bar{\mathbf{x}} - \mu_0}{s/\sqrt{n}} \tag{8.1}$$

[1] i.e., $\mathbf{X}$ is Gaussian distributed with a mean μ and a standard deviation σ.

where $\bar{\mathbf{x}}$ denotes the sample mean and

$$s^2 = \frac{1}{n-1}\sum_{i=1}^{n}(\mathbf{x}_i - \bar{\mathbf{x}})^2 \tag{8.2}$$

is an unbiased estimate of the variance σ^2. If H_0 is true, t is distributed as a Student t variable with $n-1$ degrees of freedom. Thus H_0 is accepted at the α level of significance if $|t| \leq t_{\alpha/2;n-1}$, where $t_{\alpha/2;n-1}$ is the upper $\frac{\alpha}{2}$ percentage point of the t distribution. Otherwise, if $|t| \geq t_{\alpha/2;n-1}$, H_0 is rejected[2]. Note that whether or not H_0 is accepted depends on the alternative against which it is being tested, and that the power of the test increases with the number of available samples. If the variance σ^2 is known, σ replaces s in (8.1), and the test statistic is then distributed as a $\mathcal{N}(0,1)$ variable if H_0 is verified.

To compare the means of two normal variables $\mathbf{X}$ and $\mathbf{Y}$ with the same variance and distributions $\mathcal{N}(\mu_X, \sigma^2)$ and $\mathcal{N}(\mu_Y, \sigma^2)$, respectively, from two random samples of n_X and n_Y independent observations, one considers the test statistic

$$t = \frac{\bar{\mathbf{x}} - \bar{\mathbf{y}}}{\sqrt{\frac{1}{n_X} + \frac{1}{n_Y}}\, s_p} \tag{8.3}$$

where s_p is an unbiased pooled estimate of the variance σ^2 given by

$$s_p^2 = \frac{1}{n_X + n_Y - 2}\left[\sum_{i=1}^{n_X}(\mathbf{x}_i - \bar{\mathbf{x}})^2 + \sum_{i=1}^{n_Y}(\mathbf{y}_i - \bar{y})^2\right] \tag{8.4}$$

Indeed, if the null hypothesis that $\mu_X = \mu_Y$ is true, t is distributed as a t variable with $n_X + n_Y - 2$ degrees of freedom, and we reject H_0 in favor of H_A if $|t|$ is larger than the corresponding critical value. If $\mathbf{X}$ and $\mathbf{Y}$ have different variances, the test remains valid when the number of samples is comparable for the two populations, $n_X \approx n_Y$. Otherwise, approximate solutions can be obtained (Behrens-Fisher problem).

The t-test is powerful and robust against departure from normality, and it has been widely used to evaluate the statistical significance of prescribed change experiments with atmospheric GCMs, starting with the work of Chervin and Schneider (1976). However, the tests have not always been interpreted carefully. Indeed, GCM fields are spatially correlated and different grid points do not provide independent information, as discussed below. Finally, note that a variant of the t-test based on the likelihood of recurrence in two random samples has been advocated by H. von Storch and Zwiers (1988).

[2]Of course, "rejecting" or "accepting" H_0 only implies that the observations are "unfavourable to" or "favourable to" the hypothesis.

8.2.2 Tests for Autocorrelated Variables

The standard t-test requires samples of independent observations, which are easily available when ensembles of GCM experiments are performed, providing sets of independent realizations of, say, monthly means. However, some GCM experiments are performed in "perpetual conditions", providing instead long time series that can be more effectively dealt with by taking into account their finite correlation time when estimating the standard errors of time averages.

The variance of the sample mean $\bar{\mathbf{x}}$ of a statistically stationary process $\mathbf{x}(1)\ldots\mathbf{x}(n)$ is given by

$$\sigma_{\bar{\mathbf{x}}}^2 = \frac{1}{n} \sum_{i=-(n-1)}^{n-1} (1 - \frac{|i|}{n})\gamma_x(i) \tag{8.5}$$

where $\gamma_x(i)$ is the lagged covariance of $\mathbf{X}_i$. Thus, if $\bar{\mathbf{x}}$ is approximately normal (which should always be the case for large samples, because of the central limit theorem) and if a reliable estimator $s_{\bar{\mathbf{x}}}^2$ of (8.5) could be found, the null hypothesis $\mu_X = \mu_0$ could be tested since

$$t = \frac{\bar{\mathbf{x}} - \mu_0}{s_{\bar{\mathbf{x}}}} \tag{8.6}$$

would be again distributed as a t variable if H_0 is true, with an approximate equivalent number of degrees of freedom ν given by

$$\nu = \frac{s_x^2}{s_{\bar{\mathbf{x}}}^2} - 1, \tag{8.7}$$

where s_x^2 is the sample variance. The main problem is that there is no unbiased estimator of $\gamma_x(i)$ when the true mean is not known, and that the use of traditional estimators in (8.5) results in unacceptable biases for small samples (Anderson, 1971). When samples are sufficiently large, (8.6) is asymptotically distributed as a $\mathcal{N}(0,1)$ variable if H_0 is true, and the usual t-test can be used. Guidelines for the application of the latter and alternative tests for the small sample case are given in Zwiers and H. von Storch (1995).

Alternatively, one can use spectral analysis to evaluate (8.5), since one has asymptotically for large n

$$\sigma_{\bar{\mathbf{x}}}^2 \approx \frac{1}{n}\Gamma_x(o) \tag{8.8}$$

where $\Gamma_x(0)$ is the spectral density near zero frequency. Atmospheric spectra are approximately white at low frequencies, hence $\Gamma_x(0)$ can be estimated if the time series are long enough, providing a direct estimate of the denominator in (8.6), or its generalization to the two-sample case. Many degrees of freedom are needed to get a stable estimate of $\Gamma_x(0)$, which can be obtained by averaging spectral estimates at low frequencies. However, the averaging may then include frequencies for which the spectrum is not entirely flat, resulting in an underestimation or an overestimation (depending on the sample size and the shape of the spectrum near zero frequency) of the percentage of rejections of the null hypothesis (Jones, 1976; Zwiers and Thiébaux, 1987).

An alternative way to take serial correlation into account is to fit a parametric model to the time series, and then use the fitted model to derive the variances needed in the asymptotic t-test (Katz, 1982; Thiébaux and Zwiers, 1984).

8.2.3 Field Significance

The interpretation of GCM experiments usually requires evaluating the significance of changes in a field composed of many grid points (and, possibly, variables). This raises the question of the collective significance of an ensemble of univariate tests, which requires taking into account the multiplicity of local tests and their interdependency.

Even in the simple case where all the univariate tests are independent, the overall rate of rejection of the null hypothesis at the α level should be larger than α for global or field significance, if the number of local tests is finite. The critical rejection rate can be inferred from the binomial distribution (H. von Storch, 1982; Livezey and Chen, 1983) and, for a small number of tests, the threshold for field significance (global rejection of the null hypothesis) can be large (see Figure 9.2). Furthermore, the interpretation of the results is difficult because only global decisions about the experiment are possible (H_0 rejected or accepted everywhere), without a way to discriminate between sub-regions of true or accidental change. For local decisions, one may have to adopt Madden and Julian's (1971) stringent choice of a very high significance level in the individual tests (e.g., $\alpha = 0.01$).

In the more realistic case where the GCM data, hence the local tests, are not independent but spatially correlated, one observes that H_0 tends to be rejected in "pools" of grid points, rather than at randomly distributed points, and one expects the critical rejection rate to be larger since the effective number of independent tests is smaller. This number is difficult to estimate, because the tests have complex and poorly known spatial correlations. Livezey and Chen (1983) have thus suggested to establish field significance by using permutation techniques, which is costly in computer time but can take properly into account the interdependence between the tests (see Chapter 9). An alternative is to use a multivariate approach, although, as discussed below, it may also have limitations.

Figure 8.1: *Mean sea-level pressure difference (in hPa) between anomaly and control runs in the GISS GCM Model I experiment. The region of positive and negative SST anomalies are delineated. The forward slashes indicate grid points where the null hypothesis is rejected at the 5% level using the two-sample t-test, and the backward slashes those using the test for autocorrelated variables. (From Hannoschöck and Frankignoul, 1985).*

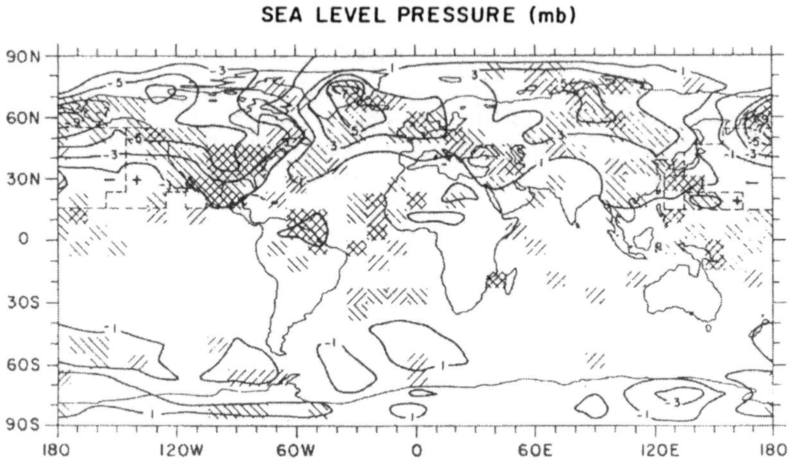

8.2.4 Example: GCM Response to a Sea Surface Temperature Anomaly

For illustration, consider the SST anomaly experiment of Hannoschöck and Frankignoul (1985). To evaluate the sensitivity of the GISS[3] GCM Model I to a North Pacific SST anomaly, three independent runs were performed in perpetual January conditions: a 15-month and an 8-month control run, and an 8-month anomaly run. The mean difference between the anomaly and the control runs is illustrated in Figure 8.1 for SLP. To evaluate its statistical significance, the two-sample t-test based on (8.3) was first applied, using 6-month averages to define the independent samples, which lead to one anomaly and three control samples. Forward slashes in Figure 8.1 indicate areas corresponding to grid points where the null hypothesis is rejected at the 5% level of significance. Although these areas are rather scattered, the averaged rejection rate is 14% in the Northern Hemisphere and 6% in the Southern Hemisphere. The binomial distribution suggests that, if all grid points were independent, global significance would be accepted for the former domain, but not for the latter.

The test (8.6) was then applied, using smoothed spectral estimates of the white noise level with 8 degrees of freedom. This corresponds to a low cut-

[3]Goddard Institute for Space Studies.

off period (200 days) so that the spectra should be fairly white and the bias discussed in Section 8.2.2, small. The backward slashes in Figure 8.1 indicate when H_0 was rejected by this test. We see that the rejection rate is substantially larger in the Northern Hemisphere (29%) than with the previous test, but not in the Southern Hemisphere (6%). Presumably, this occurs because the larger number of degrees of freedom enhances the signal-to-noise ratio. In the Southern Hemisphere the null hypothesis is frequently rejected over different areas with the two tests, while the regions of rejection coincide more often in the Northern Hemisphere. As the variance estimates used in the two tests are nearly independent, this, together with the very high rejection rates, suggests that the SST anomaly really has an influence on the northern hemisphere SLP. For some other atmospheric variables, however, Hannoschöck and Frankignoul (1985) found that the rejection rate was lower and no global assessment of significance could be made, lacking an estimate of the effective number of independent tests.

8.2.5 Analysis of Variance

An ensemble of GCM experiments can also be used to estimate the response of the atmosphere to the observed time-varying boundary conditions. In these experiments, many independent runs are performed using the same observed SST and sea-ice extents. To quantitatively estimate the seasonal predictability at each grid point, the standard analysis of variance (ANOVA, e.g., Morrison, 1976) is often used as it aims at testing the equality of several means and is thus well adapted to distinguishing between natural and forced GCM variability. Using the F-distribution, one can assess the null hypothesis that the variance caused by the boundary forcing is not significantly greater than zero (Zwiers, 1996; Rowell, 1998). Note that the technique has little power unless the sampling is large, and that field significance should be evaluated (Section 8.2.3) in regions with small potential predictability.

8.3 Multivariate Analysis

8.3.1 Test on Means of Multidimensional Normal Variables

If a sample of n independent observations $\vec{\mathbf{x}}_1 \ldots \vec{\mathbf{x}}_n$ of a m-dimensional random vector $\vec{\mathbf{X}}$ with the multinormal $\mathcal{N}(\vec{\mu}, \Sigma)$ distribution is available, with $n-1 \geq m$, the null hypothesis H_0 that $\vec{\mu} = \vec{\mu}_0$ can be tested against the alternative hypothesis H_A, say $\vec{\mu} \neq \vec{\mu}_0$, by considering the single sample Hotelling T^2 statistic

$$T^2 = n(\bar{\vec{\mathbf{x}}} - \vec{\mu}_0)^T \hat{\Sigma}^{-1} (\bar{\vec{\mathbf{x}}} - \vec{\mu}_0) \tag{8.9}$$

where

$$\bar{\vec{\mathbf{x}}} = \frac{1}{n}\sum_{i=1}^{n}\vec{\mathbf{x}}_i \tag{8.10}$$

and

$$\hat{\mathbf{\Sigma}} = \frac{1}{n-1}\sum_{i=1}^{n}(\vec{\mathbf{x}}_i - \bar{\vec{\mathbf{x}}})^T(\vec{\mathbf{x}}_i - \bar{\vec{\mathbf{x}}}) \tag{8.11}$$

are unbiased estimates of the mean $\vec{\mu}$ and the covariance matrix $\mathbf{\Sigma}$. Indeed, when the null hypothesis is true, the quantity $F = \frac{n-m}{m(n-1)}T^2$ has the F distribution with m and $m - n$ degrees of freedom (e.g., Morrison, 1976). Thus, H_0 is rejected at the α-level if

$$T^2 > \frac{m(n-1)}{n-m}F_{\alpha;m,n-m} \tag{8.12}$$

where $F_{\alpha;m,n-m}$ denotes the upper α quantile of the F distribution with m and $m - n$ degrees of freedom, and accepted otherwise. T^2 is the direct multivariate analogue of the univariate t-ratio (8.1), and it reduces to t^2 when $m = 1$. Except for skewed distributions, the test is rather robust against departures from normality. However, the T^2 test has little power unless the sample size n is much larger than m, which is seldom the case in GCM experiments. Note that, as n tends to ∞, T^2 become asymptotically distributed as χ^2_m when H_0 is true. When the true covariance matrix $\mathbf{\Sigma}$ is known, $\mathbf{\Sigma}$ should replace $\hat{\mathbf{\Sigma}}$ in (8.9); the test statistic is then also distributed as a χ^2_m variable when H_0 is true.

The generalization to the two-sample case is straightforward. If n_X and n_Y are independent observations of two random m-dimensional normal variables $\vec{\mathbf{X}}$ and $\vec{\mathbf{Y}}$ with the same covariance matrix $\mathbf{\Sigma}$ and distribution $\mathcal{N}(\vec{\mu}_X, \mathbf{\Sigma})$ and $\mathcal{N}(\vec{\mu}_Y, \mathbf{\Sigma})$, respectively, the null hypothesis $\vec{\mu}_X = \vec{\mu}_Y$ can be tested when $m < n_X + n_Y - 2$ by considering the two-sample Hotelling T^2 statistic

$$T^2 = \frac{n_X n_Y}{n_X + n_Y}(\bar{\vec{\mathbf{x}}} - \bar{\vec{\mathbf{y}}})^T\hat{\mathbf{\Sigma}}_p^{-1}(\bar{\vec{\mathbf{x}}} - \bar{\vec{\mathbf{y}}}) \tag{8.13}$$

where $\hat{\mathbf{\Sigma}}_p$ is a pooled estimate of the covariance matrix with $n_X + n_Y - 2$ degrees of freedom, defined as in (8.4) The rejection rule (8.12) applies by replacing n by $n_X + n_Y - 1$, and the test is only powerful if $n_X + n_Y - 2 \gg m$.

The effect of unequal dispersion matrices is discussed e.g. in Seber (1984). When n_X and n_Y are very large and $n_X \approx n_Y$, the effects of the differences in the covariance matrices on the significance level and the power of the T^2 test are minimal. If $n_X \neq n_Y$ and $\mathbf{\Sigma}_X \neq \mathbf{\Sigma}_Y$, the effects are serious. An appropriate test statistic is then

$$T^2 = (\bar{\vec{\mathbf{x}}} - \bar{\vec{\mathbf{y}}})^T\left[\frac{1}{n_X}\hat{\mathbf{\Sigma}}_X + \frac{1}{n_Y}\hat{\mathbf{\Sigma}}_Y\right]^{-1}(\bar{\vec{\mathbf{x}}} - \bar{\vec{\mathbf{y}}}) \tag{8.14}$$

where $\hat{\mathbf{\Sigma}}_X$ and $\hat{\mathbf{\Sigma}}_Y$ are conventional estimates of the covariance matrices $\mathbf{\Sigma}_X$ and $\mathbf{\Sigma}_Y$. When n_X and n_Y are very large, this statistic becomes asymptotically distributed as χ^2_m when H_0 is true. When the sample sizes are not large, (8.14) is approximately distributed as Hotelling's T^2, where the degrees of freedom can be estimated from the data (Seber, 1984).

Hasselmann (1979) has pointed out that, because of the very large dimension of the GCM fields, the atmospheric response will normally fail a multivariate significance test, even if $\mathbf{\Sigma}$ is known, since too many (noisy) parameters (grid-points) are needed to describe the circulation patterns. Hence, a signal will be very difficult to recognize from noise, unless filtering technics are used to increase the signal-to-noise ratio. In addition, GCM samples are always limited, so that the covariance matrice estimates have strongly reduced rank: information about the "noise" is only available in a subspace of much lower dimension.

To apply multivariate tests, the dimensionality must thus be strongly reduced, which severely limits the amount of model details that can be investigated in practice. Because of the subjective choice of a highly truncated representation, the data reduction must be done a priori. This can be a stringent limitation for response studies, as discussed below.

8.3.2 Application to Response Studies

To analyze sensitivity experiments with atmospheric GCMs, Hasselmann (1979) has suggested using a hypothesis testing strategy, where the anticipated GCM response is represented by an a priori sequence of guessed patterns derived from former knowledge or simpler dynamical models. The guesses are characterized by only a few parameters, and they must be ordered a priori in a sequence reflecting their anticipated contribution to the total response. Suppose that n_X independent realizations of a GCM field $\vec{\mathbf{X}}$ in control conditions are available, and n_Y independent realizations of the same field in anomaly conditions (denoted by $\vec{\mathbf{Y}}$). It is then assumed that the GCM response, described by the mean differences between anomaly and control runs, $\bar{\vec{\mathbf{x}}} - \bar{\vec{\mathbf{y}}}$, can be represented by a linear combination of K guess vectors $\vec{p}^{\,k}$; $k = 1, 2, ..., K (K \ll m)$:

$$\bar{\vec{\mathbf{x}}} - \bar{\vec{\mathbf{y}}} = \sum_{k=1}^{K} \alpha_k \vec{p}^{\,k} + \vec{n} \tag{8.15}$$

where the α_k are scalar parameters estimated by minimizing the residual error $\| \vec{n} \|$, and where the cut-off K must be determined (see also Section 13.2.1). The null hypothesis that there is no atmospheric response ($H_0 : \vec{\mu}_X = \vec{\mu}_Y$) is that the true parameters α_k $k = 1 \dots K$ are all equal to zero, $\vec{\alpha} = 0$. In the limited sample case, the null hypothesis should be tested by using the statistic (8.13) or (8.14) (Hannoschöck and Frankignoul, 1985); the χ^2 test of Hasselmann (1979) and H. von Storch and Kruse (1985) is only applicable for very large samples. The selection criteria for K in (8.15) may be sequential or not, with slightly different significance levels, as discussed in Barnett et al. (1981). One often chooses the highest value of K for which the null hypothesis is rejected. Alternatively, all the K guess vectors may be prescribed a priori, thereby defining a reduced base of fixed dimension for the analysis.

As reviewed below, three kinds of guess patterns have been considered: empirical guesses, theoretical predictions and patterns based on observations (see also Section 13.2.3). In each case the main limitation of the approach is that it requires a priori insights on the expected response. The multivariate test of significance can tell us whether the GCM response is consistent with our assumptions, but it does not indicate whether there is any significant response at all.

Hasselmann (1979, 1993) has also suggested using a pattern recognition (optimal fingerprint) method to optimize the statistical significance of the response, but the results may become biased. Thus, much caution is required to apply the optimization procedure to the small sample case (Hannoschöck and Frankignoul, 1985; H. von Storch and Kruse, 1985).

Venzke et al. (1999) have discussed an interesting generalization of the fingerprint method to the situation in which the spatio-temporal characteristics of the GCM response pattern are not known. They considered the case of an ensemble of GCM integrations forced by the same time-varying SST and sea-ice distribution. To optimize pattern detection in each season, they worked in a truncated space defined by the main modes of internal variability (Empirical Orthogonal Functions or EOFs, see Chapter 13) of the model, so that the sample noise becomes spatially white - it was already white in time. The spatio-temporal characteristics of the dominant forced response was then defined by performing an EOF analysis in the pre-whitened space, and transformed back to ordinary coordinates. The sensitivity to truncation was shown to be small.

a) Empirical Guesses

In their study of the GISS GCM Model I response to the North Pacific SST anomaly above, Hannoschöck and Frankignoul (1985) have assumed that the response is largest at the largest scales and used as guess vectors an ordered sequence of spherical harmonics Y_i^j of decreasing spatial scales, where j is the zonal wavenumber and i the total wavenumber. A more truncated expansion was used by Frankignoul and Molin (1988a) to investigate the response of the more realistic GISS GCM Model II to the same SST anomaly. The mean monthly changes between 5 anomaly and 5 control runs is illustrated in Figure 8.2 (left) for the 320 hPa geopotential height. The hatched areas show that the rejection rate for the t-test (at the 5% level) was only 4% in the Northern Hemisphere, giving no indication that the SST anomaly had any significant impact on the atmosphere. On the other hand, a significant large scale response was found in the Northern Hemisphere with the multivariate

Figure 8.2: *Left: Mean change in 320 hPa geopotential height between anomaly and control runs. Contour interval is 20 m. The grid points where H_0 is rejected at the 5% level are indicated by a hatched area.*
Right: Significant response at the 5% level (see text). Contour interval is 10 m.

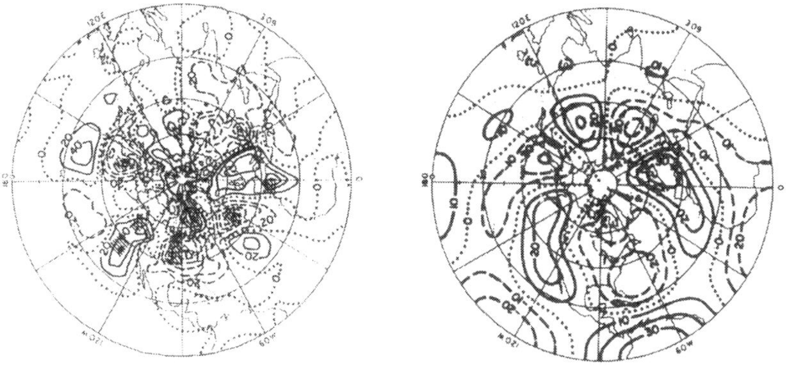

method. The a priori sequence of spherical harmonics had increasing zonal wavenumbers $j = 1, 2 \ldots$ and, for each zonal wavenumber, the meridional wavenumber was assumed to be that which dominated the climatology at the same zonal scale. Although this severely constrained the anticipated response, a significant signal was detected; using a mild selection criterion for the optimal model, which was taken as the maximum-order model satisfying the significance test at the 5% level, the sequence of guesses was interrupted at $K = 4$. The "significant 320 hPa height response" is primarily at zonal wavenumbers 3 to 5, and its projection onto the original grid shows several maxima of more than 40m (Figure 8.2 right). However, the side-lobes found in the tropics are clearly an artifact of the guess patterns, and the signal should not be interpreted locally.

This example illustrates the power of the multivariate testing approach at detecting significant changes in GCM response studies. However, it also stresses that, at least in the small sample case, the pattern of the "significant signal" is very guess-dependent.

b) Theoretical Predictions

The strong constraint inherent to having to formulate a priori guesses is not limiting for testing theoretical predictions, like whether the GCM response can be interpreted in part as a forced linear wave response, since in this case the anticipated response is well-defined a priori. We illustrate this approach with another small sample sensitivity study performed with the GISS GCM Model II (Frankignoul and Molin, 1988b). Here a positive SST anomaly was considered in the subtropical Pacific in order to test the applicability of stationary Rossby wave models to the atmosphere. The guesses were constructed using a barotropic model which was linearized about both the GCM mean zonal state and its mean wavy state, assuming for simplicity a local relationship between SST, diabatic heating, upper level divergence, and anticyclonic vorticity anomalies. The two basic states were tested separately, so there was only one guess vector in (8.15), and the test statistic (8.9) reduced to a (t-statistic)^2. Figure 8.3 (top) shows the mean monthly changes in 320 hPa geopotential height between 5 anomaly and 5 control runs, after "triangular truncation 12 (T12)". Hatched areas indicate the grid points where the null hypothesis was rejected at the 5% level, using standard univariate t-tests (8.3) (before T12 truncation). The rejection rate in the Northern Hemisphere was about 5%, consistent with the absence of a true signal. Also shown is the barotropic model predition, using the 320 hPa mean zonal (middle) and wavy (bottom) GCM flows as basic state. At the 10% level, the prediction is significant when the basic state is wavy, but not otherwise, and in the former case its magnitude is consistent with the GCM signal. However, the signal-to-noise ratio is low and only a small fraction of the GCM anomaly variance is explained by the barotropic model. Although this is due in part to the differences in the numerics of the two models (it would have been preferable to use a linearized version of the GCM to predict the linear wave response), the limited agreement was expected, since transient forcing, baroclinicity, nonlinearities and the divergent flow had been neglected in the wave model.

From a statistical viewpoint, the use of model predictions as guess vectors is more satisfactory than the use of empirical guesses, since the data reduction is much more effective, possibly leading to detecting very weak signals that could not be seen otherwise. Furthermore, the results can provide interesting theoretical clues (in the example above, on the role of the mean zonal asymmetries and the equivalent barotropic level). The multivariate method is thus a useful tool both for hypothesis testing and for interpreting sensitivity studies. However, the implications for the GCM response are again guess-dependent, and caution is required: another model might have explained a larger part of the differences between anomaly and control runs, or the use of an ill-adapted prediction might have improperly suggested that there was no significant GCM response.

c) Observed Anomaly Patterns

In some cases, one may want to verify the consistency of a GCM response with observed anomaly patterns. Of course, this requires the expected response to the prescribed anomaly to be well-documented. To a large extent, this is the case for the atmospheric response to the warm SST anomaly appearing in the equatorial Pacific during El Niño. Thus, H. von Storch and Kruse (1985) have used the 500 hPa height anomaly observed in January 1983 (Figure 8.4 left), during the most intense El Niño on record, as guess

Figure 8.3: a) *Mean change in 320 hPa geopotential height (in m) between anomaly and control runs for the subtropical SST anomaly experiment with the GISS GCM Model II, in T12. Contour interval is 20 m.*
b) *Geopotential height response estimated from the barotropic model using the 320 hPa mean zonal GCM control flow as basic state. Contour interval is 10 m.*
c) *As b) but for the 320 hPa mean wavy flow. The hatched areas indicate grid points where the null hypothesis is rejected at the 5% level. (From Frankignoul and Molin, 1988b).*

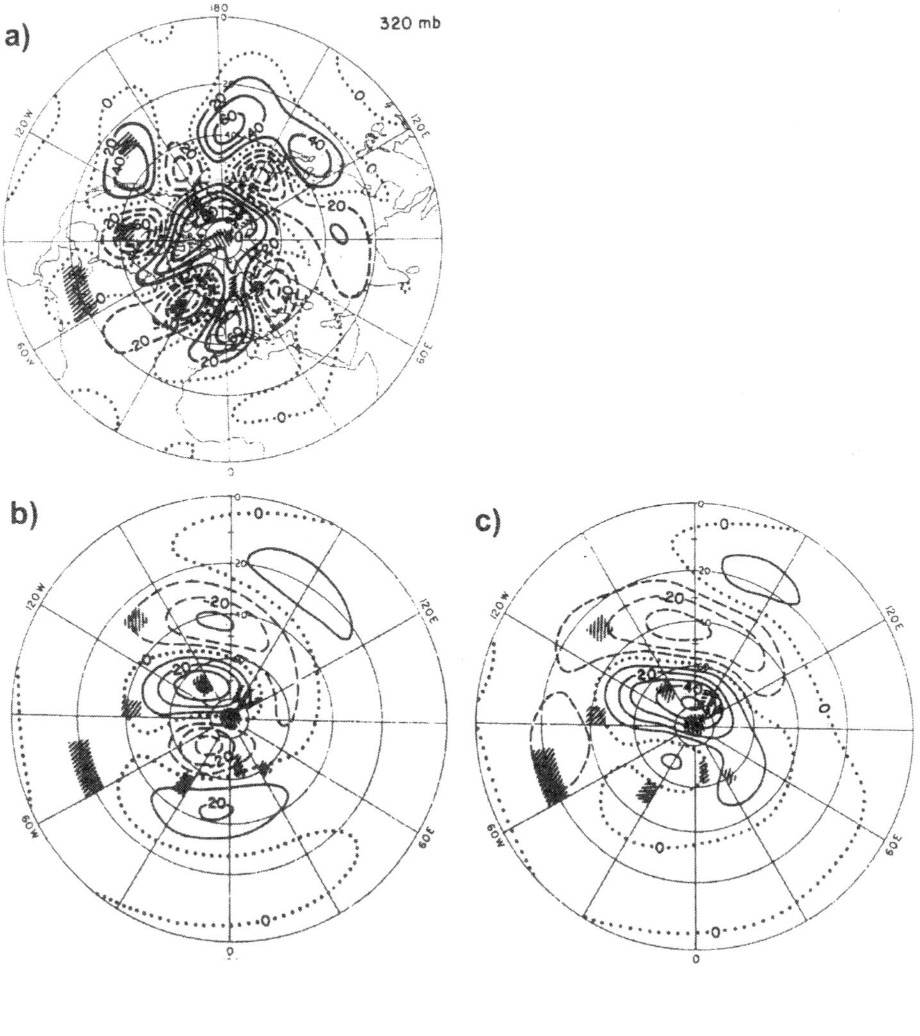

Figure 8.4: *Left: The mean 500 hPa geopotential height anomaly (in tens of m) observed in January 1983. Contour interval is 40 m.*
Right: Three-January mean northern-hemispheric response pattern to the warm anomaly composite simulated by the ECMWF GCM. Contour interval is 10 m. (From H. von Storch and Kruse, 1985).

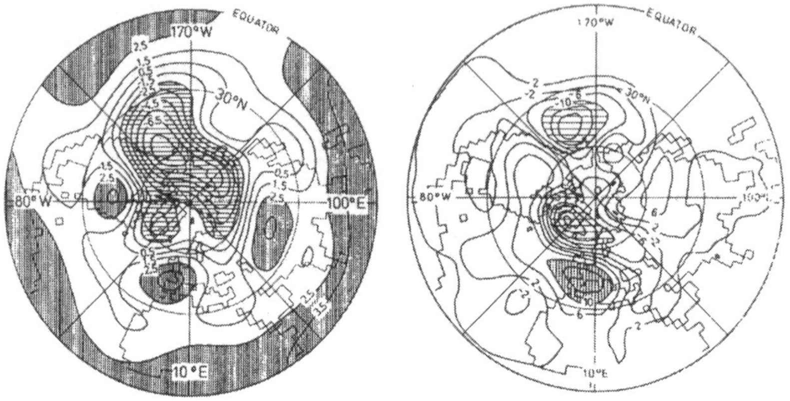

vector for the response of the T21 ECMWF[4] GCM to the composite warm El Niño SST anomaly of Rasmusson and Carpenter (1981), as well as to an anomaly of reverse sign. Nine independent January means were available, plus three anomaly Januaries in each SST anomaly case. The mean anomaly response of the model to the positive SST anomaly (Figure 8.4 right) seems rather similar to the observed pattern. This was confirmed by the multivariate test, although the guessed pattern was shown to only explain part of the GCM signal. Note that H. von Storch and Kruse (1985) used permutation procedures to establish significance, but similar results would have been found with a parametric test. On the other hand, the null hypothesis was accepted in the cold anomaly case (i.e., the amplitude of the guess vector was not significantly different from zero), suggesting that the atmospheric response is not linear. This is of interest, but the test gives no other clues on the GCM response to the cold anomaly.

Although useful, the use of observed patterns as guess vectors also has limitations, which are easily discussed in the context of this experiment. Indeed, the "observed signal" may have been strongly affected by the noise associated with the natural variability of the atmosphere, especially since it was derived from a single monthly mean. Furthermore, the prescribed SST anomaly was not exactly that observed in 1983 (nor could it be, because of observational uncertainties), and SST anomalies outside the tropical Pacific, or other external forcings, may have contributed to the observed change. Thus, the GCM

[4]European Centre for Medium Range Weather Forecast.

forcing was an approximation to the true one, and the guess vector only a noisy, and possibly biased, guess of the true signal. An improved procedure would be to base the guess vector on observed anomaly composites, thereby decreasing its noise level, and furthermore to take into account its statistical uncertainty in the testing procedure. This could be done by specifying not only the guess vector but also its error covariance matrix, as discussed below in the context of ocean model testing.

d) Summary

The multivariate approach is a very powerful tool to interpret the results of response and sensitivity studies with atmospheric GCMs. It is easy to implement, but its usefulness is limited by the difficulty, or the subjectivity, of constructing adequate guess vectors. This may explain in part why the standard univariate t-test has remained more popular in the analysis of GCM response studies.

8.3.3 Application to Model Testing and Intercomparison

A necessary step in the development of atmospheric and oceanic models is to test whether they are consistent with the available observations. Although they are well-suited for this purpose, the methods based on multivariate analysis have been little applied to atmospheric GCMs. Visual comparisons between simulated and observed climatologies remain so far the standard procedure in the atmospheric case, even though they can lead to an incorrect assessment of their agreement. On the other hand, multivariate model testing is being used more frequently in the oceanic context. The ocean is driven by the air-sea fluxes of momentum, heat and freshwater, which are only known with large uncertainties. This induces uncertainties in oceanic model response that have large correlation scales. In addition, the oceanic fields used for validation may be analyzed fields with highly correlated errors. Hence, distinguishing between ocean model inadequacies and data uncertainties requires a multivariate viewpoint, even more than in the atmospheric context where external forcing is well-documented and the observations often accurate.

As before, the multivariate approach requires a strong reduction in the dimensionality. However, the main difficulty encountered in the interpretation of GCM sensitivity studies, namely the arbitrariness of the choice of the guess vectors, disappears, since the obvious choice of the reduced base becomes that which allows the best representation of the main features of both the modeled and the observed fields. On the other hand, the observational errors need to be taken into account, which can be tedious.

a) Testing Atmospheric GCMs

Of interest is an early study of H. von Storch and Roeckner (1983), who compared four individual January simulations with the Hamburg University GCM to observations from 15 Januaries. The data compression was done by means of EOFs, and the covariance matrix estimated from the observed data. The multivariate tests were based on the χ^2 distribution, although the Hotelling's T^2 would have been more accurate, and univariate tests were used subsequently to try detecting which structures might have caused the significant model-observations disagreement that was found.

b) Testing Ocean Models

Frankignoul et al. (1989) have developed a multivariate model testing procedure which compares the main space and time structures of the model-observation differences to those of the data uncertainties. The procedure is illustrated here by an investigation of the ability of the LODYC[5] nonlinear 2-layer model at simulating the 1982-1984 variations of the tropical Atlantic thermocline depth around its 3-year mean (Frankignoul et al., 1995).

The tropical oceans are primarily wind-driven, and the thermocline variations could be simulated deterministically for the most part if the wind stress were accurately known. However, wind observations are inaccurate with frequent gaps, and the bulk formulae used to estimate the wind stress are uncertain. To represent these uncertainties, three independent, equally plausible monthly wind stress fields were constructed by Monte Carlo method and used to force the ocean model (Braconnot and Frankignoul, 1993). Although their dispersion only represents part of the forcing uncertainties, the corresponding thermocline depth uncertainty has a standard deviation of several meters, nearly as large as its observed interannual variability. The mean model response $\vec{\bar{\mathbf{m}}}(x,t)$ at location x and time t was calculated in the space-time intercomparison domain, as well as its sample error covariance matrix $\hat{\mathbf{\Sigma}}_M$. The observed thermocline depth $\vec{\mathbf{d}}(x,t)$, as determined by the depth of the 20^oC isotherm, was derived from temperature measurements by Reverdin et al. (1991). The data coverage was very sparse (Figure 8.5), so that the field was mapped monthly using a function fitting algorithm and a cut-off period of 3 months. The covariance matrix $\mathbf{\Sigma}_D$ of the analysed field was calculated, so that the non-locality of the interpolation procedure is accounted for in the error field. Although details can not be trusted, $\mathbf{\Sigma}_D$ is treated in the analysis as a true covariance matrix.

The observed and simulated thermocline depth variations are shown in Figure 8.6. The agreement is characterized by the misfit

$$T^2 = (\vec{\bar{\mathbf{m}}} - \vec{\mathbf{d}})^T(\hat{\mathbf{\Sigma}}_M + \mathbf{\Sigma}_D)^{-1}(\vec{\bar{\mathbf{m}}} - \vec{\mathbf{d}}) \qquad (8.16)$$

which is approximately distributed as Hotelling's T^2 with m (the dimension

[5] Laboratoire d'Océanographie Dynamique et de Climatologie.

Figure 8.5: *Distribution of observed temperature profiles used in the analysis for June 1984. (From Reverdin et al., 1991)*

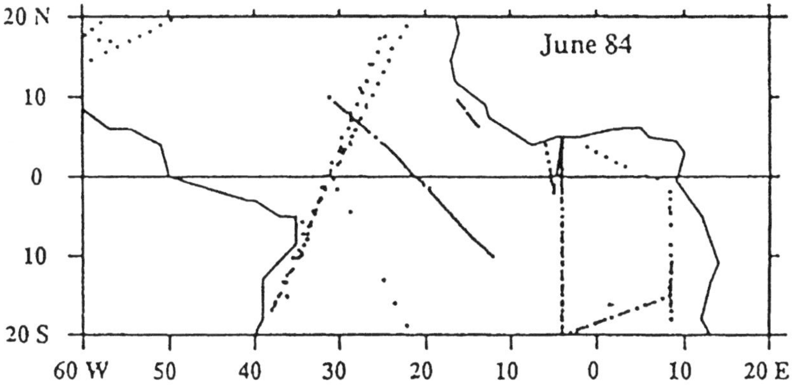

of the intercomparison space) and n (the degrees of freedom of $\hat{\Sigma}_M + \Sigma_D$) degrees of freedom, if the null hypothesis H_0 that there is no model error holds. As m is large, a strong data compression is done. Using Common Principal Component Analysis[6], an orthonormal base is first defined where the main features of simulations and observations are well represented. The first four common Empirical Orthogonal Functions (EOFs) account for more than 80% of the observed and simulated fields (Figure 8.7, left). Observations and simulations are represented in this subspace by the time series in Figure 8.7 (right), where the 95% confidence intervals already suggest that the observational and forcing uncertainties cannot explain all the model-reality discrepancies. A time compression is then done by an EOF analysis of the *differences* between observed and model time series, so with no further truncation, the dimension of the reduced space is $4 \times 4 = 16$. Due to the large estimated values of the degrees of freedom n, T^2 behaves approximately as a χ^2 variable with 16 degrees of freedom. The misfit turns out to be much larger than the critical value at the 5% level, and H_0 is rejected.

c) Model Intercomparison

Since no oceanic model is expected to be perfect and not all of the forcing uncertainties can be represented, it is expected that T^2 in (8.16) will be larger than the critical value if sufficiently accurate observations are available. However, the test statistic also indicates how closely a model is able to reproduce

[6]Generalization of principal component analysis (Chapter 13) that applies simultaneously to two or more fields. The EOFs are common to the different fields but the eigenvalues and principal components differ, because of sampling variability and/or true differences. The transformation can be viewed as a rotation yielding variables that are as uncorrelated as possible simultaneously in k groups (see Flury, 1989).

Figure 8.6: *Left: Observed departures (in m) of the 20°C isotherm depth from the 1982-1984 mean for March and July, from 1982 to 1984.*
Right: Corresponding simulations with the LODYC nonlinear 2-layer model.
(From Frankignoul et al., 1995).

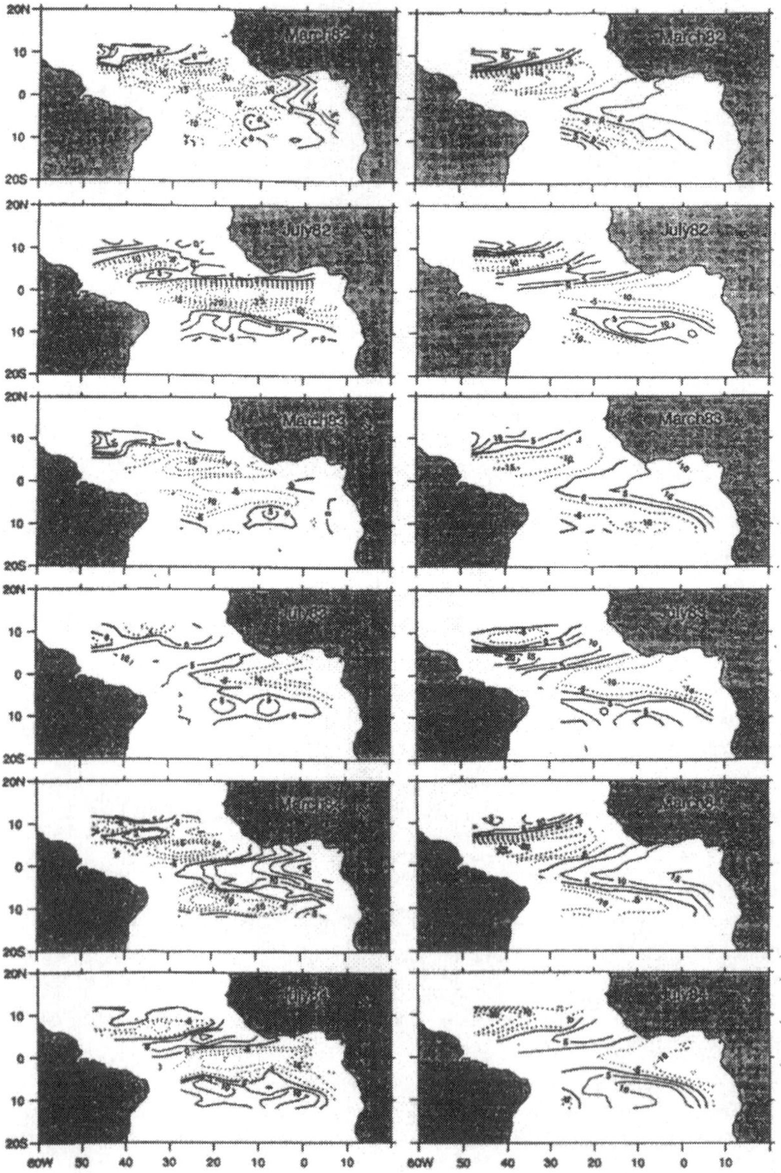

Figure 8.7: *Left: basis vectors for the spatial reduction.*
Right: projections of observed (continuous middle line) and modeled (dotted middle line) fields onto the spatial base, with univariate 95% confidence intervals. (From Frankignoul et al., 1995)

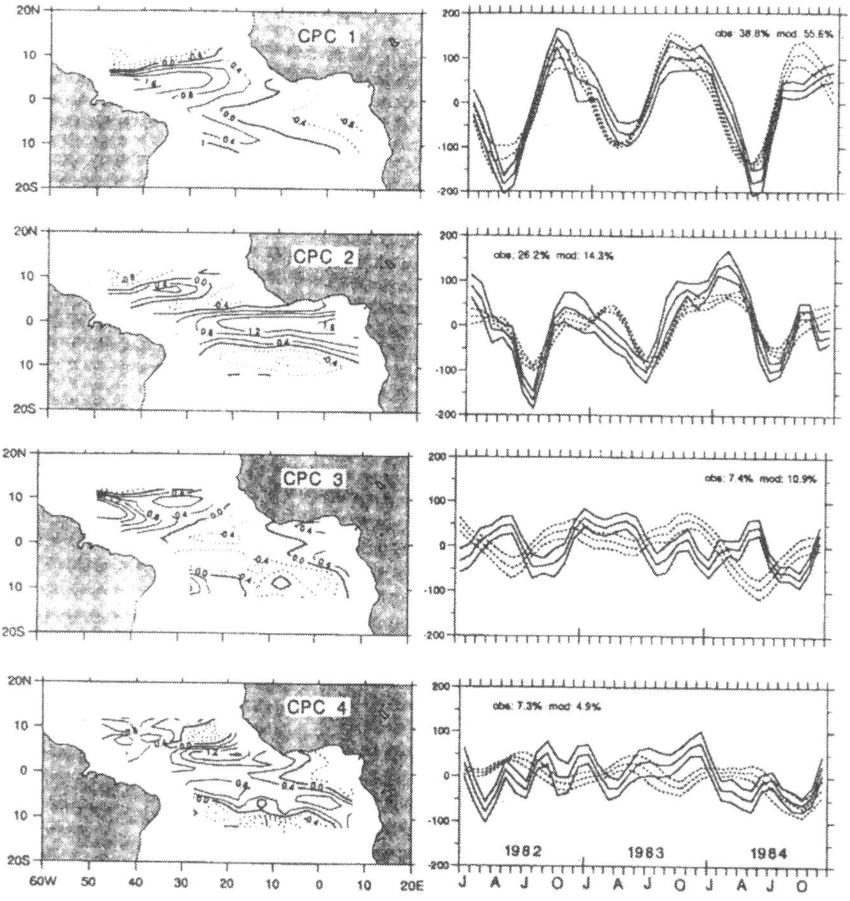

the observations and is thus handy for model intercomparisons. Figure 8.8 summarizes an intercomparison between four tropical Atlantic ocean models of increasing complexity: a linear model, the LODYC nonlinear 2-layer model, and two GCMs of increasing resolution (see Frankignoul et al., 1995, for details of the various models), in the same context of 1982-1984 thermocline depth fluctuations. The approximate error bars on the misfits are based on the non-central T^2, and they should only be taken into account once in the intercomparison since the observation and the forcing errors are the same in each case. Although H_0 is rejected in all the cases, the differences in model performance are significant, and consistent with our expectations that more complex models perform generally better than simpler ones.

Figure 8.8: *Misfit between four tropical Atlantic models and observations in the four testing regions indicated on the map below, for the 1982-1984 thermocline depth variations around the mean. The error bars represent the 95% confidence intervals, and the dotted line the critical value for rejecting the null hypothesis of no model error at the 5% level. (From Frankignoul et al., 1995).*

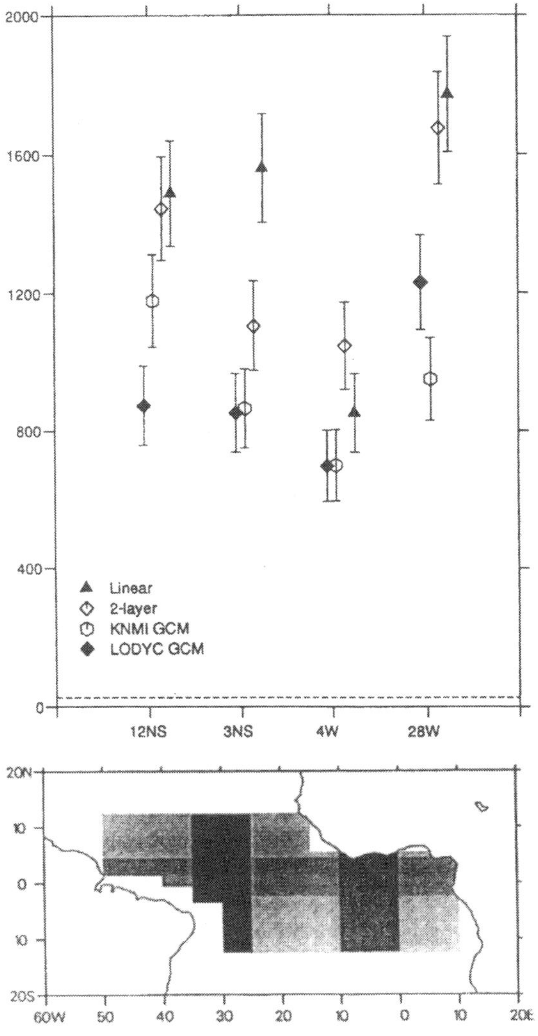

Chapter 9

Field Intercomparison

by Robert E. Livezey

9.1 Introduction

Frequently analyses of climate data, whether model-generated or observed, require the simultaneous assessment of the significance of multiple statistics. A commonly occurring situation is the test of a hypothesis for the difference in means of two fields of data, e.g. the average wintertime temperature anomaly pattern over a net of European stations in each of two sets of GCM simulations. Regardless of whether the problem is approached with the multivariate methods described in Chapter 8 or as a collection of individual or *local* significance tests, the collective or *field* significance of the results depends crucially on the number of data points or tests and their interdependence.

In those situations where the sample size is at least several (three to five) times as large as the effective dimensions of the analyzed fields, the prescriptions discussed in Chapter 8 are appropriate. However, often in climate studies or studies of interannual variability this is not the case. In these instances the permutation and Monte Carlo techniques described in this chapter are generally effective alternatives for hypothesis testing. They also have the advantage that they can be used when reference distributions of test statistics cannot be analytically derived or are otherwise unknown, thereby also permitting the use of innovative statistics (although the design basis for such statistics should still be sound statistical principles). Generally these procedures are not difficult to apply.

When, in addition to relatively small sample sizes, *a priori* expectations of experimental outcomes are missing (i.e. "fishing expeditions"), then the methods of this chapter may be the only effective alternatives for objective assessment of the statistical significance of the results. Here "fishing expedition" refers to the entire collection of statistics examined without *a priori* expectations. For instance, following the example above, if more than two sets of GCM simulations are searched two-by-two for differences in the average wintertime temperature anomaly pattern over Europe then significance tests must account for the multiple difference maps as well as the multiple locations on each map. Incorrect inferences are more likely if such tests are conducted for only the "best" difference map.

These ideas will be illustrated in Section 9.2 through an extended but relatively simple example taken from Livezey and Chen (1983). Two strategies for conducting permutation techniques and some of their properties in a multivariate test environment without serial correlation with and without interdependence among the variables will be introduced in Section 9.3, which relies heavily on the work of Zwiers (1987). Next, a discussion of the effects of serial correlation on permutation techniques and strategies to account for the effects will be presented in Section 9.4. This last material is mainly a synthesis and condensation of ideas developed by F.W. Zwiers and colleagues (Zwiers, 1990; Thiébaux and Zwiers, 1984; Zwiers and H. von Storch, 1995) and K. E. Trenberth (1984)), but significant recent contributions of Ebisuzaki (1997) and Wilks (1997) are noted as well.

Climate analysts who must design and/or evaluate studies of variability on interannual and longer time scales should find the already cited papers enormously useful if not mandatory sources of information and examples pertinent to the methods of this chapter. Further examples can be found in Livezey (1985), Zwiers and Boer (1987), Santer and Wigley (1990), Preisendorfer and Barnett (1983), and Mielke et al. (1981). The latter two references offer two formal approaches to application of permutation methods. In addition, Wigley and Santer (1990) present a detailed discussion of the relative merits of measures available for a variety of experimental situations. Measures specifically intended for evaluation of forecasts and forecast methodologies and special considerations for the conduct of these assessments are presented in Chapter 10.

9.2 Motivation for Permutation and Monte Carlo Testing

This author's introduction to the type of problem towards which permutation techniques are directed forms the central focus of Livezey and Chen (1983) and will be used to illustrate the effects of multiplicity and interdependence in multivariate hypothesis testing.

The experimental environment of an analysis by W. Y. Chen satisfies all of the conditions mentioned in Section 9.1 that suggest the use of permutation or Monte Carlo hypothesis testing: Correlations between a SOI and 700 hpa DJF heights from 29 years of data were computed at every point of a 936-point extratropical Northern Hemisphere grid without an *a priori* expectation of the result. The height data is at least cross-correlated[1], the ratio of sample size to dimension of the height field is small (even if the bulk of its meaningful variance is compressed into EOFs [Chapter 13] this dimension would still be at least 15 to 20), and the calculations are part of a "fishing expedition" (i.e. *a posteriori*). Correlation maps were also produced with the SOI leading the winter heights by one and two seasons and at all three leads (0-2 seasons) for JJA as well, thereby increasing the number of correlations and interdependencies between them to be evaluated. All correlations were tested at the 5% level with a Students-t test.

9.2.1 Local vs. Field Significance

The results of the lag-0 winter correlations are shown in Figure 9.1. The most basic problem facing the analyst given these results is whether or not a null hypothesis that the field of correlations arose by chance can be rejected with confidence, i.e. to decide whether or not there is sufficient evidence to believe the relationship between SOI and winter DJF extratropical Northern Hemisphere 700 hpa heights was non-accidental.

The fact that there are a number of areas where the correlations have local significance and that they account for more than 5% of the grid area is insufficient in itself to reject the null hypothesis at the 5% level and presume there is field significance. First, if the null hypothesis was everywhere true there would still be multiple opportunities for local significance to occur by chance. Moreover, the expected number of these occurrences would be 5% of all of the tests so the probability would be 0.5 that more than this number would occur by accident.

In fact even if all of the tests were independent there would still be a substantial probability (from binomial considerations) that noticeably more or less than 5% of the 5% local tests would pass by accident (under the null hypothesis). The per cent of area locally significant at the 5% level that is equaled or exceeded 5% of the time under the null hypothesis is given by the curve in Figure 9.2, and for 936 points is equal to more than 6%. Thus, to reject the null hypothesis at the 5% level in the absence of spatial correlation requires local significance over more than 6% of the grid. In Figure 9.1 11.4% of the area is covered by local tests significant at the 5% level.

The effect of spatial correlation in the height data is to lower the "effective" number of tests and to broaden the distribution of outcomes, leading to more stringent criteria for field significance. The winter lag-0 result would be field significant only if this unknown effective number of tests exceeded about 52 (the intercept of 11.4% with the binomially-derived curve in Figure 9.2). In

[1]All of the annual time series used in this analysis have some serial correlation that was treated heuristically in the significance tests. For the present this will be ignored. In Section 9.4 more solidly-based strategies for serial correlation will be presented.

Figure 9.1: *Correlation of winter season averages of a Southern Oscillation Index (SOI) and 700mb heights in hundreds. Negative and positive isopleths are shown as thin dashed lines and solid lines, respectively, with the two separated by heavy solid zero lines. A subarea approximating the Pacific basin, used for the experiment summarized in Section 9.2.2 is enclosed by a heavy dashed line. Points on a* $5^o \times 5^o$ *latitude-longitude grid at which correlations are 5% significant are indicated by solid dots: open dots represent additional points significant with a more liberal test. (From Livezey and Chen, 1983).*

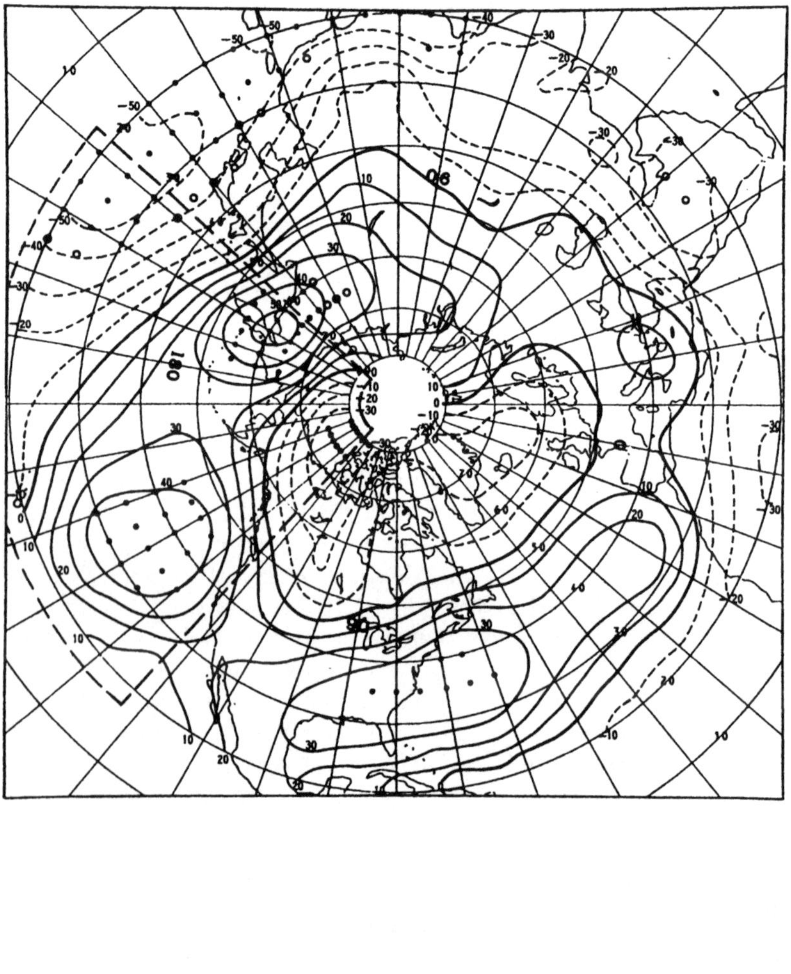

Figure 9.2: *Estimated percent of independent 5% (p = 0.05) significance tests passed that will be equaled or exceeded by accident 5% (P = 0.05) of the time versus the number of independent tests N (labeled "DOF" for "degrees of freedom"). The curve is based on the binomial distribution. The plotted point and coordinate lines and points refer to the significance test of Chen's experiment described in the text. (From Livezey and Chen, 1983).*

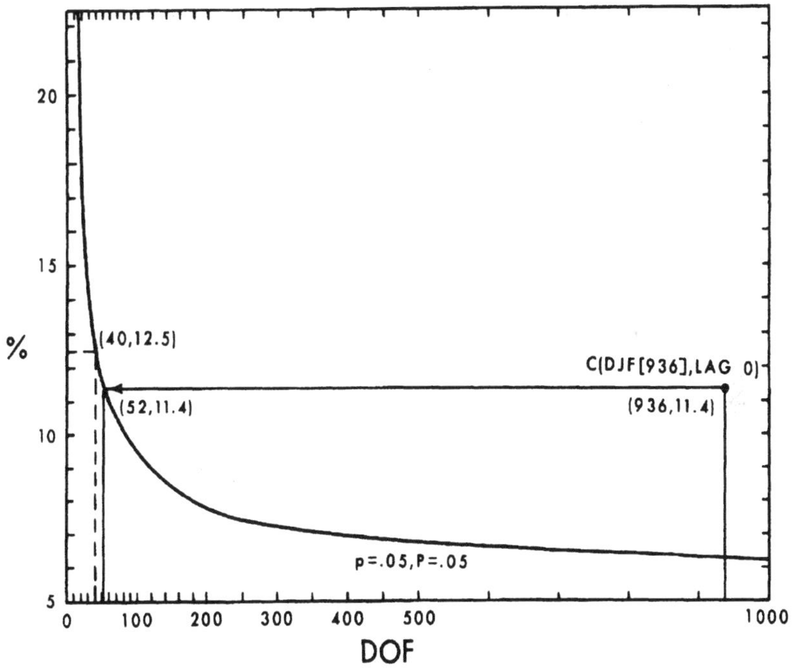

the present case there is some uncertainty that this is the case, therefore a Monte Carlo or permutation test is needed to resolve the question of field significance of the correlation map.

9.2.2 Test Example

In this situation two approaches are viable. Livezey and Chen (1983) first substituted a random series generated from Gaussian white noise for the SOI series, recomputed correlations and reperformed the local t-tests, and counted the number of rejections of the null hypothesis. An alternative approach would have been to use a random reordering (resampling without replacement; a permutation procedure as described in Section 9.3) of the SOI instead of the Gaussian noise.

Two hundred proxy SOI series were produced and all of the calculations repeated for each series. Finally, the experimental result (11.4%) was tested against the 5% tail of the empirical distribution function of the Monte Carlo test. This distribution function and the lag-0 result are shown in Figure 9.3a. It is clear that the null hypothesis cannot be rejected at the 5% level for the DJF lag-0 result. Note, however, that the lag-1 and -2 results may be significant at the 5% level, suggesting predictability of the DJF height field from the SOI, although so far the multiplicity of maps generated (different seasons, different lags) has not been taken into account.

In summer, it is known that the structural scale of the low-frequency height variability is smaller than in winter. This implies a larger number of effective tests and a narrower distribution of random outcomes. This is reflected in Figure 9.3b which also shows that none of the lag correlation maps has field significant signals.

The comparison of summer to winter is analogous to sampling a signal with different grid resolutions; because structures are generally larger in the winter they are sampled more densely than summer structures.

Instead, a comparison can be made in which spatial scales are comparable but the grid coverage is reduced. In the context of Figure 9.2 this amounts to the criterion for field significance moving along the solid curve from right to left and increasing, in other words a broadening of the empirical distribution function. This is illustrated in Figure 9.3c, which shows the empirical distribution function and experimental results for the Pacific/North America (PNA) subdomain. Only the lag-2 results are possibly field significant for this subdomain. The relative difficulty in detecting signals in the two contrasting situations (winter vs. summer and full vs. PNA domain) illustrated in Figure 9.3 was first pointed out by Hasselmann (1979).

Incidentally, Livezey and Chen (1983) did not account for the multiplicity and interdependence of seasons and lags considered, but easily could and should have by use of randomizations of the DJF SOI series. The same randomizations could be used simultaneously with the four preceding seasonal mean SOIs (DJF,MAM,JJA,SON) to consistently produce all of the lag cor-

Figure 9.3: *Histograms of percent of area with correlations of 700mb heights and Gaussian noise statistically significant at the 5% level (p = 0.05) in 200 Monte Carlo trials for: (A) the winter hemisphere; (B) the summer hemisphere; and (C) the winter north Pacific basin (outlined in heavy dashed line in Figure 9.1) The abscissa is percent of area while the ordinate is number of cases for each one percent interval. The 5% tail is schematically shown by shading the 10 of 200 largest percents. The results of correlations i-th seasonally averaged SOI's, with heights lagging by the indicated number of seasons, are shown by vertical arrows. (From Livezey and Chen, 1983).*

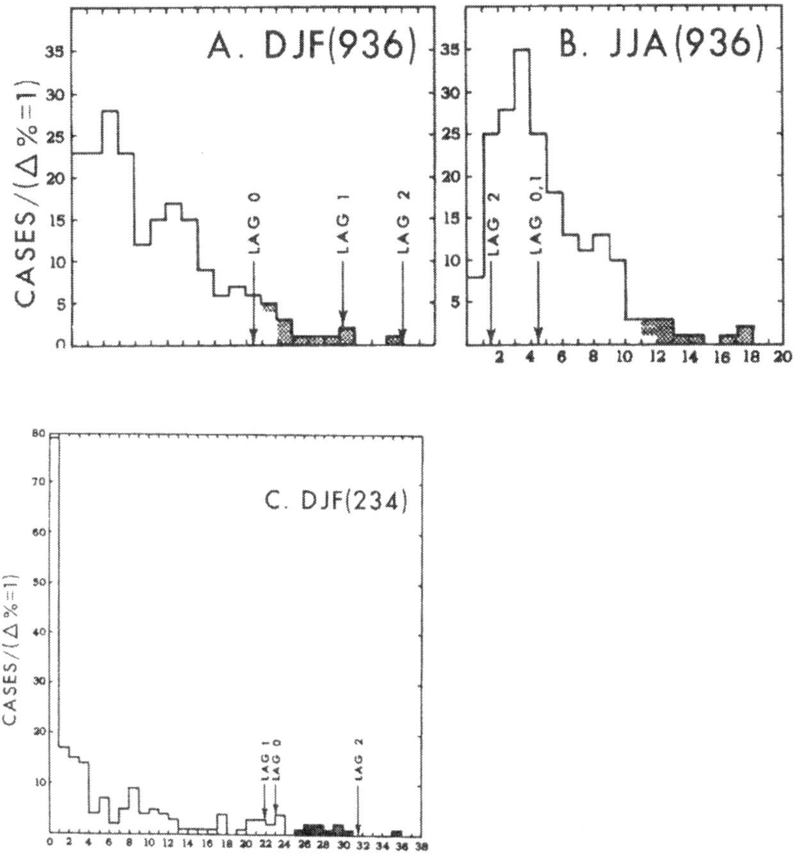

relations for both summer and winter, thereby preserving season to season serial correlation in both the SOI and height data. In this context, the permutation technique ultimately would have been superior to the Monte Carlo approach actually used. In the next section two approaches to resampling with permutation techniques will be described.

9.3 Permutation Procedures

To illustrate the application and properties of the two most commonly employed permutation methods, examples of tests of null hypotheses concerning the difference of means of two fields and the analyses of Zwiers (1987) will be used. The discussion will begin with consideration of two fields within which there is no cross-correlation. This will be subsequently added to study its impact on the procedures. Serial correlation is not considered in this section.

9.3.1 Test Environment

The analysis of Zwiers (1987) is set up so that exact parametric tests are also available as benchmarks to measure the performance of the permutation approaches. Samples of size n (=10) of each element of two m-dimensional ($m = 2, 4, 8, 12, 24$) vector fields $\vec{\mathbf{X}}$ and $\vec{\mathbf{Y}}$ are generated separately (and thus independently) from identically distributed Gaussian noise, thus there is no underlying serial or cross (either between or within each vector) correlation. Likewise, the null hypothesis that there is no difference between the means of the two vectors is true. Note that situations where the ratio of n to m varies from 5 to 5/12 are considered.

Two exact parametric approaches are available to test the null hypothesis. The first involves the statistic, $D^2 = \| \bar{\vec{\mathbf{x}}} - \bar{\vec{\mathbf{y}}} \|^2$, which is proportional to a chi-squared distribution with m degrees of freedom (DOF). Here overbars denote the sample means. The second consists of first computing the statistics, $T_j^2 = \frac{n}{2}(\bar{\mathbf{x}}_j - \bar{\mathbf{y}}_j)^2$, $j = 1\ldots, m$, each of which have chi-squared distributions with 1 DOF, then counting the number k (out of m possible) local rejections of the null hypothesis (called the *counting norm*) and using it as the test statistic. The latter is the approach used in Livezey and Chen (1983) with the Students-t statistic as the test statistic at each grid point. Refer to Chapter 8 for more details about the parametric tests.

Thus two different permutation techniques (see the next subsection), two different test statistics (D^2 and the counting norm), and six different sample size-to-dimension ratios will be considered below. This will provide some appreciation for the performance of permutation methods in the most commonly encountered field significance problem, differences in means. In all instances nominal significance levels of 5% are used.

9.3.2 Permutation (PP) and Bootstrap (BP) Procedures

For both PP and BP the null hypothesis of $D^2 = 0$ allows the two samples to be pooled, i.e. $(\vec{\mathbf{x}}_1 \ldots \vec{\mathbf{x}}_n, \vec{\mathbf{y}}_1 \ldots \vec{\mathbf{y}}_n)$. Two new n samples of $\vec{\mathbf{X}}$ and $\vec{\mathbf{Y}}$ are created by random resampling from the pooled sample, either without replacement in the case of PP or with replacement in the case of BP.

For PP this amounts to reordering the 20 vectors and relabeling the first 10 as $\vec{\mathbf{x}}_1$ to $\vec{\mathbf{x}}_n$ and the second 10 as $\vec{\mathbf{y}}_1$ to $\vec{\mathbf{y}}_n$; in the Livezey and Chen (1983) example in the last section this amounts to reordering the SOI series. In contrast, for BP the original samples may not be used proportionately, indeed parts may not be used at all. However, this should not be construed as a deficiency, because BP is equivalent to random sampling from a monotonically nondecreasing step function with steps at the observations. As the number of observations increases, this empirical distribution function converges to the true distribution.

Empirical distribution functions for each case considered (PP or BP, test statistic, m) are built up with 200 different resamplings of the original samples to be tested. The original sample is then tested against the 5% tail of the empirical distribution function and the results noted.

9.3.3 Properties

The performance of PP and BP under various conditions (test statistic, m) was studied by generating 1000 sets of paired samples $(\vec{\mathbf{x}}_1 \ldots \vec{\mathbf{x}}_n, \vec{\mathbf{y}}_1 \ldots \vec{\mathbf{y}}_n)$ for every m and then conducting field tests using both procedures described above with both test statistics (D^2 and k) on every one of these twin samples. The true 5% value of both chi-squared and the counting norm are known and the expected rejection rate for the exact parametric tests is 5%, so two summary quantities that are of particular interest are the proportion of test statistic estimates that exceed the true values (i.e. the bias in estimating the true values) and the rate at which the null hypothesis is rejected in the 1000 tests. Perfect performance implies values of 0.50 and 0.05 for these two quantities respectively. Results are displayed in Table 9.1.

Generally the procedures perform well for most m, but overall BP performs a bit better in terms of rejection rates and PP in terms of bias sizes (the bias for BP/D^2 tests is particularly large for large m). There really are no significant differences between the performance of the D^2 and counting norm tests for PP.[2]

[2]The counting norm has the advantage of highlighting the locations in the field that lead to field significance when it is present. Of course, there is still no automatic guarantee in this case that any particular location that is locally significant is not so by accident. The confidence that can be placed on local significance increases with both its nominal value and the level of the field's significance.

Table 9.1: *The characteristics of 1000 PP or BP estimates of the 5% critical points of either D^2 or k based on samples of size 10 from multivariate normal distributions of dimension m. The second column contains the proportion of estimates greater than the true critical point. The last column contains the proportion of times the null hypothesis was erroneously rejected using the indicated procedure. (From Zwiers, 1987).*

BP/D^2	Probability to		PP/D^2	Probability to	
	estimate	reject		estimate	reject
m	$> \chi^2_{0.05}$	H_0	m	$> \chi^2_{0.05}$	H_0
2	0.50	0.05	2	0.50	0.04
4	0.47	0.04	4	0.40	0.05
8	0.64	0.06	8	0.38	0.08
12	0.79	0.05	12	0.39	0.07
24	0.98	0.02	24	0.39	0.05

BP/k	Probability to		PP/k	Probability to	
	estimate	reject		estimate	reject
m	$> k_{0.05}$	H_0	m	$> k_{0.05}$	H_0
2	0.35	0.05	2	0.49	0.05
4	0.02	0.09	4	0.47	0.09
8	0.61	0.06	8	0.37	0.07
12	0.17	0.07	12	0.43	0.08
24	0.54	0.05	24	0.38	0.07

In Zwiers (1987) the power to detect known uniform signals was also examined with three sample sizes, n =5, 10, and 20. This is therefore a test situation where the effects of sampling resolution can also be studied. The most important result is that the permutation methods were at least as good (within estimation error) at detecting the signals (rejecting the null hypothesis) as the exact methods, with PP again slightly superior to BP and the use of D^2 slightly superior to the counting norm. The other notable result is that the power of BP and PP both were substantially lower for $n = 5$, perhaps because of severe estimation error in the empirical distribution functions. In this case $\frac{n}{m}$ was as little as $\frac{5}{24}$, a level at which no method would be expected to perform well, except with a very strong signal.

Obviously, these results are not directly transferable to situations where the reference distributions are different or unknown, but they lend considerable confidence to the application of PP and BP in these instances.

9.3.4 Interdependence Among Field Variables

If the m vector elements of $\vec{\mathbf{X}}$ and $\vec{\mathbf{Y}}$ respectively are intercorrelated the situation is analogous to testing differences in means of maps of climate data where there is spatial correlation between grid locations. Because BP and PP build up empirical distribution functions from samples in which the covariance structures of the two fields are preserved the capability of the tests to estimate appropriate significance levels is unaffected. What is affected is the power of signal detection as has already been implied in the discussion in Section 9.2.2.

Zwiers (1987) examines these effects in a modified experiment in which he imposes cross-correlation on the randomly generated m-dimensional vectors in two different ways, one that mimics an expanding network (recall the PNA vs. hemispheric domain example) and one that mimics increasing resolution (the summer vs. winter example) as m increases. To estimate power a strong uniform signal is added to the m components of the sample of $\vec{\mathbf{X}}$. Only PP with D^2 is used for the permutation tests, otherwise the experiments are the same as before, and the principal result of interest is the rate of rejections of the null hypothesis. In this experiment, however, the best performing tests are those with high rates because the null hypothesis is false in all instances. The exact test becomes a Hotelling-T^2 test applied with known covariance structure.

The power of the permutation tests is first compared to the exact tests and the case with independent vector variables. Results of these tests are presented in Table 9.2 under "Multivariate test". First note the loss of power of both the PP and exact tests with the addition of spatial correlation. This is comparable to a reduction in network density.

Second, with spatial correlation the power of the PP test is only slightly inferior overall to the exact test in the case of the expanding network and is

Table 9.2: *A comparison of the power of several scalar and multivariate tests in the case of two types of spatial correlation structure. The scalar tests look for departures from the null hypothesis in the direction of the first EOF, in the direction of the true departure μ_0 from the null hypothesis and in the direction of ν of maximum power. The multivariate tests are PP/D^2 and an exact chi-square test. Power of multivariate tests with independent variables are indicated in brackets. (From, Zwiers, 1987).*

Spatial correlation structure	m	Scalar test: First EOF	μ_0	ν	Multivariate test: PP/D^2	χ^2
Geographically expanding network	2	0.31	0.31	0.31	0.25 [0.30]	0.24 [0.42]
	4	0.36	0.36	0.54	0.29 [0.63]	0.34 [0.60]
	8	0.48	0.60	0.83	0.34 [0.83]	0.51 [0.83]
	12	0.05	0.88	0.95	0.57 [0.87]	0.63 [0.94]
	24	0.05	>0.99	>0.99	0.77 [1.00]	0.86 [0.99]
Network with increasing density	2	0.42	0.42	0.42	0.33	0.33
	4	0.48	0.48	0.50	0.42	0.31
	8	0.49	0.51	0.54	0.38	0.25
	12	0.50	0.51	0.55	0.45	0.22
	24	0.50	0.51	0.56	0.46	0.20

superior for increasing density. Moreover, the test environment here is unrepresentative in that the covariance structure of the data is known precisely. In practice this is rarely the case, the covariance matrix must be estimated with the practical result that the exact Hotelling-T^2 test is either ineffective or cannot be used at all here except for $m = 2$.[3] This implies that the permutation procedure is a competitive alternative to the exact test for all situations examined here.

Third, note the diminishing returns with increasing density of the network. This illustrates the point made by Hasselmann (1979) that with large-scale signals denser networks often add little non-redundant information about the signal but a considerable amount of noise.

The powers of three exact scalar tests are also compared to the permutation tests. These scalar tests look for signals in three pre-specified directions respectively: the first EOF, the known signal μ_0 and the maximum signal-to-noise ratio ν. All utilize a test statistic that is normally distributed and all require the use of the known covariance structure. They are related to several of the tests covered in Chapter 8 which are alternatives in situations with small n and spatial correlation. For further details refer to Zwiers (1987).

It is clear from the left half of Table 9.2 that all three scalar tests are at least as powerful as the permutations tests in all situations examined, with the exception of the first EOF test for the larger geographically expanded networks. For these same cases analogous to the larger spatial domains the μ_0 and ν tests are considerably superior to the permutation procedure. However, as emphasized above, in practice covariance matrices must be estimated, thereby requiring the use of different test statistics and less powerful tests. Nor are signals known perfectly a priori in practice. For both of these reasons the powers of the scalar tests shown in Table 9.2 are likely to be overestimates, especially for the μ_0 and ν tests.

Overall, the results in Table 9.2 and the stated caveats suggest that for $\frac{n}{m}$ less than about five (the range studied here) permutation methods are viable test alternatives in situations with spatial correlation. Further, in the absence of *a priori* knowledge of the signal to be detected they are practically the only currently available choice for $\frac{n}{m}$ less than about one.

9.4 Serial Correlation

The discussion of serial correlation has been deferred to the end of this chapter despite the fact that it is perhaps the most serious difficulty the analyst faces in the application of permutation or Monte Carlo techniques. Without attention to its effects it is completely debilitating to the effectiveness of the procedures. If two samples to be compared each have serial correlation the process of resampling destroys this property of the data and leads to

[3] With an estimated covariance matrix the Hotelling T^2 test requires that $n-1=9 \geq m$ and is ineffective unless $n \gg m$ (see Chapter 8).

Table 9.3: *Rejection rate using the 5% critical point of tests using the PP together with the Euclidean distance statistic D^2, when samples are taken from Gaussian Markov processes. (From Zwiers, 1987).*

	size	
m	$\rho_1 = 0.3$	$\rho_1 = 0.75$
2	0.12	0.55
4	0.28	0.81
8	0.32	0.91
12	0.40	0.98
24	0.72	1.00

empirical distribution functions that are not representative of the population of the observed statistic to be tested. The severity of the problem is apparent in Table 9.3 from Zwiers (1987) which shows the results of modifying the tests described in Sections 9.3.1 – 9.3.3 so that the elements of $\vec{\mathbf{X}}$ and $\vec{\mathbf{Y}}$ are independently generated from first-order Markov AR(1) processes with lag-1 autocorrelations of respectively 0.3 and 0.75. Clearly the tests are now useless.

Zwiers (1990) examined the effects in more detail with univariate ($m = 1$) difference of means tests using PP based on 1000 resamplings each with samples of $n = 20$, 40, and 100 taken from two random series generated from AR(1) processes with a range of lag-1 autocorrelations, ρ_1. To assess its performance PP was performed on 2080 paired series for every combination of n and ρ_1. Overall rejection rates are summarized in Figure 9.4 for nominal levels of 1, 2.5, 5, and 10%. Several things are very clear from the curves shown, including the insensitivity of the effects of serial correlation to the sample sizes considered and a universal tendency for liberalization of the tests, i.e. rejection rates higher than their nominal values making it more likely that the null hypothesis will be erroneously rejected with use of the tests. This is expected because empirical distribution functions will be narrower than the populations from which the original samples were drawn because of the destruction of serial correlation by the resampling.

Solutions to the problem of accounting for serial correlation in permutation and Monte Carlo test situations are offered below. Some are available only under special circumstances while others can be tried at the discretion of the analyst. The list is almost certainly not all-inclusive but does represent a synthesis of the ideas presented in the six papers mentioned in the introduction and the author's experience:

Figure 9.4: *The actual significance level of a permutation procedure-based difference of means test applied to pairs of samples obtained from a simulated AR(1) process. The actual significance level was estimated by replicating the process of generating the samples and conducting the test 2080 times. The curves display the proportion of the 2080 tests which were rejected in each simulation experiment as a function of the serial correlation of the observations. Panels (a)-(d) correspond to tests conducted nominally at the 1%, 2.5%, 5% and 10% significance levels. Within each frame, the solid line, the long dashed line, and the short dashed line record the performance of the test with samples of size* n *= 20, 40, and 100, respectively. (From Zwiers, 1990).*

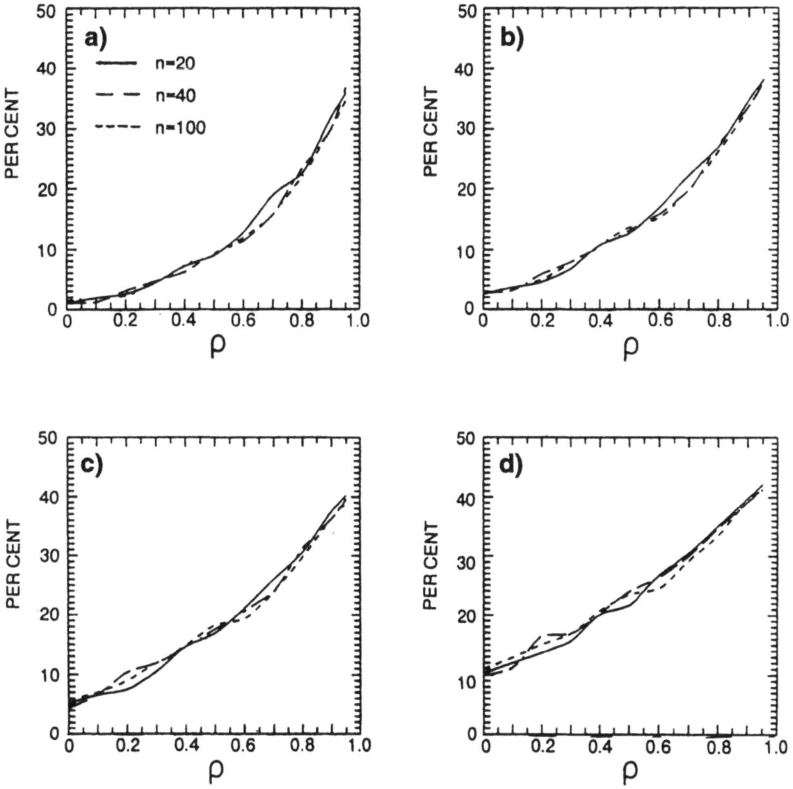

9.4.1 Local Probability Matching

If the counting norm statistic is used with PP or BP and it is possible to account properly for serial correlation in the local tests such that the local probabilities of false rejections of the null hypothesis are the same for both the original field samples to be tested and all of the samples constructed by resampling, then the analyst can proceed as before. This is what Livezey and Chen (1983) attempted to do heuristically for their local tests.

A considerable amount of progress has been made by Zwiers and von Storch (1995) in solving the problem of dealing with serial correlation in local tests of the difference in means and by Ebisuzaki (1997) in local tests of correlations. Thus, matching the local probabilities mentioned above should be possible in many test situations through use of local tests proposed by these authors for the original samples (which are serially correlated).

9.4.2 Times Series and Monte Carlo Methods

In field test situations that result from a single time series operating on a full grid or field of series, like for example the field of correlations between the SOI and extratropical Northern Hemisphere 700 hpa heights in Livezey and Chen (1983), then a viable strategy is to model the single series with autoregressive techniques and use the resulting model to generate random Monte Carlo samples which preserve the autocorrelation structure of the original series.

Both Thiébaux and Zwiers (1984) and Trenberth (1984) describe procedures for the autoregressive modeling which rely heavily on the work of Katz (1982). Zwiers and H. von Storch (1995) suggest that a simple AR(1) process is an adequate approximation for many meteorological and climatological time series with the annual cycle removed. This approximation would obviously be incorrect for interannual series (like the SOI) in which the quasi-biennial oscillation (QBO) is important.

Simultaneous modeling of both spatial and temporal correlation for generation of Monte Carlo samples is considerably more challenging, so the use of this approach for field vs. field problems generally will often be intractable.

9.4.3 Independent Samples

The most direct remedy for dealing with serial correlation is to eliminate it. For GCM experiments careful experimental design can ensure that samples are temporally uncorrelated. In the case of real data studies it is also possible to eliminate temporal correlation but at the price of smaller samples.

The procedure for obtaining test samples without serial correlation is simple: Estimate an "effective time between independent samples", T_0, and prune the data accordingly. For example, if T_0 is approximately twice the sampling interval elimination of all of the even or odd numbered samples will ensure a test series with inconsequential serial correlation in terms of the statistic to be tested.

The formula for T_0 depends on the tested statistic (Thiébaux and Zwiers, 1984; Trenberth, 1984). For tests of the mean,

$$T_0 = 1 + 2\sum_{\Delta=1}^{N}(1 - \frac{\Delta}{n})\hat{\rho}_\Delta \tag{9.1}$$

where $N \leq n - 1$ is the maximum of lags Δ of the estimated autocorrelation function $\hat{\rho}_\Delta$. For tests of the standard deviation, $\hat{\rho}_\Delta$ is replaced by $\hat{\rho}_\Delta^2$ in (9.1), and for tests of the correlation of series 1 and 2, $\hat{\rho}_\Delta$ is replaced by $\hat{\rho}_{\Delta;1} \cdot \hat{\rho}_{\Delta;2}$.

The autocorrelation function, ρ_Δ, in all cases must be estimated and this is often difficult to do well. Generally, the maximum lag N must be set well below n so that the sample size for all N lag estimates is reasonably large ($>$ 30 or so). Then all ρ_Δ can be estimated with relatively small estimation error. If not, then a time series model should be fitted to the data as outlined in the previous section and used to specify the $\hat{\rho}_\Delta$. In some cases an AR(1) model will be appropriate, giving simply

$$\hat{\rho}_\Delta = \hat{\rho}_1^\Delta \tag{9.2}$$

Using (9.1) and (9.2), $T_0 \sim 2$ for both a single AR(1) process with $\hat{\rho}_1 = 0.3$, and two AR(1) processes, $\hat{\rho}_{1;1} = 0.6$ and $\hat{\rho}_{1;2} = 0.6$, for tests of the mean and correlation respectively. Consequently, the amount of data that must be pruned depends on the statistic being estimated and tested.

9.4.4 Conservatism

In the example of Figure 9.4, note that PP tests of the univariate difference in means at the nominal 2.5% level for $\hat{\rho}_1 = 0.2$ and at the nominal 1% level for $\hat{\rho}_1 = 0.3$ are both operating at about the 5% level. If serial correlation is modest but consequential a simple expedient to account for it in testing is to reject the null hypothesis at a much lower level than would ordinarily be appropriate.

9.4.5 The Moving-Blocks Bootstrap

As noted above, conventional resampling procedures which randomly pick one datum at a time cannot reproduce the time correlation in the underlying data whose properties one would like to simulate: independent resampling destroys the autocorrelation structure. Recently, however, an extension to conventional bootstrap resampling called the *moving-blocks bootstrap* has been proposed in the statistical literature, allowing bootstrap resampling to be used with serially correlated data (see for example Liu and Singh 1992). As the name implies, the moving-blocks bootstrap is a procedure that resamples with replacement, but the resampling operates on comparatively short consecutive lengths (blocks) of values of the time series rather than on single observations as would the conventional bootstrap. To illustrate, consider a time series of length n. The ordinary bootstrap resamples the series with replacement, so for each resample the algorithm has available n objects from which to choose with equal probability. The block bootstrap operates in the same way, except that the objects that are resampled are blocks of consecutive values of length L. There are therefore $n - L + 1$ overlapping blocks from which the resampling algorithm can choose with equal probability: the first of these is the string $\{\mathbf{x}_1, \mathbf{x}_2, \ldots, \mathbf{x}_L\}$, the second is $\{\mathbf{x}_2, \mathbf{x}_3, \ldots, \mathbf{x}_{L+1}\}$, and the $(n - L + 1)$th of these is $\{\mathbf{x}_{n-L+1}, \mathbf{x}_{n-L+2}, \cdots, \mathbf{x}_n\}$. Each synthetic sample is then assembled by concatenating n/L of these resampled blocks into an ersatz time series of length n, and an ensemble of these synthetic samples can then be used in the usual way to estimate confidence intervals, conduct hypothesis tests, etc.

The critical choice that must be made in employing the moving-blocks bootstrap is choice of the blocklength, L. Choices for this parameter in early work have been ad hoc and qualitative. Wilks (1997) presents prescriptions for the block lengths as a function of the sample size, for several simple time-series forms. For example, for AR(1) series, choosing the block length according to

$$L = (n - L + 1)^{\frac{2}{3}(1-exp(-2T_0/n)/T_0)}, \tag{9.3}$$

where T_0 is given by (9.1), yields bootstrap hypothesis tests for the mean that reject true null hypotheses with the correct probability, and are nearly as powerful as the best available parametric tests for univariate time series. Since (9.3) is implicit it is evaluated iteratively.

The moving-blocks bootstrap is especially attractive for use in field intercomparison. Wilks (1997) has shown that multivariate tests of this kind for differences in vector mean perform properly and with good power, even in difficult circumstances where the series lengths are small in comparison to the dimensionality of the fields being compared.

9.5 Concluding Remarks

A knowledge of the material in this chapter, consultation of the listed references when needed, and some creativity will permit the application of permutation techniques in a wide variety of test situations. Their application in turn permits a degree of objectivity in assessment of results of exploratory studies that is often lacking in current practice. Obviously, difficulties remain that will pose obstacles for some problems, some of which have been highlighted here. For example, work on the treatment of serial correlation in tests of the standard deviation or correlation has not progressed as far as that for the mean. However, some of these problems will be solved as a result of heightened awareness of the need for proper testing.

Chapter 10

The Evaluation of Forecasts

by Robert E. Livezey

10.1 Introduction

The ultimate goal of climate dynamics is the prediction of climate variability on interannual to interdecadal and longer timescales. Two necessary, but not sufficient, conditions for the usefulness of such forecasts is that they have real skill in discriminating one forecast situation from another and that this skill is known in advance by the user. Two "no-skill" forecast schemes would be random forecasts or constant (e.g. fixed climatology) forecasts. An additional factor relevant to a scheme's usefulness is its efficiency – in other words its skill with respect to more simple forecast schemes (like persistence of existing conditions).

The evaluation of these two crucial factors to the utility of forecasts, absolute and relative skill, are the concerns of this chapter. The information contained here, combined with some of the considerations in preceding chapters (especially Chapter 9) regarding tests of null hypotheses (here hypotheses of no skill or no difference in skill), should be sufficient to design and conduct a wide range of forecast assessments. One exception is the evaluation of probability forecasts, for which the reader is referred to Daan (1985) and Wilks(1995). Also, ultimately forecast usefulness depends on the cost/benefit environment of individual users. Existing analyses of user-dependent utility are few in number (see Mjelde et al., 1993, for a recent example), often quite complex, and beyond the scope of this chapter.

The remainder of the chapter will highlight three related topics. First, basic ingredients and considerations for objective forecast verification will be discussed in Section 10.2, which borrows heavily from Livezey (1987). This is followed in Sections 10.3 and 10.4 by overviews of skill measures in modern use and some of their important properties and interrelationships for both categorical and continuous forecasts respectively. Virtually all of the material in these sections is synthesized from papers that have appeared in the last dozen years. Finally, the difficult problem of evaluating statistical prediction schemes in developmental situations where it is not feasible to reserve a large portion of the total sample for independent testing is examined in Section 10.5. Cross-validation (Michaelson, 1987) will be the emphasized solution because of its versatility, but references to Monte Carlo and other alternatives for evaluation of multiple (including screening) regression models will be provided. Complementary sources rich in examples for much of the material in this chapter are Wilks (1995) and the forthcoming book by H. von Storch and Zwiers (1999).

10.2 Considerations for Objective Verification

The skill of a forecaster or forecast system is easy to overstate and this has frequently been done both in and out of the refereed literature. Such overstatement inevitably leads to overstatement about the usefulness of forecasts, though as pointed out above skill alone is not a sufficient condition for forecast utility. Although the analysis of utility is beyond the scope of this chapter, it is possible to describe four elements whose consideration should enhance the possibility of objective assessment of the level of skill, and thereby of utility. Some of these components may strike some as obvious or trivial. Nevertheless, it has been the experience of the author that they need to be continually restated.

10.2.1 Quantification

Forecasts and verifying observations must be precisely defined, so that there is no ambiguity whatsoever whether a particular forecast is right or wrong, and, for the latter, what its error is. Also, it is important that it be clear what is not being predicted. For example, predictions of the amplitude of a particular atmospheric structure could be highly accurate but of little value for most users if the variance of that structure represents only a small fraction of total variance of the field.

10.2.2 Authentication

Verifications of forecast methodologies are conducted on one or another of two types of prediction sets, often referred to as forecasts and *hindcasts*. In the former prognostic models or experience are based only on information preceding independent events that are predicted, while in the latter they are based on information either following or both preceding and following the target forecast period.

In operational situations only forecasts are possible and their issuance by firm deadlines preceding the forecast period is sufficient guarantee that they are *true* forecasts. In the laboratory, either type is possible and special attention must be paid to ensure that no skill inflation takes place. For forecasts, this means that absolutely no use is made of future information in formulating prognostic procedures or in making actual forecasts. A frequently encountered and subtle example of a violation of this condition would be a set of zero-lead forecasts made using predictor series subjected to a symmetric filter; in this instance the forecasts would not be possible in practice because future information would be necessary to determine current values of the filtered predictors. Skill estimates for such laboratory forecasts will invariably be inflated.

In the case of laboratory hindcasts, uninflated estimates of forecast skill are impossible unless it can be assumed that the climate is reasonably stationary. If the assumption is appropriate then hindcasts can be considered true forecasts if the target forecast period is statistically independent of the model developmental sample. Much more will be said about these matters in Section 10.5.

10.2.3 Description of Probability Distributions

Because there is generally a non-trivial probability that at least some number of random categorical forecasts can be made correctly or that the mean error of some number of random continuous forecasts will be less than some amount, the notion of forecast skill in the absence of information about the joint probability distribution of forecasts and observations is meaningless (e.g. a statement like "the forecasts are 80% correct" is a statement about the accuracy not skill). Thus it is not possible to make objective assessments of forecast skill without some information about this joint distribution.

A very instructive example occurs for the prediction of monthly mean precipitation. The distribution of this quantity for a wide range of climates is negatively skewed, i.e. the median is less than the mean, as schematically illustrated in Figure 10.1. For most U. S. locations in all seasons this skewness is such that about 60% of cases have below normal precipitation and 40% above. Thus the probability of randomly making a correct forecast of above or below normal precipitation is not 0.5 (i.e. $2 \cdot 0.5^2$), but is 0.52 ($0.6^2 + 0.4^2$). An equally unintelligent (and useless) forecast procedure would be to forecast below normal precipitation all of the time, ensuring an even higher 0.6 ($0.6 \cdot 1.0 + 0.4 \cdot 0.0$) probability of success. In Livezey (1987) two examples of published monthly precipitation forecasts are cited for which extravagant claims of success were made, but that had no skill statistically distinguishable

Figure 10.1: *Schematic of skewed (upper) and symmetric (lower) probability density functions p of some continuous predictand f. (From Livezey, 1987).*

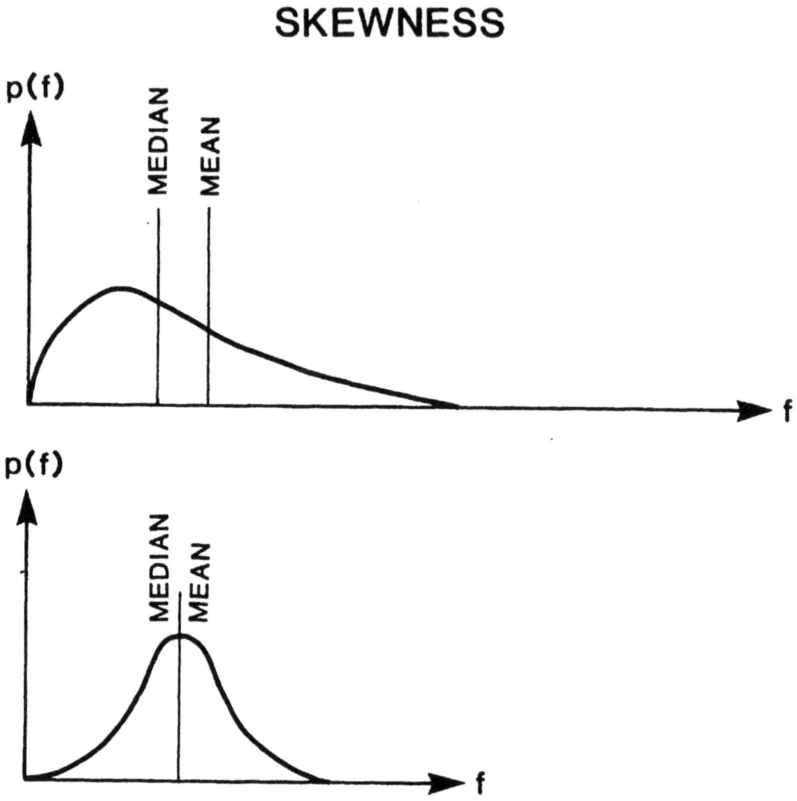

from random forecasts. In both instances skills in forecasting signs of departures from the *mean* were inappropriately compared to the expected skill of random forecasts of signs of departures from the *median*, namely 50%, rather than the higher value implied by the skewness of the distribution.

A second situation in which misinformation about the distribution of observations and/or forecasts leads to misstatements about forecast skill occurs for locations with pronounced anthropogenic urban heat-island trends. A schematic example is shown in Figure 10.2. Suppose a statistical forecast model with a trend term to predict temperature terciles (three equally probable classes) is constructed over a developmental period which also defines the temperature classes. This model's independent period forecasts will be dominantly biased towards the above normal category and in terms of the developmental period climatology will generally be correct.

But how skillful is the model? Because of its design it obviously has no

Figure 10.2: *Schematic of effect of trend on appropriateness of reference levels determined over developmental period for an independent forecast period. (From Livezey, 1987).*

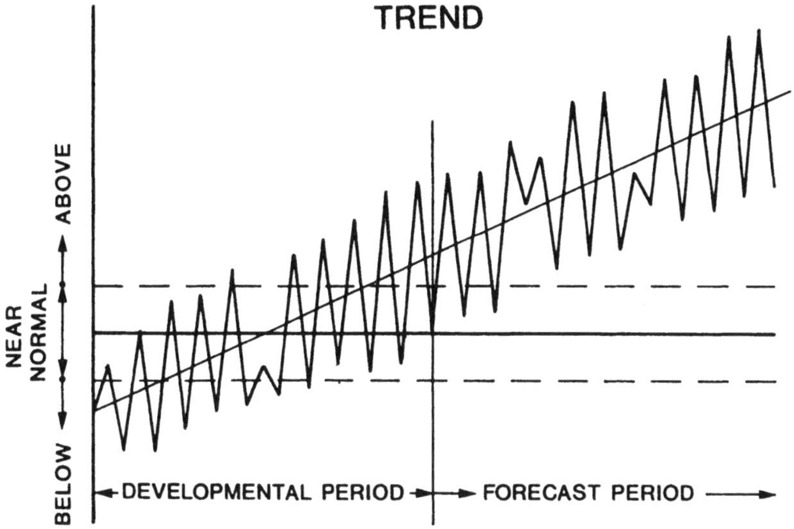

ability to discriminate interannual climate fluctuations nor does it have any applicability over the broader region of which the city for which it was developed is a part. Its skill should not be compared to forecasts drawn randomly from the probability distribution for the developmental period. These random forecasts would be expected to be correct only 33% of the time.

Instead, just as in the precipitation example, the expected number of correct random forecasts should be computed from the skewed distribution, which, for example, might have frequencies of above, near, and below normal classes of 0.9, 0.1, and 0.0 respectively. In this instance the expected percent of correct random forecasts is 82% ($0.9^2 + 0.1^2$). Also, again following the precipitation example, it is possible to guarantee (uselessly) that 90% of the forecasts will be correct by simply forecasting above normal all of the time.

Lest the reader think the trend example is too blatant an error to be committed in practice, the author has encountered it in prediction studies in which pooled verifications over many stations were reported. Because heat-island trends were severe at only a small proportion of stations their inflation of skill was not readily apparent, but in fact this inflation led the forecaster to conclude that his methodology had skill when without it skill was nonexistent. See Livezey (1987) for details.

Both of these examples show how crucial it is in verification studies to have appropriate references or yardsticks against which to measure forecast system performance. This topic will be expanded in the next subsection.

Figure 10.3: *Schematic of time series to illustrate three types of control forecasts. The horizontal line denotes the climatological mean while the vertical dashed line denotes the most recent observation time t_0. The dots denote forecasts at $t_0 + \Delta t$. (From Livezey, 1987).*

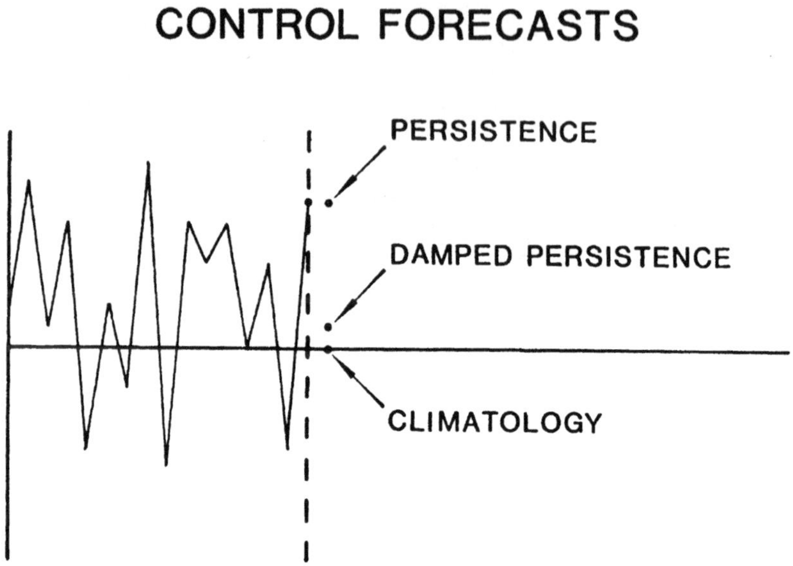

Also, scores for categorical forecasts that are immune to skill inflation from biased forecasting will be presented in Section 10.3.

10.2.4 Comparison of Forecasts

As emphasized above, statements about forecast success are not meaningful unless placed in the context of what is possible or expected from simple "strawmen" or *control* forecast schemes. In the previous section the most basic of these, random forecasting, was described. Three other common controls are illustrated schematically in Figure 10.3: *climatology* is a fixed forecast of zero departure from a suitable climate mean, *persistence* is a forecast that current anomalous conditions will continue, *damped persistence* is an AR(1) model forecast. Higher order autoregressive models frequently also are useful as control forecasts. Biased forecasts described in the last subsection will not be emphasized here as controls because the use of measures immune to bias effects will be advocated in Section 10.3.

Generally, the control selected should be the one that has the most apparent skill for the measure used in the evaluation. For example, if the measure used is mean squared error (MSE), random or persistence forecasts should not be used as no-skill standards because their MSE is quite high compared

Figure 10.4: *Schematic illustrating the difference between persistence and "operational" persistence forecasts. The long horizontal line is a time line. (From Livezey, 1987).*

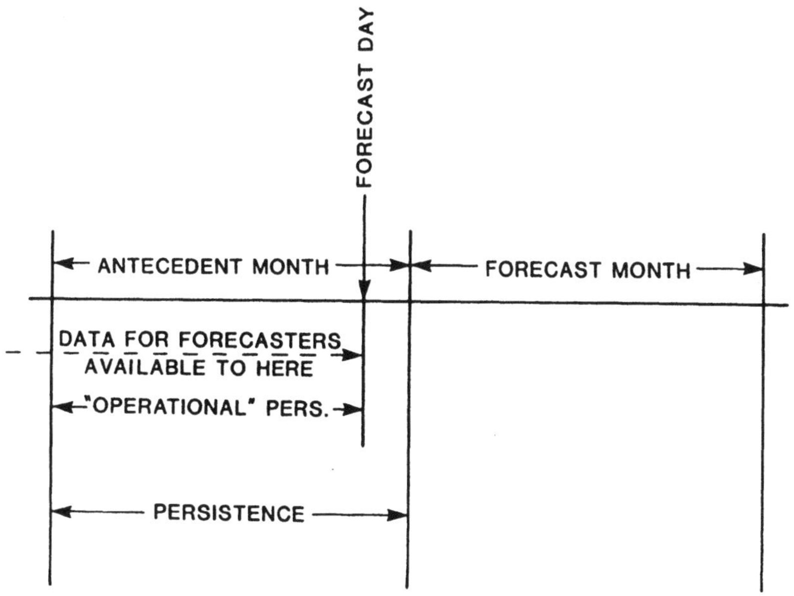

to damped persistence or climatology and is easy to improve upon. In fact, damped persistence is the right choice because it minimizes MSE on the developmental sample. Properties of commonly-used measures and relationships between them will be discussed in the next two sections.

When a forecast system is being compared to other competing systems, apart from the controls a number of additional considerations are important. First, *homogeneous* comparisons are preferred, i.e. on the same forecast set, otherwise the comparisons may be unfair if inhomogeneities are not taken into account. This is because some situations are easier to predict than others. Second, none of the forecasts being compared should use data that are unavailable or arbitrarily withheld from any of the others.

For example, a comparison between a neural net time series model that looks at antecedent data at four lags and an AR(1) model would be unfair. The fair comparison would be between the former and higher order autoregressive models that also use up to four lags. Another example is shown in Figure 10.4 in which monthly forecasters are operationally constrained to produce their forecasts several days before the forecast periods begin. Therefore, it would be unfair to compare their performance to a persistence forecaster who utilizes data over entire antecedent months. Instead, "operational" persistence should be used as the control.

Table 10.1: *Contingency table for forecasts/observations in three categories.*

Fore-	Observed		
cast	B	N	A
B	f_{BB}	f_{BN}	f_{BA}
N	f_{NB}	f_{NN}	f_{NA}
A	f_{AB}	f_{AN}	f_{AA}

10.3 Measures and Relationships: Categorical Forecasts

Under ideal circumstances continuous forecasts (i.e. quantitative forecasts of the predictand) are preferred to categorical forecasts (forecasts of a range of values of the predictand). However, when predictability is modest (as it is in most climate predictions) the specificity and detail of continuous forecasts are often unwarranted. Additionally, it is often much easier to cast probability forecasts in terms of probabilities of categories than to specify the full forecast probability distribution. The information in the first subsection below is partially drawn from Murphy and Winkler (1987), and that in the second condensed from Barnston (1992) and Ward and Folland (1991).

10.3.1 Contingency and Definitions

A contingency table summarizes outcomes of a categorical forecast scheme and contains an enormous amount of information. Consider the case of forecasts in three categories, below (B), near (N), and above (A) normal. The contingency table (Table 10.1) is constructed by first summing all forecasts for each of the nine possible forecast/observation outcomes and then dividing each sum by the total number of forecasts, T, to get outcome probabilities, f_{ij}, $(i,j) = (B,N,A)$, where the first subscript refers to forecast category (table row) and the second observed (table column).

From Table 10.1 the following quantities are frequently defined:

Hits:	$T(f_{BB} + f_{NN} + f_{AA}) = H$
One-Class Misses:	$T(f_{NB} + f_{AN} + f_{BN} + f_{NA})$
Two-Class Misses:	$T(f_{AB} + f_{BA})$
Distribution of Observations:	$f_{\cdot B} = f_{BB} + f_{NB} + f_{AB}$, etc.
Distribution of Forecasts:	$f_{B\cdot} = f_{BB} + f_{BN} + f_{BA}$, etc.
Marginal Probabilities,	
Hit Rate:	$f_B(o\|p) = f_{BB}/f_{B\cdot}$, etc.
Probability of Detection:	$f_B(p\|o) = f_{BB}/f_{\cdot B}$, etc.

In the last two lines o and p refer to observation and forecast respectively. Note also that H plus one- and two-class misses sum to T. A number of these quantities will be used below, but those that are not are useful in other contexts (see Murphy and Winkler, 1987, Wilks, 1995 and H. von Storch and Zwiers, 1999).

Table 10.2: *Four versions of Heidke credit/penalty scoring systems for three equally likely categories, progressing from*
(a) the original system in which all misses and hits count equally but differently,
(b) the system in which penalties depend only on the class of error (no longer equitable),
(c) system (b) adjusted for E value by forecast category to restore equitability, and
(d) system (c) recalibrated for expected hit mean of 1 and grand mean of 0. (From Barnston, 1992).

(a) The original Heidke scoring system.

Fcst.	Observed Category		
cat.	B	N	A
B	1	0	0
N	0	1	0
A	0	0	1

(b) Heidke system modified to account for error classes.

Fcst.	Observed Category		
cat.	B	N	A
B	1	0	-1
N	0	1	0
A	-1	0	1

(c) Matrix (b) made equitable by adjusting E by forecast category.

Fcst.	Observed Category		
cat.	B	N	A
B	1	0	-1
N	$-\frac{1}{3}$	$\frac{2}{3}$	$-\frac{1}{3}$
A	-1	0	1

(d) Matrix (c) recalibrated for $(0, 1)$ random vs. perfect levels.

Fcst.	Observed Category		
cat.	B	N	A
B	1.125	0.000	-1.125
N	-0.375	0.750	-0.375
A	-1.125	0.000	1.125

10.3.2 Some Scores Based on the Contingency Table

a) Heidke

The *Heidke* score is defined

$$S = \frac{H - E}{T - E} \tag{10.1}$$

where E is the expected number of correct random forecasts. For three equally-likely classes, $E = (1/9 + 1/9 + 1/9)\ T$, and, further, if the T forecasts are random and independent, the standard deviation of S is $\sigma = \frac{1}{\sqrt{2T}}$

(Radok,1988). Table 10.2a, which is reproduced from Barnston (1992) is a template for the assignment of weights for each forecast outcome to determine H.

In this specific example the Heidke score is *equitable* (Gandin and Murphy, 1992) because no advantage can be gained by forecasting one class or another all of the time. Thus an equitable skill score is one in which the biased forecasting described in the examples of Section 10.2.3 would not result in higher skills. Suppose forecast and observed classes are not equiprobable, say with (A, N, B) probabilities of (0.3, 0.4, 0.3) and $E = (0.3^2 + 0.4^2 + 0.3^2)T = 0.34T$. In this instance, the Heidke is inequitable, because all forecasts of near normal lead to $H = 0.4T$ and a positive score, but all above or below normal lead to $H = 0.3T$ and a negative score.

b) Refinements

The Heidke score can be thought of as a "poor man's" correlation score (the correlation score is defined in Section 10.4) so it is of some interest to examine the relationship between these respective categorical and continuous forecast scores. The results of Monte Carlo simulations by Barnston (1992) in Figures 10.5a and b show that the Heidke is generally smaller than the correlation and that this difference increases with increasing number of equi-probable classes for the categorical forecasts. In other words, for a given level of forecastability (as measured by correlation between forecasts and observations) the largest Heidke scores will be for categorical forecasts with the fewest classes. This undesirable property is related to the source of a more serious deficiency of the Heidke score, namely its inability to distinguish between two forecasts with the same number of hits but different numbers of one-, two-, etc. -class misses. Both problems are caused by the failure to account for the severity of misses.

An adjusted scheme that linearly discriminates between zero-, one- and two-class misses is shown in Table 10.2b, in which a two-class miss is now regarded as a loss of a hit. To use (10.1) with this scheme requires the redefinition of E as the expected number of hits with possible loss of hits from two-class misses taken into account, and corresponding redefinition of H. To accommodate this and more general redefinitions with more complex reward/penalty matrices, let the entries for boxes (i, j) in any of the panels in Table 10.2 be denoted by s_{ij} and n_{ij} the number of cases for the box. Now,

$$E = \frac{T}{9} \sum_{i,j=1}^{3} s_{ij} \quad \text{and} \quad H = \sum_{i,j=1}^{3} s_{ij} n_{ij} \qquad (10.2)$$

Note from Figures 10.5c and d (again from Monte Carlo simulations conducted by Barnston, 1992) that this scheme is an improvement on the Heidke score in two respects: The score is generally closer to the correlation score and scores for different numbers of classes are quite similar.

Unfortunately, this new score is inequitable, because it is higher for all forecasts of near normal ($n_{2j} = \frac{T}{3}$, $n_{1j} = n_{3j} = 0$, $j = 1,2,3$) than for all of either of the other two categories. To fix this the second row in Table 10.2b is adjusted so that the row elements sum to zero (as they already do for the first and third rows) while preserving the differences among them, i.e. by subtracting

$$\frac{1}{3}\sum_{j=1}^{3} s_{2j} = 0.33$$

from each s_{2j}. This ensures that H in (10.2) will sum to zero when every forecast is for near normal, for below or above. The result of this operation is shown in Table 10.2c. Note, however, that now for perfect forecasts ($n_{ij} = \frac{T}{3}$ for $i = j$ and $= 0$ for $i \neq j$)

$$H = \frac{T}{3}\sum_{i=1}^{3} s_{ii} \neq T$$

and therefore, $S < 1$. This is caused by the reduction of the near normal hit reward.

Thus, in a final transformation (Table 10.2d) diagonal elements in the weighting table are proportionately adjusted to sum to 3 so that $H = T$ for perfect forecasts, and then rows are readjusted so

$$\sum_{j=1}^{3} s_{ij} = 0$$

for all i to ensure $H = 0$ for single category forecast sets. This weighting scheme preserves Heidke's equitability and scaling as well as the linearly weighted scheme's smaller difference from the correlation score and relative insensitivity to the number of categories.

c) Linear Error in Probability Space (LEPS)

While the refinements of the Heidke score described above were well motivated, they were nevertheless implemented arbitrarily. In contrast, the family of LEPS scoring rules (Ward and Folland, 1991, and Potts et al., 1996) all have a rational basis and follow from a score

$$S = 1 - |F_p - F_0|$$

where $|F_p - F_0|$ denotes the difference in cumulative probabilities of a forecast and verifying observation. This leads eventually to scoring schemes for continuous forecasts (*LEPSCONT*), probability forecasts (*LEPSPROB*), and categorical forecasts (*LEPSCAT*). Only the latter will be presented here.

The scoring rules for LEPSCAT with three equi-probable classes are displayed in Table 10.3 (again from Barnston, 1992, who also has worked out LEPSCAT rules for two, four, and five equiprobable classes) and are used in the same way as the rules in Table 10.2 to determine a value for S in (10.1). The differences between Table 10.3 and Table 10.2d are subtle and include triangular symmetry in the former but not in the latter. From Figures 10.5e-f it is clear that the LEPSCAT scheme is slightly superior to the refined Heidke rules in several respects. Because it is also equitable and has a more fundamental genesis, LEPSCAT is the preferred score for categorical forecasts in the view of the author.[1]

Table 10.3: *The equitable credit/penalty scoring system based on LEPS (Ward and Folland 1991) for three equally likely categories, calibrated such that random forecasts have an expected score (E) of zero and perfect forecasts unity. (From Barnston, 1992).*

Fore-cast	Observed B	N	A
B	1.35	-0.15	-1.20
N	-0.15	0.29	-0.15
A	-1.20	-0.15	1.35

10.4 Measures and Relationships: Continuous Forecasts

In this section the verification of a sample of forecasts of a single variable and the verification of a single forecast of a field of variables will be separately treated in the two subsections. Barnston (1992), Wilks (1995), and H. von Storch and Zwiers (1999) are excellent supplements to Section 10.4.1. Murphy and Epstein (1989) is the principal source for Section 10.4.2, while Livezey and Hoopingarner (1995) provides graphic examples of its application. An exhaustive and unified treatment of error decompositions like those in Section 10.4 is provided by Murphy (1996).

[1]The relative merits of LEPSCONT scoring and other continuous measures will not be treated in Section 10.4, so LEPSCONT may be a viable alternative to the approaches detailed there. It will be noted, however, that the ranked probability score (RPS; cf. Daan, 1985) is still preferred to LEPSPROB.

Figure 10.5: *Score as a function of correlation score for three and five equally likely categories, respectively, based on a large number of simulations using a random number generator. The solid curve represents mean results, the short-dashed curves the plus and minus-one standard deviation interval, and the long-dashed curves the maximum and minimum results. Panels (a) and (b) are for the Heidke, (c) and (d) for the modified Heidke in Table 10.2b, and (e) and (f) for LEPSCAT. (From Barnston, 1992).*

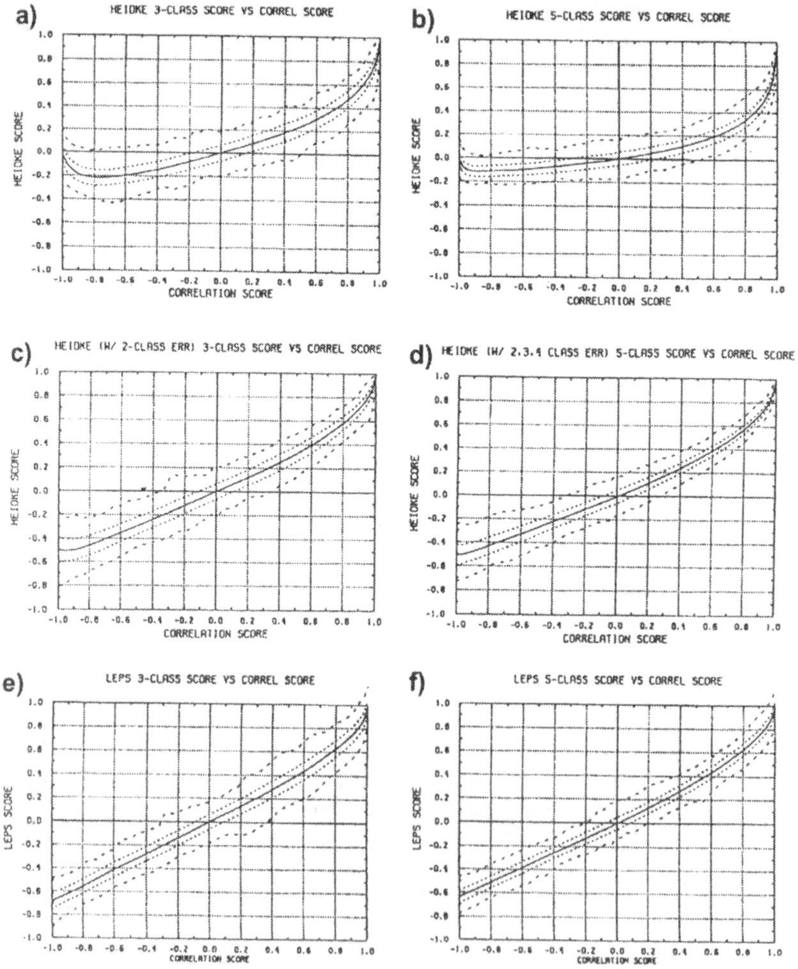

10.4.1 Mean Squared Error and Correlation

Let $(f_i, x_i), i = 1 \dots n$, denote a sample of forecasts and verifying observations respectively of the variable x taken at different times. The *mean squared error* is defined

$$MSE_{fx} = \frac{1}{n}\sum_{i=1}^{n}(f_i - x_i)^2 \tag{10.3}$$

For convenience suppose that the forecasts and observations can and have been temporally standardized (to means of zero and unit standard deviations). This standardization is not possible for climatology forecasts. The *correlation* is then

$$\rho_{fx} = \frac{1}{n}\sum_{i=1}^{n} f_i x_i \quad \text{and} \tag{10.4}$$

$$MSE_{fx} = 2(1 - \rho_{fx}). \tag{10.5}$$

If the forecasts or observations are imperfectly standardized so that the samples do not both have zero means and unit standard deviations (i.e. standardization is based on a different sample's statistics, e.g. a different period's climatology), the comparable expression to (10.5) contains additional terms to account for bias and amplitude errors in the forecasts. These terms are analogous to those developed in the next subsection for verification of fields. However, ρ_{fx} remains insensitive to bias and amplitude errors.

The relationship (10.5) is plotted in Figure 10.6 from Barnston (1992). Note from the figure or (10.5) that for random forecasts ($\rho_{fx} = 0$) $MSE_{fx} = 2$, but from (10.3) for climatology forecasts ($f_i \equiv 0$) $MSE_{fx} = 1$, so that random forecasts have double the squared error of climatology forecasts. Note further that for $MSE_{fx} < 1$, the level for climatology, $\rho_{fx} > 0.5$. The latter is a commonly used criterion in medium to long range prediction for usability of circulation forecasts.

MSE_{fx} can be reduced by damping the x_i by an estimate of ρ_{fx}. The result of this is represented by the lower curve in Figure 10.6.

An additional useful skill measure called the *Brier-based* score is the percent reduction of MSE from some reference or control forecast system. If the control is climatology, c, then from (10.3) the Brier-based score is

$$\beta = \frac{MSE_{cx} - MSE_{fx}}{MSE_{cx}} = 1 - \frac{MSE_{fx}}{MSE_{cx}} = 1 - \frac{\sigma_E^2}{\sigma_x^2} \tag{10.6}$$

where the last term is the ratio of the error variance to the observed variance. Thus the Brier-based score with climatology as the reference is also the *proportion of explained variance.*

If the forecasts and observations are standardized, the denominator in (10.6) becomes one and, from (10.5)

$$\beta = 2\rho_{fx} - 1. \tag{10.7}$$

Figure 10.6: *Root-mean-square error (RMSE) as a function of correlation for standardized sets of forecasts and observations (curve A), and for same except that the forecasts have been damped and possibly sign reversed by multiplying by* ρ_{fx} *- i.e., the correlation between forecasts and observations (curve B). (From Barnston, 1992).*

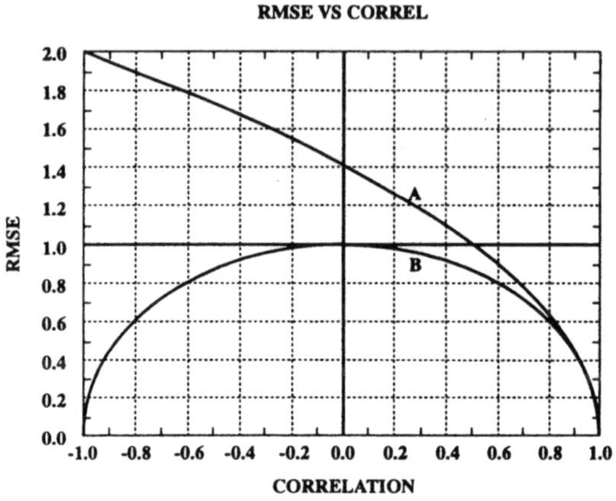

Thus, for $\beta > 0, \rho_{fx} > 0.5$, and for random forecasts $\beta = -1$.

To close this subsection, it is worth recalling from Section 10.2.4 that it is not appropriate to compare a forecast scheme to persistence if MSE_{fx} is used as the measure of accuracy, because MSE_{fx} is relatively large for persistence and easy to beat. This is likewise true of β. Recall also from above that MSE_{fx} can be reduced by damping the forecasts by an estimate of ρ_{fx}. For persistence this is simply the lag-one autocorrelation, so a more competitive control is an AR(1) forecast model, i.e. damped persistence. The correlation score, ρ_{fx}, remains unaffected by this damping.

10.4.2 Pattern Verification (the Murphy-Epstein Decomposition)

In contrast to (10.3) define the MSE between a map of gridded forecasts f_i and a field of observations x_i with i denoting a gridpoint,as

$$MSE(f,x) = < (f_i - x_i)^2 > \qquad (10.8)$$

where the angle brackets denote a spatially weighted mean (not a sum over grid points). Below the subscript i will be deleted when the angle brackets are used.

If climate anomalies are denoted by $()' = () - c_i$, where c_i is the climate mean at gridpoint i and (for future reference) σ_{oi} is the climate standard deviation, then the following represent moments of the map anomalies with map means removed:

$$\begin{aligned} s_{f'}^2 &= < f'^2 > - < f' >^2, \quad s_{x'}^2 = < x'^2 > - < x' >^2 \\ s_{f'x'} &= < (f' - < f' >)(x' - < x' >) >, \quad \text{and} \\ \rho_{f'x'} &= \frac{s_{f'x'}}{s_{f'} s_{x'}} \end{aligned} \tag{10.9}$$

The quantities in the first line of (10.9) can be thought of as estimates of map variance (for forecast and observed maps respectively) and in the second line as an estimation of the covariance between the forecast and observed maps. The last line of (10.9) is called the ***anomaly correlation***, a commonly used measure of spatial correspondence of forecast and observed anomaly features.

The scores (10.8) and (10.9) are linked through the decomposition by Murphy and Epstein (1989) of the spatial analog to (10.6), the Brier-based score with respect to climatology,

$$\beta = \frac{A^2 - B^2 - C^2 + D^2}{1 + D^2} \tag{10.10}$$

where

$A^2 = \rho_{f'x'}^2$	Phase association between forecast and observed anomalies; $1 - A^2$ is the degree of phase error,
$B^2 = \left[\rho_{f'x'} - s_{f'}/s_{x'}\right]^2$	Conditional bias, reduction in skill due to amplitude error,
$C^2 = \left[(< f' > - < x' >)/s_{x'}\right]^2$	Unconditional bias, reduction in skill due to map mean error,
$D^2 = \left[< x' > /s_{x'}\right]^2$	Normalization factor, adjustment due to the observed map mean.

To help in interpreting B^2 consider a least-squares linear regression fit of the observations to the forecasts, namely

$$E(x'|f') = a + bf' \quad \text{where} \quad b = \frac{s_{x'}}{s_{f'}} \rho_{f'x'} \quad \text{and} \quad a = < x' > -b < f' > \tag{10.11}$$

(The expression $E(x'|f')$ represents the expectation of x' conditional upon f'.) Then, from (10.10) and (10.11), $B^2 = \left[(b-1)s_{f'}/s_{x'}\right]^2$ and vanishes, i.e. the forecast anomalies are not conditionally biased, only if $b = 1$. The link between conditional bias and amplitude error becomes clearer with the consideration of the case in which $< x' > = < f' > = 0$ and (10.11) reduces to $E(x'|f') = bf'$. In this instance amplitudes of observed anomalies tend to be larger than, smaller than, or of the same magnitude as their forecast values depending on the value of b. In the event there is no conditional bias ($b = 1$), if the forecasts are also perfectly phased with the observations ($\rho_{f'x'} = 1$), then $s_{x'} = s_{f'}$ and the forecast must be perfect.

The term D^2 in (10.10) acts as a normalization factor for β such that for perfect forecasts ($A^2 = 1, B^2 = C^2 = 0$) $\beta = 1$, and for climate forecasts ($A^2 = B^2 = 0, C^2 = D^2$) $\beta = 0$. With a reasonable climatology, C^2 and D^2 are usually small and substantially offset each other. They become important terms for small domains.

It is possible to reduce both B^2 and C^2 by postprocessing of forecasts if the forecast (e.g. model) climatology, $\bar{f}_i$ and $\sigma_{fi} = \overline{(f_i - \bar{f}_i)^2}$ are well estimated. Here the overbar denotes the sample mean corresponding to the period being forecast (e.g. January, winter, etc.). Redefine

$$f_i' = (f_i - \bar{f}_i)\frac{\sigma_{oi}}{\sigma_{fi}}$$

This amounts to removal of systematic bias and amplitude errors.

A warning is in order here: There are at least two other simple ways to inflate β that usually do not imply any meaningful gain in forecast information content. Inspection of (10.9) and (10.10), for instance, reveals that general smoothing of less skillful forecasts, i.e., ones with small anomaly correlations relative to the ratio of forecast to observation map standard deviations or average anomalies that are greater in magnitude or opposite in sign to the average observational anomaly, leads to reductions in B^2 and C^2 respectively. Likewise, it is clear from (10.6), the definition of the Brier-based score, that the use of some inappropriate climatologies can lead to an increase in MSE_{cx} and thereby an increase of β. This can occur if estimates of the climatology are based on a sample that is either too small in size (generally it should be around 30 years) or substantially out of date. The problem is exacerbated when interdecadal variability is present in the data, as is often the case in climate studies.

With these cautions in mind, it will turn out in many applications that the separation of conditional and unconditional forecast biases from the skill in phase prediction in (10.10) will provide useful diagnostic information about the forecast system which may lead to its improvement. In any event, it is clear that a multi-faceted perspective like the Murphy-Epstein decomposition is necessary to properly assess pattern prediction skill.

10.5 Hindcasts and Cross-Validation

Often the skill of a statistical climate prediction model is estimated on the sample used to develop the model. This is necessitated by a sample size that is too small to support both approximation of model parameters and accurate estimation of performance on data independent of the subsample on which the parameter estimation was based. Such a dependent-data based estimate is referred to as ***hindcast skill*** and is an overestimate of skill that is expected for future forecasts. The difference between hindcast and true forecast skill is called ***artificial skill.***

There are two notable sources of artificial skill. The first is non-stationarity of the climate over the time scale spanning the developmental sample period and the expected period of application of the statistical forecast model. If it cannot be reasonably assumed that this non-stationarity is unimportant, then a source of error is present for independent forecasts that is not present during the developmental period. The assumption of stationarity is usually reasonable for prediction of interannual variability.

Artificial skill also arises in the statistical modeling process because fitting of some noise in the dependent sample is unavoidable. The ideal situation of course is to have a large enough sample that this overfitting is minimized, but this is usually not the case for the same reasons that reservation of an independent testing sample is often not feasible.

There are two approaches available to deal with small sample estimation of true skill. Both require that the climate is negligibly non-stationary on the time scales of the particular forecast problem. The first is to attempt to estimate artificial skill simultaneously with hindcast skill, either through analytical or Monte Carlo means. Techniques for some problems have been presented in the literature. In particular, a thorough discussion of the considerations involved in artificial skill estimation for linear regression modeling (including screening regression) can be found in the series of papers by Lanzante (1984), Shapiro (1984), Shapiro and Chelton (1986), and Lanzante (1986). Much of their thrust is summarized by Michaelson (1987), who simultaneously examines the powerful alternative of *cross-validation*, the subject of the remainder of this section.

Cross-validation is arguably the best approach for estimating true skill in small-sample modeling/testing environments; in most instances it permits independent verification over the entire available sample with only a small reduction in developmental sample size. Its key feature is that each forecast is based on a different model derived from a subsample of the full data set that is statistically independent of the forecast. In this way many different (but highly related) models are used to make independent forecasts. The next two subsections, which consist of a general description of the procedure and a discussion of two crucial considerations for application respectively, should clarify how and why cross-validation works. To augment the material here, Michaelson (1987) is highly recommended.

10.5.1 Cross-Validation Procedure

For convenience the narratives here and in the next subsection are cast in terms of a typical interannual prediction problem (for example seasonal prediction) for which predictor and predictand data are matched up by year. The possibility of other correspondences is understood and easily accommodated.

The cross-validation begins with deletion of one or more years from the complete sample of available data and construction of a forecast model on the subsample remaining. Forecasts are then made with this model for one or more of the deleted years and these forecasts verified.

The deleted years are then returned to the sample and the whole procedure described so far is repeated with deletion of a different group of years. This new group of deleted years can overlap those previously withheld but a particular year can only be forecast once during the entire cross-validation process.

The last step above is repeated until the sample is exhausted and no more forecasts are possible.

10.5.2 Key Constraints in Cross-Validation

For each iteration of model building and prediction during the course of cross-validation, every detail of the process must be repeated, including recomputation of climatologies from the subsample and recalculation of anomalies (i.e. departures from climatology). Otherwise, inflation of skill or certain degeneracies will occur in the scoring (see Van den Dool, 1987, and Barnston and Van den Dool, 1993, for examples).

Finally, forecast year data in each iteration must be independent of the developmental sample. To ensure this a forecast year predictor or predictand should not be serially correlated with itself in the years reserved for model building. Suppose, for example, that for a particular predictand adjacent years are serially correlated but every other year is effectively not. In this instance three years must be set aside in each iteration and a forecast made only for the middle withheld year. To determine the effective time between independent samples, refer to the procedures outlined in the previous chapter.

Chapter 11

Stochastic Modeling of Precipitation with Applications to Climate Model Downscaling

by Dennis Lettenmaier

11.1 Introduction

Probabilistic models of precipitation are mathematical representations of the probability distribution of precipitation without regard to time, that is

$$prob\,(\mathbf{x} < \mathbf{X} < \mathbf{x} + d\mathbf{x}) = \int_x^{x+dx} f_X(r)dr, \tag{11.1}$$

where $\mathbf{X}$ might be, for instance, the total annual (or monthly) precipitation occurring at a given station, or the maximum annual precipitation occurring in a 24 hour period. Probabilistic models are useful for assessing risk, which, in its simplest form, is the probability of an undesirable outcome. For instance, in an unirrigated agricultural area, it may be important to know the

Acknowledgments: The assistance of James P. Hughes and Larry W. Wilson in assembling the review materials is greatly appreciated. Portions of this material have previously appeared in a paper by the author included in the proceedings of the Conference on Stochastic and Statistical Methods in Hydrology and Environmental Engineering (June, 1993, University of Waterloo).

probability that the growing season precipitation will be less than the threshold required for crop survival. Likewise, the design of small flood protection structures, particularly in urban areas, requires knowledge of what engineers term a probability intensity duration relationship. A probability intensity duration relationship is simply the family of probability distributions of the annual maximum precipitation for duration **D** where **D** might take on, for example, values of 1, 2, 3, 6, 12, and 24 hours.

Stochastic models represent the joint probability density function of precipitation, where (discrete) time t_i, $i = 1 \dots n$ is the index, that is

$$prob\,(\mathbf{x}_{t_1} < \mathbf{X}_{t_1} < \mathbf{x}_{t_1} + d\mathbf{x}_{t_1}; \dots \mathbf{x}_{t_n} < \mathbf{X}_{t_n} < \mathbf{x}_{t_n} + d\mathbf{x}_{t_n}) = \quad (11.2)$$

$$\int_{x_{t_1}}^{x_{t_1}+dx_{t_1}} \cdots \int_{x_{t_n}}^{x_{t_n}+dx_{t_n}} f_{X_{t_1} \dots X_{t_n}}(r_1, \dots r_n) dr_1 \dots dr_n \quad (11.3)$$

Stochastic models of precipitation were originally developed to address practical problems of data simulation, particularly for water resource systems design and management in data-scarce situations, and to aid in understanding the probabilistic structure of precipitation. However, they have important applications for local interpretation of large area climate simulations, such as are produced by General Circulation Models. Some of these applications are discussed in Section 11.5. First, however, some important probabilistic and stochastic characteristics of precipitation are described.

11.2 Probabilistic Characteristics of Precipitation

The probabilistic characteristics of precipitation are strongly dependent on time and space scales. At short time scales (e.g., hourly to daily), precipitation is intermittent, that is, precipitation amounts do not have a continuous probability distribution. Instead, there is a finite probability of zero precipitation, and a continuous probability distribution for non-zero amounts (in practice, there is a minimum measurement threshold, typically 0.25 mm, and larger measurements are in increments of the threshold; nonetheless, non-zero amounts are usually approximated with a continuous distribution). At longer time scales (for instance, monthly or annual), precipitation is usually not intermittent, except in some arid areas. In addition to the tendency away from intermittency, the probability distribution of precipitation amounts tends to become more symmetric as the time scale increases. For instance, Figure 11.1 shows empirical probability distributions of daily, monthly, and annual precipitation for West Glacier, Montana, USA. In Figure 11.1, the precipitation amounts have been normalized by the mean for each time scale, and the abscissa has a normal probability scale, so that a straight line corresponds to a normal distribution. In addition, the slope of the plot is related to the

Figure 11.1: *Empirical probability distributions of daily, monthly, and annual precipitation for West Glacier, Montana, USA. Precipitation amounts have been normalized by the mean for each time scale, and are plotted on a normal probability scale.*

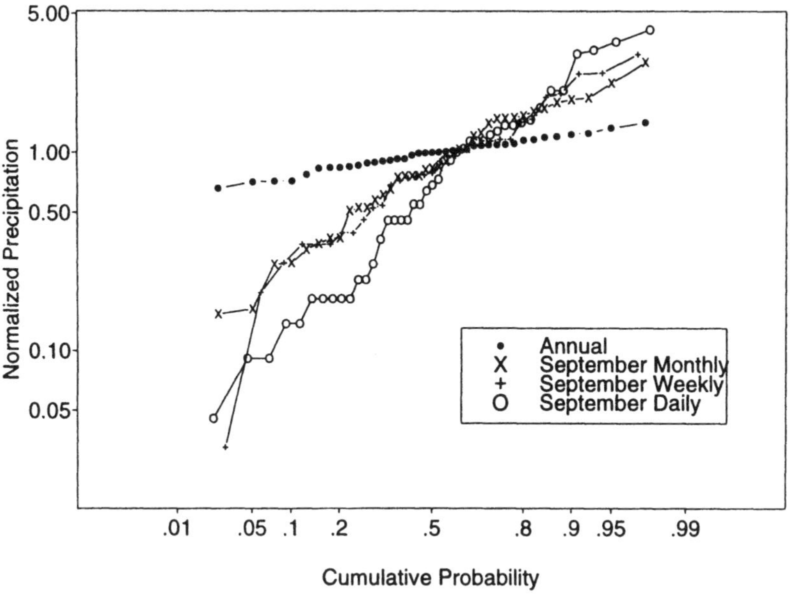

variance of the distribution; variance increases with slope. Clearly, the distributions become less variable, and more closely approximate the normal distribution, as the time scale increases. The reason for the tendency toward normality is explained by the central limit theorem, which states that, under fairly general conditions, the sum of independent random variables approaches normal.

At time scales where the normal distribution is not appropriate, a number of other distributions have been fit to precipitation amounts. At the daily time scale, the exponential, mixed exponential, gamma, and Kappa distributions have been used (Foufoula-Georgiou, 1985). Woolhiser and Roldhan (1982) applied several candidate distributions for daily precipitation to five stations located throughout the U.S., and found that the mixed exponential performed the best. At longer time scales, distributions such as the log-normal or gamma, which can be obtained from the normal distribution via transformation, have often been applied.

The spatial correlation of precipitation is also strongly dependent on the time scale. Figure 11.2 shows schematically typical variations in the spatial correlation with time period, and by season, which would be appropriate to much of North America and northern Eurasia. The correlation length

Figure 11.2: *Typical variations in the spatial correlation with time period, and by season, which appropriate to much of North America and northern Eurasia.*

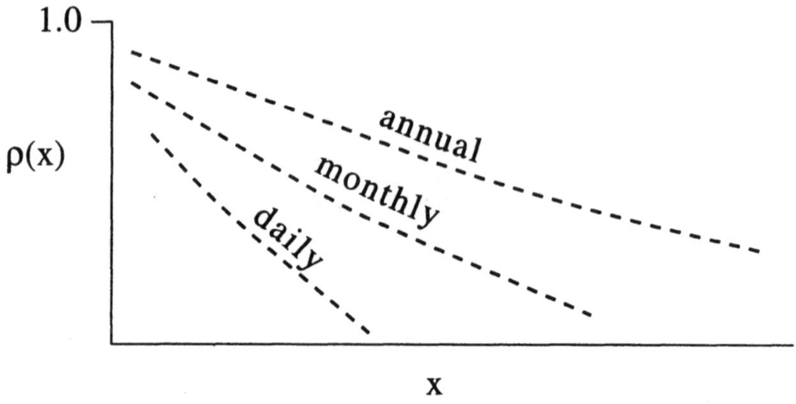

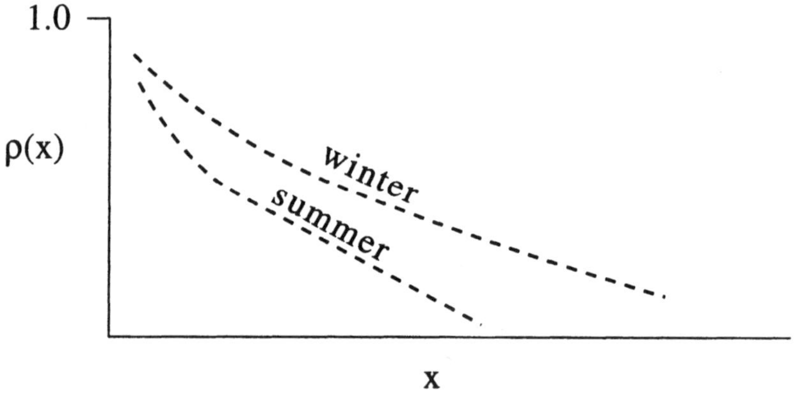

increases with the accumulation interval, and is longer in winter, when frontal storms of large spatial extent dominate, than in summer. The zero-intercept for the spatial correlation is less than one due to measurement error. An extensive literature exists that deals with precipitation of spatial correlation, the random component of the measurement error tends to average out at long time scales, resulting in a zero intercept closer to one. However, it should be emphasized that any bias present in the measurements is passed through from the short to the long time scales.

11.3 Stochastic Models of Precipitation

11.3.1 Background

As noted above, the main characteristic that distinguishes precipitation time scales in the context of stochastic modeling is intermittency. Our interest here is primarily in short time scales, at which distinct wet and dry periods are apparent in the record. Early attempts to model the stochastic structure of the precipitation arrival process (wet/dry occurrences) were based on first order homogeneous Markov chains (e.g., Gabriel and Neumann 1957; 1962), which essentially state that the precipitation state (wet or dry) probability on day t depends only on the state on days $t-i,\ i=1\ldots k$, where k is the order of the Markov chain (often one). The probabilities of state $p,\ p=1,2$ (wet or dry) occurring on day t given the state on day $t-i,\ i=1\ldots k$ are the transition probabilities.

Various extensions of Markov models have been explored to accommodate inhomogeneity (such as seasonality) of the transition probabilities (e.g., Weiss, 1964; Woolhiser and Pegram, 1979; Stern and Coe, 1984) and to incorporate and discrete amounts intervals (e.g., Khanal and Hamrick, 1974; Haan, 1976). Markov models have generally fallen from favor, however, because they are unable to reproduce the persistence of wet and dry spells (that is, they underestimate the probability of long runs of wet or dry sequences, also known as clustering) that is observed in rainfall occurrence series at daily or shorter time intervals (Foufoula-Georgiou, 1985).

Much of the recent activity in stochastic precipitation modeling, starting with the work of Kavvas and Delleur (1981), has sought to develop and/or apply more advanced point process models that represent clustering [see, for example, Rodriguez-Iturbe et al. (1987), Foufoula-Georgiou and Lettenmaier (1987), Smith and Karr (1985), and others]. Most of this recent work, which is reviewed in Georgakakos and Kavvas (1987), is based on a branch of probability known as point process theory (e.g., LeCam, 1961). Markov chain and point process models are similar to the extent that they are restricted to single-station applications, and are not easily generalizable to multiple station applications, at least without (in the case of Markov chain models) explosive growth in the number of parameters. In addition, all of the above models describe the precipitation process unconditionally, that is, they do not incorporate cause-effect information, such as descriptors of large-scale meteorological conditions that might give rise to wet, or dry, conditions.

11.3.2 Applications to Global Change

Recent interest in assessments of the hydrologic effects of climate change has placed different demands on stochastic precipitation models. Much of the concern about global warming has been based on simulations of climate produced by global general circulation models of the atmosphere (GCMs). These

Figure 11.3: *Long-term average rainfall predicted by the CSIRO GCM for a grid cell in northern Australia as compared with an average of long-term precipitation gauges located in the grid cell.*

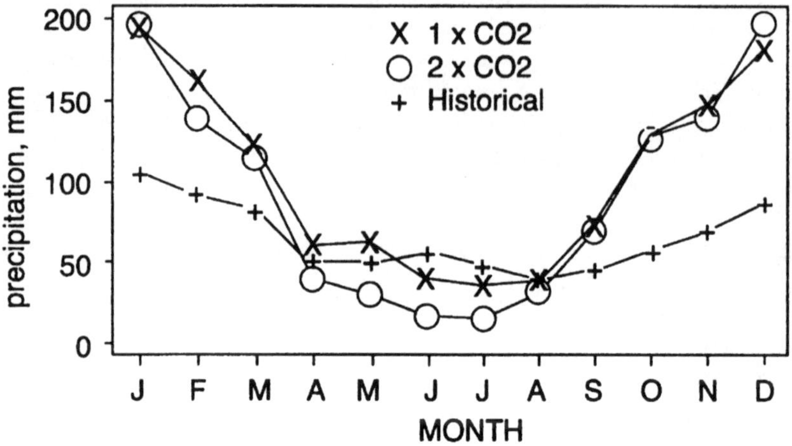

models operate at spatial scales of several degrees latitude by several degrees longitude, and computational time steps typically ranging from several minutes to several tens of minutes. GCMs are fully self-consistent produce predictions of both free atmosphere variables as well as surface fluxes.

In principle, the surface fluxes could be used directly to drive land surface (e.g., hydrologic) models which could serve to spatially disaggregate the GCM surface fluxes to predict such variables as streamflow. However, this approach is not at present feasible for two reasons. First, there is a scale mismatch between the GCM grid mesh and the catchment scale of interest to hydrologists (typically $10^2 - 10^4\ km^2$). Second, GCM surface flux predictions are notoriously poor at scales much less than continental. Figure 11.3 shows, as an example, long-term average rainfall predicted by the CSIRO GCM (Pittock and Salinger, 1991) for a grid cell in northern Australia as compared with an average of several long-term precipitation gauges located in the grid cell. Although the seasonal pattern (winter-dominant precipitation) is the same in the observations and model predictions, the model underpredicts the annual precipitation by a factor of about two. Such differences in GCM predictions of precipitation are not atypical (see, for instance, Grotch, 1988), and in fact the predictions shown in Figure 11.3 might in some respects be considered a "success" because the seasonal pattern is correctly predicted by the GCM.

Giorgi and Mearns (1991) review what they term "semi-empirical approaches" to the simulation of regional climate change. These are essentially stochastic approaches, which rely on the fact that GCM predictions of free atmosphere variables are usually better than those of surface fluxes. Stochastic approaches therefore attempt to relate local (e.g., catchment-scale) surface variables (especially precipitation) to GCM free atmosphere variables. Among these methods are Model Output Statistics (MOS) routines, which are essentially regressions that adjust numerical weather predictions (produced on grid meshes smaller than those used for GCM climate simulations, but still "large" compared to the scale required for local assessments). MOS adjustments to precipitation are used, for instance, in quantitative precipitation forecasts (QPFs) for flood forecasting. One drawback of these routines is that they attempt to produce "best estimates" in the least squares sense. While this may be appropriate for forecasting applications, least squares estimates underestimate natural variability (especially when the forecasts are most accurate), which can be critical for climate effects assessments.

Other semi-empirical approaches have been developed to relate longer term GCM simulations of free atmosphere variables to local precipitation. Among these are the canonical correlation approach of von Storch et al. (1993), and the regression approach of Wigley et al. (1990). The disadvantage of these approaches is that they are most appropriate for prediction of local variables at a time scale much longer than the catchment response scale (e.g., monthly or seasonal). Therefore, there is no practical way to incorporate the predictions within a rainfall-runoff modeling framework from which effects interpretations might be made. Moreover, in the case of the regression approach of Wigley et al. (1990), the input variables for local precipitation predictions include large-scale (e.g., GCM) precipitation, which is responsible for much of the predictive accuracy. Unfortunately, as indicated above, GCM precipitation predictions are often badly biased, and this bias would be transmitted to the local predictions.

11.4 Stochastic Precipitation Models with External Forcing

Several investigators have recently explored stochastic precipitation models that operate at the event scale (defined here as daily or shorter) and incorporate, explicitly or implicitly, external large-area atmospheric variables. The motivation for development of these methods has been, in part, to provide stochastic sequences that could serve as input to hydrologic (e.g., precipitation-runoff) models. Most of the work in this area has utilized, either directly or via summary measures, large scale free atmosphere variables rather than large area surface fluxes. In this respect, their objective has been to simulate stochastically realistic precipitation sequences that incorporate external drivers, rather than to disaggregate large area predictions, for instance, of precipitation.

11.4.1 Weather Classification Schemes

Weather classification schemes have been the mechanism used by several authors to summarize large-area meteorological information. The general concept of weather classification schemes (see, for example, Kalkstein et al., 1987) is to summarize measures of similar atmospheric conditions. Most of the externally forced stochastic precipitation models can be classified according to whether the weather classification scheme is subjective or objective, and whether it is unconditional or conditional on the local conditions (e.g., precipitation occurrence).

Subjective classification procedures include the scheme of Baur et al. (1944), from which a daily sequence of weather classes dating from 1881 to present has been constructed by the German Federal Weather Service (Bárdossy and Caspary, 1990) and the scheme of Lamb (1972), which has formed the basis for construction of a daily sequence of weather classes for the British Isles dating to 1861. The subjective schemes are primarily based on large scale features in the surface pressure distribution, such as the location of pressure centers, the position and paths of frontal zones, and the existence of cyclonic and anticyclonic circulation types (Bárdossy and Caspary, 1990).

Objective classification procedures utilize statistical methods, such as principal components, cluster analysis, and other multivariate methods to develop rules for classification of multivariate spatial data. For instance, McCabe (1990) utilized a combination of principal components analysis and cluster analysis to form classifications of daily weather at Philadelphia. The statistical model was compared to a subjective, conceptual model, which was found to give similar results. Briffa (Chapter 7 in this book) describes an objective approximation to the Lamb scheme, and uses it for regional validation of selected GCM control runs. Wilson et al. (1992) explored classification methods based on K-means cluster analysis, fuzzy cluster analysis, and principal components for daily classification of weather over a large area of the Pacific Northwest. All of the above methods are unconditional on local conditions, that is, no attempt is made to classify the days in such a way that local precipitation, for instance, is well-described by the weather classes.

Hughes et al. (1993) used an alternative approach that selected the weather classes so as to maximize the discrimination of local precipitation, in terms of joint precipitation occurrences (presence/absence of precipitation at four widely separated stations throughout a region of dimensions about 1000 km). The procedure used was CART (Breiman et al., 1984), or Classification and Regression Trees. The large area information was principal components of sea level pressure. Figure 11.4 shows the discrimination of the daily precipitation distribution at one of the stations modeled, Forks, according to weather class. As expected, because the classification scheme explicitly attempts to "separate" the precipitation (albeit occurrence/absence rather than amount) by the selected classes, the resulting precipitation distributions are more distinguishable than those obtained by Wilson et al. Hughes et al. also simu-

Figure 11.4: *Discrimination of daily precipitation distribution at Forks, WA for Fall using CART procedure.*

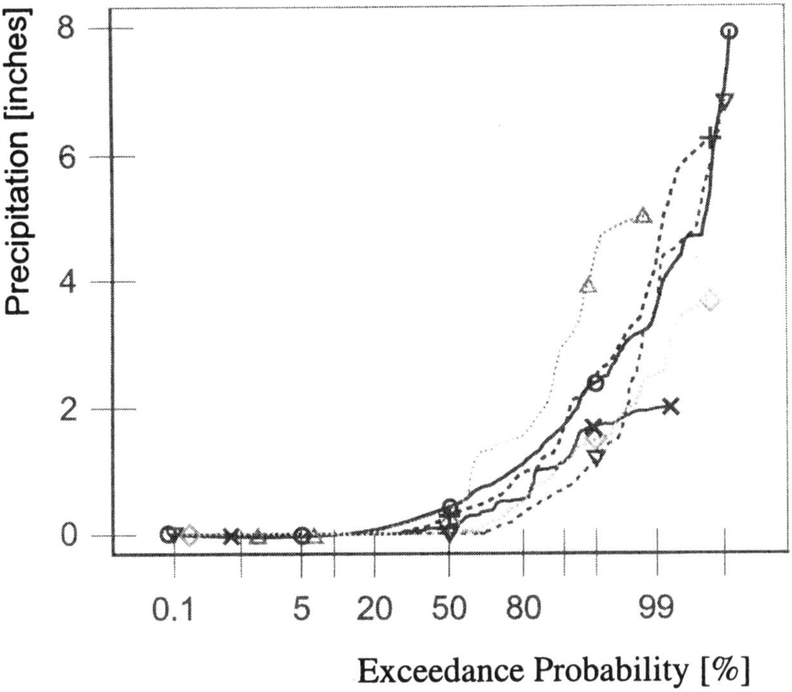

lated daily temperature minima and maxima. For this purpose, they used a Markov model conditioned on the present and previous days' rain state.

A final method of weather class identification is implicit. Zucchini and Guttorp (1992) describe the application of a set of models known as hidden Markov to precipitation occurrences. The objective of their study was to model the (unconditional) structure of the precipitation arrival process, and the properties of the hidden states, which could be (although do not necessarily need to be) interpreted as weather states, were not explicitly evaluated. Hughes (1993) and Hughes and Guttorp (1994) explored a larger class of nonhomogeneous hidden Markov models (NHMM), of which the model of Zucchini and Guttorp is a special case. He explored models of the precipitation occurrence process in which the atmospheric states were explicit, but were inferred by the NHMM estimation procedure. In this model, therefore, the weather state and stochastic precipitation structure are completely integrated. For this reason, further comments on the NHMM model are deferred to the next section.

11.4.2 Conditional Stochastic Precipitation Models

Hay et al. (1991) used a classification method (McCabe, 1990) based on wind direction and cloud cover which was coupled with a semi-Markov model to simulate temporal sequences of weather types at Philadelphia. Semi-Markov models (Cox and Lewis, 1978) with seasonal transition probabilities and parameters of the sojourn time distribution, were used to simulate the evolution of the weather states. This step is not strictly necessary if a lengthy sequence of variables defining the large-area weather states (the classification used required daily wind direction and cloud cover data) is available. Where such sequences are not available (sometimes the case for GCM simulations) fitting a stochastic model to the weather states has the advantage that it decouples the simulation of precipitation, and other local variables, from a particular GCM simulation sequence. The method of simulating daily precipitation conditional on the weather state used by Hay et al. was as follows. For each weather state and each of 11 weather stations in the region, the unconditional probability of precipitation was estimated from the historic record. Then, conditional on the weather state (but unconditional on precipitation occurrence and amount at the other stations and previous time) the precipitation state was selected based on the unconditional precipitation occurrence probability. Precipitation amounts were drawn from the product of a uniform and exponential distribution. Retrospective analysis of the model showed that those variables explicitly utilized for parameter estimation (conditional precipitation occurrence probabilities, mean precipitation amounts) were reproduced by the model. An analysis of dry period lengths suggested that the length of extreme dry periods was somewhat underestimated.

Bárdossy and Plate (1991) also used a semi-Markov model to describe the structure of the daily circulation patterns over Europe, with circulation types based on synoptic classification. They developed a model of the corresponding rainfall occurrence process that was Markovian within a weather state (circulation type), but independent when the weather state changed. Precipitation occurrences were assumed spatially independent. Bárdossy and Plate (1991) applied the model to simulate the precipitation occurrences at Essen, Germany. For this station, they found that the persistence parameter in the occurrence model was quite small, so that the model was almost conditionally independent (that is, virtually all of the persistence in the rainfall occurrence process was due to persistence in the weather states). The model reproduced the autocorrelations of the rainfall occurrences, as well as the distributions of dry and wet days, reasonably well. This is somewhat surprising, since other investigators (e.g., Hughes et al., 1993) have found that conditionally independent models tend to underestimate the tail of the dry period duration distribution. However, this finding is likely to depend on both the structure of the weather state process, and the precipitation occurrence process, which is regionally and site-specific. Bárdossy and Plate (1992) extended the model of Bárdossy and Plate (1991) to incorporate spatial persistence in the rainfall occurrences, and to model precipitation amounts explicitly. The weather state classification procedure was the same as in Bárdossy and Plate (1991), and they retained the assumption of conditional independence under changes in the weather state. Rather than modeling the occurrence process explicitly, they modeled a multivariate normal random variable W, negative values of which corresponded to the dry state, and (a transform of) positive values are the precipitation amount. Within a run of a weather state, W was assumed to be lag-one Markov. Spatial correlation in the occurrence process, and in the precipitation amounts, is modeled via the first two moments of W, which were weather state-dependent. The model was applied to 44 stations in the Ruhr River catchment. The model was able to reproduce the first two unconditional moments of rainfall amounts, and precipitation probabilities, as well as the dry day durations, reasonably well at one of the stations (Essen, also used in the 1991 paper) selected for more detailed analysis.

Wilson et al. (1991) developed a weather classification scheme for the Pacific Northwest based on cluster analysis of surface pressure and 850 mb temperature over a 10 degree by 10 degree grid mesh located over the North Pacific and the western coast of North America. A ten-year sequence of weather states (1975-84) was formed, and was further classified according to whether or not precipitation occurred at a station of interest. The partitioned weather state vector was then modeled as a semi-Markov process. For wet states, precipitation amounts were simulated using a mixed exponential model. The wet and dry period lengths were simulated quite well, although some of the weather state frequencies were mis-estimated, especially in summer. The authors noted that the semi-Markov model used a geometric distribution for the lengths-of-stay, they suggested that a heavier tailed distribution might be necessary. The above model is somewhat limited in that its generalization to multiple stations results in rapid growth in the number of parameters.

Wilson et al. (1992) explored a slightly different multiple station model, based on a Polya urn structure. Rather than explicitly incorporating the wet-dry state with the weather state, they developed a hierarchical modified model for the rainfall state conditioned on the weather state and the wet-dry state of the higher order station(s). In a Polya urn, the wet-dry state is obtained by drawing from a sample, initially of size N + M, a state, of which N are initially wet, and M are initially dry. For each wet state drawn, the sample of wet states is increased by n, likewise for each dry state drawn, the sample of dry states is increased by m, and the original state drawn is "replaced". Thus, the Polya urn has more persistence than a binomial process, in which the state drawn would simply be replaced, and the probability of the wet or dry state is independent of the dry or wet period length. The modification to the Polya urn (employed by others as well, e.g., Wiser, 1965) is to replace the persistent process with a binomial process once a given run (wet or dry period) length w has been reached. In addition, the parameters of the model (N, n, M, m, and w) are conditioned on the weather state and the wet-dry status of the higher stations in the hierarchy, but the memory is "lost" when the state (combination of weather state and wet-dry status of higher stations in the hierarchy) changes. The model was applied to three precipitation stations in the state of Washington, using a principal components-based weather classification scheme for a region similar to that used by Wilson et al. (1991). The precipitation amounts were reproduced reasonably well, especially for the seasons with the most precipitation. The dry and wet period lengths were also modeled reasonably well, although there was a persistent downward bias, especially for the lowest stations in the hierarchy. The major drawback of this model is that the number of parameters grows rapidly (power of two) with the number of stations. Also, the model performs best for the highest stations in the hierarchy, but there may not be an obvious way of determining the ordering of stations.

All of the above models define the weather states externally, that is, the selection of the weather states does not utilize station information. Hughes et al. (1993) linked the selection of the weather states with observed precipitation occurrence information at a set of index stations using the CART procedure described above. Precipitation occurrences and amounts were initially modeled assuming conditional independence, by simply resampling at random from the historical observations of precipitation at a set of target stations, given the weather states. They found that this model tended to underestimate the persistence of wet and dry periods. Model performance was improved by resampling precipitation amounts conditional on the present day's weather state and the previous day's rain state. Unlike the models of Bárdossy and Plate (1991; 1992) the Markovian persistence was retained regardless of shifts in the weather state. Inclusion of the previous rain state reduced the problem with simulation of wet and dry period persistence. However, by incorporating information about the previous day's rain state the number of parameters grows rapidly with the number of stations.

A somewhat different approach is the hidden Markov model (HMM), initially investigated for modeling rainfall occurrences by Zucchini and Guttorp (1991). The hidden Markov model is of the form

$$prob\left(\mathbf{R}_t|\vec{\mathbf{S}}_1^t, \vec{\mathbf{R}}_1^{t-1}\right) = prob\left(\mathbf{R}_t|\mathbf{S}_t\right) \tag{11.4}$$

$$prob\left(\mathbf{S}_t|\vec{\mathbf{S}}_1^{t-1}\right) = prob\left(\mathbf{S}_t|\mathbf{S}_{t-1}\right) \tag{11.5}$$

where $\mathbf{R}_t$ is the rainfall occurrence (presence-absence) at time t, $\mathbf{S}_t$ is the value of the hidden state at time t, and the vectors $\vec{\mathbf{S}}_1^t$ comprise all states $\mathbf{S}_1 \ldots \mathbf{S}_t$ of the unobserved process $\mathbf{S}_t$. Similarly does the vector $\vec{\mathbf{R}}_1^{t-1}$ represent all rainfall events $\mathbf{R}_1 \ldots \mathbf{R}_{t-1}$. Essentially, the model assumptions are that the rainfall state is conditionally independent, that is, it depends only on the value of the hidden state at the present time, and the hidden states are Markov. $\mathbf{R}_t$ can be a vector of rainfall occurrences at multiple stations, in which case a model for its (spatial) covariance is required. The shortcoming of the HMM is that the hidden states are unknown, and even though they may well be similar to weather states, they cannot be imposed externally. Therefore, only unconditional simulations are possible, and in this respect the model is similar to the unconditional models of the precipitation arrival process discussed in Section 11.3.

Hughes (1993) explored a class of nonhomogeneous hidden Markov models (NHMM) of the form

$$prob\left(\mathbf{R}_t|\vec{\mathbf{S}}_1^t, \vec{\mathbf{R}}_1^{t-1}, \vec{\mathbf{X}}_1^t\right) = prob\,(\mathbf{R}_t|\mathbf{S}_t) \tag{11.6}$$

$$prob\left(\mathbf{S}_t|\vec{\mathbf{S}}_1^{t-1}, \vec{\mathbf{X}}_1^t\right) = prob\,(\mathbf{S}_t|\mathbf{S}_{t-1}, \mathbf{X}_t) \tag{11.7}$$

where $\mathbf{X}_t$ is a vector of atmospheric variables at time t. In this model, the precipitation process is treated as in the HMM, that is, it is conditionally independent given the hidden state $\mathbf{S}_t$. However, the hidden states depend explicitly on a set of atmospheric variables at time t, and the previous hidden state. As for the HMM, if $\mathbf{R}_t$ is a vector of precipitation states at multiple locations, a model for the spatial covariances is required. Also, $\mathbf{X}_t$ can be (and in practice usually will be) multivariate. Hughes (1993) explored two examples in which $\mathbf{X}_t$ was a vector of principle components of the sea level pressure and 500 hPa pressure height, and the model for $prob\,(\mathbf{S}_t|\mathbf{S}_{t-1}, \mathbf{X}_t)$ was either Bayesian or autologistic. The Bayes and autologistic models are similar in terms of their parameterization; the structure of the autologistic model is somewhat more obvious structurally and is used for illustrative purposes here. It is of the form

$$prob\,(\mathbf{S}_t|\mathbf{S}_{t-1}, \mathbf{X}_t) = \frac{e^{a_{s_{t-1},s_t} + \mathbf{X}_t \cdot b_{s_{t-1},s_t}}}{\sum_k e^{a_{s_{t-1},k} + \mathbf{X}_t \cdot b_{s_{t-1},k}}} \tag{11.8}$$

where s_t denotes the particular values of $\mathbf{S}_t$. In this model, if there are m hidden states, and w atmospheric variables (that is, $\mathbf{X}_t$ is w-dimensional) the logistical model has $m(m-1)(w+1)$ free variables. Note that the model for the evolution of $\mathbf{S}_t$ conditioned on $\mathbf{S}_{t-1}$ and $\mathbf{X}_t$ is effectively a regional model, and does not depend on the precipitation stations. Hughes (1993) explored two special cases of the model:

1: $a_{s_{t-1},s_t} = a_{s_t}$ and $b_{s_{t-1},s_t} = b_{s_t}$, and
2: $b_{s_{t-1},s_t} = b_{s_t}$.

In the first case, the Markovian property of the NHMM is dropped, and the evolution of the hidden states depends only on the present value of the atmospheric variables. In the second model, the "base" component of the hidden state transition probabilities is Markov, but the component that depends on the atmospheric variables is a function only of the present value of the hidden state, and not the previous value. In one of the two examples explored,

Hughes found, using a Bayes Information Criterion to discriminate between models, that the second model was the best choice.

In the other example, the full Markov dependence was retained.

In the two examples, Hughes evaluated the means of the weather variables corresponding to the hidden states. He found that the large area characteristics were reasonable. For winter, the states with the most precipitation on average corresponded to a low pressure system off the north Pacific coast, and the cases with the least precipitation corresponded to a high pressure area slightly inland of the coast. In the second example, with modeled precipitation occurrences at 24 stations in western Washington, transitional states with differences in the surface and 500 hPa flow patterns were shown to result in partial precipitation coverage (precipitation at some stations, and not at others). These results suggest that the NHMM may offer a reasonable structure for transmitting the effects of large area circulation patterns to the local scale.

11.5 Applications to Alternative Climate Simulation

Although the development of most of the models reviewed above have been motivated in part by the need for tools to simulate local precipitation for alternative climate scenarios, there have only been a few applications where climate model (GCM) scenarios have been downscaled using stochastic methods. Hughes et al. (1993) estimated parameters of semi-Markov models from five-year $1 \times CO_2$ and $2 \times CO_2$ GFDL simulations of surface pressure and 850 hPa temperature. From these five-year sequences, they computed the daily weather states using algorithms developed from historical sea level pressure observations (see Section 11.4.1), and fit semi-Markov models to the weather states. The semi-Markov models were used to simulate 40-year weather state sequences corresponding to the $1 \times CO_2$ and $2 \times CO_2$ runs. Daily precipitation (and temperature maxima-minima, using the model described in Section 11.4.2) were then simulated for the 40-year period, and were used as input to a hydrologic model which was used to assess shifts in flood risk that might be associated with climate change.

Zorita et al. (1995) used a model similar to that of Hughes et al. (1993) to simulate daily precipitation for four sites in the Columbia River basin. They found that the model performed reasonably well in winter, but there were difficulties in application of the CART procedure to determine climate states in summer. When stations relatively far from the Pacific Coast were used in definition of the multistation rain states in the CART algorithm, no feasible solutions resulted. This problem could be avoided by restricting the index stations to be relatively close to the coast, but when the model was applied to the middle Atlantic region, CART weather states could be obtained only when the index stations were quite closely spaced, and then only in winter. The difficulty appeared to be the shorter spatial scale of summer precipitation, and weaker coupling of local precipitation with the regional circulation patterns.

One of the motivations for development of stochastic models that couple large area atmospheric variables with local variables, such as precipitation, is to provide a means of downscaling simulations of alternative climates for effects assessments. However, as noted above, most of the applications to date have been to historic data, for instance, local precipitation has been simulated using either an historic sequence of weather states (e.g., Hughes et al., 1993) or via a stochastic model of the historic weather states (e.g., Hay et al., 1991, Bárdossy and Plate, 1992; Wilson et al., 1992).

For most of the models reviewed, it should be straightforward to produce a sequence of weather states corresponding to an alternative climate scenario (e.g., from a lengthy GCM simulation). There are, nonetheless, certain complications. Selection of the variables to use in the weather state classification is problematic. Wilson et al. (1991) classified weather states using sea level pressure and 850 hPa temperature. However, if this scheme is used with an alternative, warmer climate, the temperature change dominates the classification, resulting in a major change in the stochastic structure of the weather class sequence that may not be physically realistic. Although this problem is resolved by use of variables, such as sea level pressure, that more directly reflect large area circulation patterns, elimination of temperature from consideration as a classifying variable is somewhat arbitrary. A related problem is the effect of the strength of the linkage between the weather states and the local variables. In a sense, the problem is analogous to multiple regression. If the regression is weak, i.e., it does not explain much of the variance in the dependent variable (e.g., local precipitation), and changes in the independent variables (e.g., weather states) will not be evidenced in predictions of the local variable. Therefore, one might erroneously conclude that changes in, for instance, precipitation would be small, merely because of the absence of strong linkages between the large scale and local conditions (see, for example, Zorita et al., 1995).

Application of all of the models for alternative climate simulation requires that certain assumptions be made about what aspect of the model structure will be preserved under an alternative climate. All of the models have parameters that link the large area weather states with the probability of occurrence, or amount of, local precipitation. For instance, in the model of Wilson et al. (1992) there are parameters that control the probability of precipitation for each combination of weather state and the precipitation state at the higher order stations. In the model of Bárdossy and Plate (1991) there is a Markov parameter that describes the persistence of precipitation occurrences given the weather state. These parameters, once estimated using historical data, must then be presumed to hold under a different sequence of weather states corresponding, for instance, to a GCM simulation. Likewise, many of the models (e.g., Bárdossy and Plate, 1992; Hughes, 1993) have spatial covariances that are conditioned on the weather state. The historical values of these parameters likewise must be assumed to hold under an alternative climate.

Another complication in application of these models to alternative climate simulation is comparability of the GCM predictions with the historic observations. For instance, McCabe (1990) used a weather classification scheme based on surface wind direction and cloud cover. The resulting weather classes were shown to be well related to precipitation at a set of stations in the Delaware River basin. Unfortunately, however, GCM predictions of cloud cover and wind direction for current climate are often quite biased as compared to historic observations, and these biases will be reflected in the stochastic structure of the weather class sequence.

11.6 Conclusions

The coupling of weather state classification procedures, either explicitly or implicitly, with stochastic precipitation generation schemes is a promising approach for transferring large-area climate model simulations to the local scale. Most of the work reported to date has focused on the simulation of daily precipitation, conditioned in various ways on weather classes extracted from large-area atmospheric features. The approach has been shown to perform adequately in most of the studies, although there remain questions as to how best to determine the weather states. Further, no useful means has yet been proposed to determine the strength of the relationship between large-area weather classes and local precipitation, and to insure that weak relationships do not result in spurious downward biases in inferred changes in local precipitation at the local level. This is an important concern, since at least one of the studies reviewed (Zorita et al., 1995) found conditions under which weather classes well-related to local precipitation could not be identified.

There have been relatively few demonstration applications of these procedures for climate effects interpretations. One of the major difficulties is accounting for biases in the GCM present climate, or "base" runs. In addition, few of the models reviewed presently simulate variables other than precipitation needed for hydrological studies. Temperature simulation is especially important for many hydrological modeling applications, but methods of preserving stochastic consistency between local and large-scale simulations are presently lacking.

Part IV

Pattern Analysis

Chapter 12

Teleconnection Patterns

by Antonio Navarra

12.1 Objective Teleconnections

Walker and Bliss devoted their entire life to a pure subjective search for significant statistical relation among the myriads of correlation values that their limited data set could generate. Modern times do not require such enormous efforts, but the two papers by Mike Wallace in collaboration with D. Gutzler and J. Horel (Wallace and Gutzler, 1981; Horel and Wallace, 1981) realized a significant breakthrough.

In modern terms, teleconnections relations can be visualized by computing the correlation matrix $\mathcal{R}$. The elements of $\mathcal{R}$ are the correlation coefficient r_{ij} between the time series of the variable of interest (for instance, geopotential or surface pressure) at the grid point i with the temporal series at the grid point j. The data usually consist of anomaly time series of fields, as for instance monthly mean fields, since the correlation coefficients give more stable results for anomalies. The grid points are assumed to be ordered using only one index, rather than the more familiar two-indices ordering. Considerable information is contained in the matrix $\mathcal{R}$. The i-th column represents the correlation information for the i-th grid-point and it is usually called a *one-point teleconnection map* for the ***basis point*** i.

These patterns represents how each grid-point is connected with its neighbors. The patterns are characterized by a roughly elliptical area of positive correlations around the basis point. This area is the scale of the local

Acknowledgments: Many thanks go to Neil Ward and Mike Wallace for kindly allowing to reproduce the pictures from their papers. This work was partially supported by the EEC contract EV5V-CT92-0101

Figure 12.1: *An example of one-point teleconnections map. The basis point is at 55N,20W, in the North Pacific. The basis point is surrounded by the ellipsoidal area of high correlation pointing to the typical spatial scale of the anomalies in that region. Teleconnection patterns appear as remote regions of large correlations, either positive or negative. (From Wallace and Gutzler, 1981).*

anomaly, namely the horizontal dimension of the patch of fluid that is presumably dynamically coherent. Each basis grid point will have one of these ellipses because the grid points are not independent of each other but they tend to be organized in features much larger than a single grid-point.

For most grid points only the local ellipses will be generated, resulting in rather uninteresting distributions, but for a few others, complicated patterns show up. Basically there are two broad classes of large scale patterns. There are patterns that exhibit large negative correlations that together with the original ellipse form a dipole structure. These patterns are most commonly found over the oceans and they present a north-south orientation of the axis of the dipole. There are also more elongated patterns, with multiple centers of positive and negative high correlation, stretching over the hemisphere.

Figure 12.1 shows an example from the Wallace and Gutzler paper. In this case the basis point is located in the North Pacific and it is easily recognizable by the value of 1.0 of the correlation coefficient. The basis point is obviously perfectly correlated with itself. The correlation drops to -.65 in the next center of action to the east, meaning that positive anomalies in that location are often simultaneous to positive anomalies in the basis point.

Testing the statistical significance of the teleconnections is rather difficult, because it is not known a priori which is the level of correlation coefficient that is relevant. A popular, but not mathematically rigorous method, is to check the reproducibility of the patterns with subsets of the original data or with different data sets. The reproducibility is then taken as a strong indication of significance for a given teleconnection pattern.

The original analysis of Wallace and Gutzler identified five principal patterns for the 500mb geopotential height field. Three patterns consisted of multiple centers, the Eastern Atlantic pattern (EA), thePacific/North American pattern (PNA) and the Eurasian pattern (EU) pattern. The remaining two represented dipole structure, the West Pacific pattern (WP) and the West Atlantic pattern (WA). Each pattern is identified by the location of the nodes and anti-nodes. The anti-nodes correspond to the *centers of action*, namely the geographical locations of the grid point more strongly correlated with each other. The centers of action can be used to define teleconnections indices by combining linearly the anomalies at such centers. For instance the index for the PNA can be defined as

$$PNA = \frac{1}{4}[(20^oN, 160^oW) - (45^oN, 165^oW) + (55^oN, 115^oW) - (30^oN, 85^oW)]$$

The index is such that very high values of the index indicate an anomaly field with highs and lows in phase with the PNA and with high amplitude. Similarly, we can define indices for the other patterns

$$\begin{aligned} EA &= \frac{1}{2}(55^oN, 20^oW) - \frac{1}{4}(25^oN, 25^oW) - \frac{1}{4}(50^oN, 40^oE) \\ WA &= \frac{1}{2}(55^oN, 55^oW) - \frac{1}{2}(30^oN, 55^oW) \end{aligned}$$

Figure 12.2: *Difference in meters between composites of high and low PNA index states, based on monthly mean 500mb heights, from Wallace and Gutzler, 1981. These are not correlations, but real anomalies. They indicate a relation between anomalies in the Pacific and Atlantic coast of the United States.*

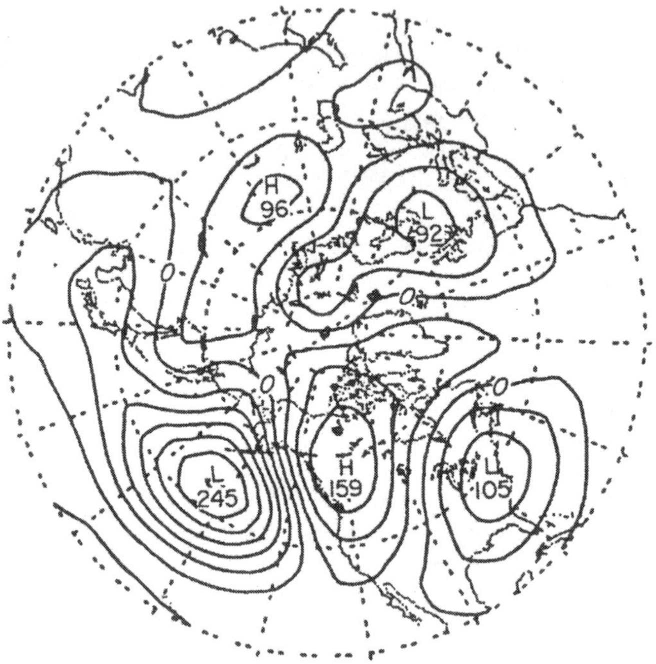

$$WP = \frac{1}{2}(60^o N, 155^o E) - \frac{1}{2}(30^o N, 155^o E)$$

$$EU = -\frac{1}{4}(55^o N, 20^o E) + \frac{1}{2}(55^o N, 75^o E) - \frac{1}{4}(40^o N, 145^o E)$$

The indices can then be used to generate *composites*, namely ensemble means of all anomalies labeled by an index value above or below some chosen threshold. The result (Figure 12.2) shows the characteristic signature of the teleconnection pattern anomalies. Composites are generated by averaging anomaly fields with similar values of the index, in the case of Figure 12.2 positive PNA-index anomalies are subtracted from negative PNA-index anomalies. The result is a characteristic string of anomalies that stretches from the Pacific to the Atlantic. The emergence of this pattern indicates a systematic tendency for high anomalies in the North Pacific to coexist with high anomalies over the Atlantic coast of the United States.

The teleconnection technique is general and in particular there is no limitation to constraint its use to fields of the same kind, the basis point time

Figure 12.3: *Correlations in hundredth between July-September sea level pressure at 10° resolution and the index of Sahelian rainfall 1949-88, from Ward, 1992. Correlations locally significant at 5% are shaded. A strong indication of a link with the equatorial Atlantic and weaker evidence of relations with Northern Europe and the Indian sub-continent are evident.*

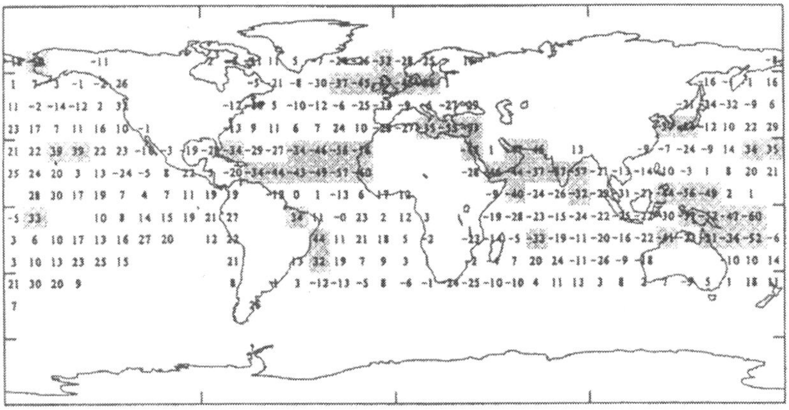

series need not be the same physical variable that is used at the other grid-point. For instance, one could use as a basis grid point time series a sea surface temperature anomaly and correlate this series with sea level pressure or upper air geopotential field. This aspect was investigated by Horel and Wallace (1980) to show the statistical relation between the tropical Pacific SST and large scale anomalies in the atmosphere. They found that positive anomalies in the Central Pacific are correlated with a system of alternating anomalies that extend from the Pacific to the North Atlantic and that bring close resemblance to the PNA. After them the relation between SST and atmospheric anomalies were investigated by Lanzante (1984), Wallace and Jiang (1987), Namias et al. (1988), Wallace et al. (1990) and Wallace et al. (1992). The teleconnections in atmosphere were further investigated by Namias (1981), Esbensen (1984), Barnston and Livezey (1987) and Kushnir and Wallace (1989) and Livezey and Smith (1999).

This technique of heterogeneous[1] teleconnections has been used extensively in the following years and the teleconnections have now been investigated in many respect. Figure 12.3, from Ward (1992) shows heterogeneous correlation in a slightly different format for a teleconnection map. In this case the values of the correlation between the Sahelian rainfall series and sea level pressure anomalies are displayed directly at 10° resolution. Large areas of

[1]Teleconnections are named *homogeneous* if the index time series with which all values in a field are correlated with, is derived from the same field. A teleconnection pattern in which an index time series and a different field are examined is *heterogeneous*.

locally significant correlations[2] correlations are visible, pointing to a possible dynamical link between precipitation in the Sahel and sea level pressure in those areas.

The teleconnection technique is one of the most elementary techniques to identify patterns. Other mathematical techniques, like the Empirical Orthogonal Function (EOF) analysis (see Section 13.3) can also reveal patterns. One of the strong points of the teleconnections maps is that there is no *a priori* assumption on the shape of the patterns to be found. EOFs require by definition the patterns to be orthogonal to each other, constructing in the solution their wave-like character to some extent. The orthogonality requirement can be relaxed if one accepts the usage of rotated EOF (see Section 13.2.4), though this option implies some other subtleties.

In a EOF analysis there is no information on the relation between grid points. The wave-like patterns obtained through EOFs do not necessarily point to a strong coherency between the centers of action (see Section 13.3.2.)

On the other hand, the EOFs provide a measure of the relative importance of a pattern by yielding the percentage of variance of the fields that can be attributed to that pattern. The information provided by EOF analysis and teleconnections maps is complementary and they should always be used together to avoid some danger of misinterpretations.

12.2 Singular Value Decomposition

The heterogeneous teleconnection maps extend to different physical fields in the investigation of teleconnections. It would be very desirable to have a technique similar to EOF analysis, but that could be applied to cross-covariances between different fields. The generalization of EOF analysis to cross-covariance is based on the Singular Value Decomposition of the cross-covariance. Eigenvalues algorithms can also be used, but they are much more computational expensive.

The *Singular Value Decomposition* (SVD) is a powerful algebraic technique to decompose arbitrary matrices in orthogonal matrices. It is based on a result from linear operator theory (Smithies, 1970; Gohberg and Krein, 1969) that proves that the SVD realize a completely orthogonal decomposition for any matrix $\mathcal{A}$.

The *SVD* of a matrix $\mathcal{A}$ of order $m \times n$ (with m columns and n rows) is its decomposition in the product of three different matrices, that is

$$\mathcal{A} = \mathcal{U}\Sigma\mathcal{V}^T \tag{12.1}$$

The $m \times m$ matrix $\mathcal{U}$ and the $n \times n$ matrix $\mathcal{V}$ obey the orthogonality relations $\mathcal{U}\mathcal{U}^T = I$ and $\mathcal{V}\mathcal{V}^T = I$. The diagonal matrix Σ is defined by $\Sigma = diag(\sigma_1, \sigma_2 \ldots \sigma_{min(n,m)})$ where, and $\sigma_1, .., \sigma_{min(n,m)}$ are nonnegative

[2]Refer to the discussion in Chapter 9 of the subtleties involved in the assessment of the over-all significance by plotting distributions of local significance.

numbers ordered in decreasing magnitude and are called ***singular values***. The columns of $\mathcal{U}$ and $\mathcal{V}$ form an orthonormal basis and they are called ***u-vectors*** and *v-vectors*, respectively.

If we put $\mathcal{U} = \left(\vec{u}_1|\vec{u}_2\ldots\vec{u}_{min(n,m)}\right)$, and $\mathcal{V} = \left(\vec{v}_1|\vec{v}_2\ldots\vec{v}_{min(n,m)}\right)$, the following relations hold

$$\left.\begin{array}{rcl}\mathcal{A}\vec{v}_i & = & \sigma_i\vec{u}_i\\ \mathcal{A}^T\vec{u}_i & = & \sigma_i\vec{v}_i\end{array}\right\} \quad i = 1\ldots min(n,m) \tag{12.2}$$

then we can relate the u-vectors $\vec{u}_i$ to the v-vectors $\vec{v}_i$ searching the corresponding i-th singular values. Sometimes, the u-vector and the v-vector are referred to as the left and right ***singular vectors***, respectively. The decomposition can be used to obtain the following representation of $\mathcal{A}$:

$$A = \sum_{i=1}^{r} \sigma_i\vec{u}_i\vec{v}_i^T \tag{12.3}$$

where $r = min(m,n)$. In the case of square matrices, i.e. $m = n$, if r is not equal to n, namely one or more of the σ_i is zero, then the matrix is singular and does not possess a complete set of eigenvectors. The relations (12.2) can be used to show that the following relations also hold

$$\left.\begin{array}{rcccl}\mathcal{A}^T\mathcal{A}\vec{v}_i & = & \sigma_i\mathcal{A}^T\vec{u}_i & = & \sigma_i^2\vec{v}_i\\ \mathcal{A}\mathcal{A}^T\vec{u}_i & = & \sigma_i\mathcal{A}\vec{v}_i & = & \sigma_i^2\vec{u}_i\end{array}\right\} \quad i = 1\ldots min(n,m) \tag{12.4}$$

Thus the $\vec{u}$ are the eigenvectors of $\mathcal{A}\mathcal{A}^T$ and the $\vec{v}$ are the eigenvectors of $\mathcal{A}^T\mathcal{A}$ and the eigenvalues of both matrices are given by the singular values squared. The SVD can be computed with a classic algorithm that is available in now easily available packages, like LAPACK and MatLab. Alternatively one can use (12.4) and compute the singular values and vectors from the eigenvalue/eigenvector analysis of $\mathcal{A}\mathcal{A}^T$ and $\mathcal{A}^T\mathcal{A}$, but this latter method is less numerically stable than the direct algorithm (Golub and Reinsch, 1970) A detailed discussion of the mathematical properties of the SVD can be found in the book by Golub and Van Loan (1996). A good description of an application of SVD to climate data can be found in Bretherton et al. (1992).

The interpretation of the left and right vectors is a natural extension of the EOF concept. They defined the couple of patterns that in the space of the left and right fields explain a fraction of the total cross-covariance given by $\sigma/\sum_{i=1}^{n}\sigma_i$.

12.3 Teleconnections in the Ocean-Atmosphere System

The Singular Value Decomposition can then be applied to the analysis of climate data to identify patterns that maximize covariation.

The technique is described in detail in Bretherton et al. (1992), but the technique will be described here on the basis of the example of the analysis of Wallace et al. (1992). In that case they applied SVD to time series of the North Pacific Sea Surface Temperature (SST) and to anomalies of upper air geopotential (Z500mb). The intent was to reveal patterns that were statistically related to each other. The SVD analysis can be difficult to interpret because of the possibility of spurious relation caused by the mathematical structure of the SVD (Cherry, 1996, 1997). However, if handled correctly it is a method that can give very useful information. The total ban on SVD that authors like Newman and Sardeshmukh (1995) seems to suggest appear excessive.

As in the teleconnection case, the starting point of the analysis is the normalized matrices of the anomalies obtained removing the time mean from the observations,

$$\mathcal{S} = (\vec{s}_1|\ldots|\vec{s}_t) \quad \text{and} \quad \mathcal{Z} = (\vec{z}_1|\ldots|\vec{z}_t) \tag{12.5}$$

The matrices $\mathcal{S}$ and $\mathcal{Z}$ are made up of the same number t of observation fields for the SST and Z500 data. The dimensions of the vectors $\vec{s}$ and $\vec{z}$ correspond to the number of grid points in the observations and need not be the same. In the following we will assume that we have n grid points for the SST and m points for the geopotential height, Z500. The vectors are obtained ordering the observations data, usually in the form of synoptic maps, as a long string of points rather than with a geographical ordering. The cross-covariance matrix between the time series of SST and Z500 can then be computed as,

$$\mathcal{C}_{SZ} = \frac{1}{t}\mathcal{S}\mathcal{Z}^T \tag{12.6}$$

since normalized anomalies were used, the elements of $\mathcal{C}_{SZ}$ are correlation coefficients.

The rows and columns of the cross-covariance matrix are teleconnection maps between the right and left fields. The interpretation is similar to the case of the covariance matrix discussed in the previous sections. In that case the matrix is symmetric, the k-th column or the k-th row can be used indifferently as the teleconnection map for the k-th grid point as basis point. The interpretation of the cross-covariance must be done with a little more caution, but is straightforward.

The rows of $\mathcal{C}_{SZ}$ are teleconnection maps for SST basis points. They are m-vectors representing the distributions of correlation coefficients between the time series of a single SST point time series and the time series of the Z500 at all grid points. Similarly the columns of $\mathcal{C}_{SZ}$ are n-vectors representing the distributions of correlation coefficients between the time series of a single Z500 point time series and the time series of the SST at all grid points.

A complete orthogonal decomposition of the cross-covariance cannot now be performed with an eigen-analysis as in the case of the EOF because the matrix is not symmetric, but we can still obtain such a decomposition if we use a SVD. Using the definition of the previous Section 12.2 the SVD of $\mathcal{C}_{SZ}$ can be written as

$$\mathcal{C}_{SZ} = \mathcal{U}\Sigma\mathcal{V}^T \tag{12.7}$$

The orthogonal matrices $\mathcal{U}$ and $\mathcal{V}$ can be used to project the SST and Z500 data vectors $\vec{s}$ and $\vec{z}$,

$$\left.\begin{array}{rcl} \vec{s}_i & = & \sum_{k=1}^{t} a_{ki}(t)\vec{u}_k \\ \vec{z}_i & = & \sum_{k=1}^{t} b_{ki}(t)\vec{v}_k \end{array}\right\} \quad i = 1 \ldots t \tag{12.8}$$

or in matrix form

$$\mathcal{S} = \mathcal{A}\mathcal{U} \quad \text{and} \quad \mathcal{Z} = \mathcal{B}\mathcal{V} \tag{12.9}$$

where the coefficient a_{ik} measures the projection of the i-th observation onto the k-th u-vector. Following Bretherton et al. (1992), the vectors $\vec{u}$ and $\vec{v}$ will be called in the following *patterns*.

The projection coefficients a_{ik} and b_{ik} can be used to generate correlations maps. For instance the map of correlations between the coefficients of the k-th SST pattern and the Z500 maps, $corr(a_{ki}, \vec{z}_j)$, is a measure of how well the anomaly pattern at Z500 can be specified from the expansion coefficients of the k-th SST pattern. These maps that involve correlation between the pattern of one field and the grid point values of the other are called heterogeneous correlation maps. Correlation maps can be created from the pattern and grid point values of the same field, i.e. $corr(a_{ki}, \vec{s}_j)$, thus providing homogeneous correlation maps, but they do not provide a direct link to the SVD modes and they are not orthogonal to each other (Wallace et al., 1992). There is no restriction on the choice of the grid point fields for which we can compute the heterogeneous correlation maps. Once the patterns have been obtained they can be correlated with any fields as long as it is composed of the same number of observations.

The total amount of squared covariance

$$||\mathcal{C}_{SZ}||_F^2 = \sum_{i,j} c_{ij} \tag{12.10}$$

can be shown that is mathematically equivalent to the *Frobenius norm* of the cross-covariance matrix. Using standard algebraic results (Golub and Van Loan, 1996), and the SVD decomposition we can show that

$$\begin{aligned} ||\mathcal{C}_{SZ}||_F^2 &= Trace\,(\mathcal{C}_{SZ}^T\mathcal{C}_{SZ}) = Trace\,(\mathcal{V}\Sigma\mathcal{U}^T\mathcal{U}\Sigma\mathcal{V}^T) \\ &= Trace(\Sigma^2) \end{aligned} \tag{12.11}$$

Figure 12.4: *Heterogeneous correlation maps for the first SVD mode from Wallace et al., 1992. The top map has been obtained correlating the SST fields against the time series of the Z500 part of the first SVD mode, the bottom map has been obtained correlating the Z500 fields against the time series of the SST part of the first SVD mode. The fraction of cross-covariance explained by this mode is 0.52, and the correlation between the time series of the SST and Z500 part of the mode is 0.81, indicating that this mode is a strongly coupled mode. (After Wallace et al., 1992).*

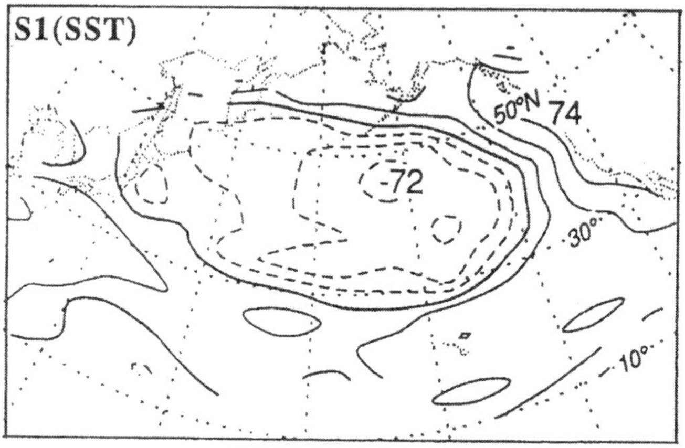

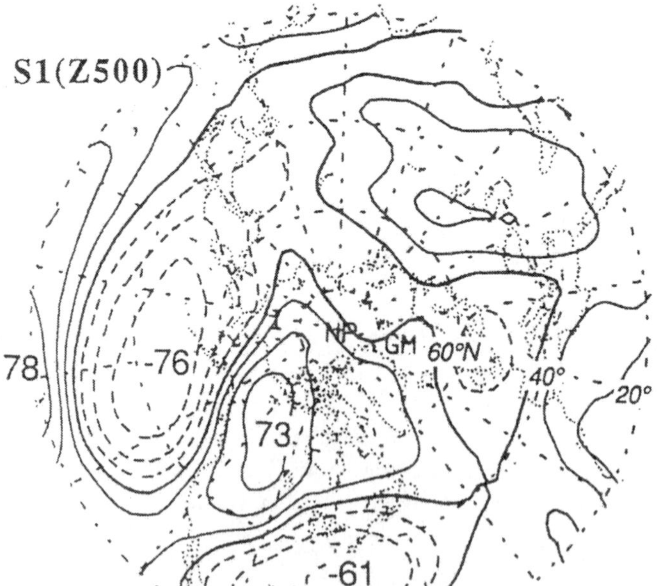

The total cross-covariance is then given by the sum of the square of the singular values, and the fraction of cross-covariance explained by the k-th pair of patterns, the k-th SST pattern and the k-th Z500 pattern is given by the ratio $\sigma_k^2 / \sum_{allpatterns} \sigma^2$. It is important to note that SVD method automatically limits the investigation to the portion of variance that is cross-correlated. The method does not give any information on the importance of these patterns for the portion of the Z500 variance that is not correlated with the SST variance.

An example of a direct SVD analysis is shown in Figure 12.4, from Wallace et al. (1992). Here the heterogeneous correlation maps for the first pattern resulting from a SVD analysis of the North Pacific SST and the Z500 geopotential field are shown. This pattern explains 52% of the cross-covariance and the time correlation between the two time-series is 0.81. The mode show that when SST anomalies in a large area in the North Pacific are connected with a complex system of anomalies extending well into the East Atlantic. The atmospheric pattern is reminiscent of the PNA teleconnections found by Wallace and Gutzler (1981) and the teleconnection pattern induced by equatorial SST (Horel and Wallace, 1981). In this case the analysis indicates that also mid-latitudes SST can generate PNA-like structures.

12.4 Reproducibility and Skill of Teleconnection Patterns

The teleconnection patterns are clearly representing a major feature of the circulation. It is then an important component of model evaluation to assess to what extent numerical models can simulate the teleconnection and with what degree of consistency. The methods of analysis of variance, and in particular the SVD, can be helpful also in this case. There are two aspects to the issue of representation of teleconnection in models. The first aspect concerns the shape of the pattern in the model is distorted. A connection is created between point on the globe that are not linked in reality, or, conversely, points that are known to variate together are independent. This is the case of a geographical distortion of the teleconnection.

The second aspect regards the existence of the teleconnection itself. The presence of the pattern may be elusive. The model can fail to realize the teleconnection itself, especially when the simulation is carried out in Monte Carlo fashion, with several realization, slightly perturbed. Some member of the ensemble can exhibit a teleconnection, whereas some others fail to show it.

The first aspect can be regarded as the skill of model in reproducing the pattern, whereas the second has to do with the ability of the model to reproduce successfully the pattern in a consistent way.

The possibility of using SVD to assess the reproducibility of teleconnection patterns has been pointed out by Ward and Navarra (1997), that used SVD to extract the common response from an ensemble of simulation. The method is based on the SVD analysis of the cross-covariance matrix among the members of the ensemble. A particular order allow to obtain a symmetric cross-covariance matrix. It can be shown that for large ensemble this modes converge to the EOF of the ensemble mean, but for relatively small number it is a more efficient way to extract the common signal.

The Ward and Navarra method yields the common variability mode of an ensemble, but to gain some information on the skill of the mode in reproducing the teleconnection another method must be used. Moron et al. (1998) have proposed a method to assess the ability of a model to simulate the teleconnection patterns based on SVD between the simulation and the observations. It is a very generous test for the model, since we are trying to extract from the best signal co-varying with the observations.

The patterns in Figure 12.5 show the results of such an analysis. The top panel shows simultaneously the pattern corresponding to an SVD between model precipitation and simulation. The shading is the "observation" part of the mode and the contours are the "model" part. There is a good agreement between positive and negative areas, indicating a relatively good skill for this teleconnection in winter (the spatial correlation is 0.35, but higher values can be obtained for selected areas in the tropics). Panel (b) shows the "common" mode. This is the "consensus" mode that explains most of the covariance within the ensemble. It is very similar to the "model" part of panel (a), indicating that when the experiments agree (panel , b) they tend to agree on the real response to SST ("model" part of panel a). The two bottom panels show the time variations of the coefficients of the modes in panels (a) and (b) indicating that there is also a good temporal correspondence. Once again the similarity between panel (c) and (d) indicates the over all skill of the model (temporal correlation is 0.85).

12.5 Concluding Remarks

The pattern identification techniques that have been described are powerful ways to identify systematic modes of variations and they have greatly enhanced our understanding of the climate system. In fact they are so intriguing and powerful that there is the danger sometimes to forget that the beautiful wave-like pattern found, for instance in correlation maps, are just that, ***maps of correlation coefficients***, and they are not physical wave-like phenomena. It must always be kept in mind that this patterns are expression of statistical relations that can be conveniently displayed over a geographic map, but they may or may not correspond to waves in the physical space. They are pointers to some unknown underlying physical mechanism that is responsible for the relations between fields that so gracefully emerge as teleconnections, but after so much time and effort there is still no convincing comprehensive dynamical explanation for the teleconnection patterns.

Figure 12.5: *Skill and reproducibility analysis for precipitation in DJFM. The simulations have been obtained from three experiments with prescribed SST for the period 1961-1994. (a) SVDA between GCM simulated and observed precipitation. Observed singular vector 1 indicated by shading, dark shading > 2, light shading < −2. GCM singular vector one indicated by contours at +/-2, 4, 6. (b) SVDA between the three GCM simulations. Contours at +/-2, 4, 6, dark shading > 2, light shading < −2. (c) SVDA time series from (a). Thin solid shows the three GCM simulations, thick solid is the ensemble mean, thick dashed is the observed. (d) SVDA time series from (b). Thin solid shows the three GCM simulations, thick solid is the ensemble mean. (After Moron et al., 1998).*

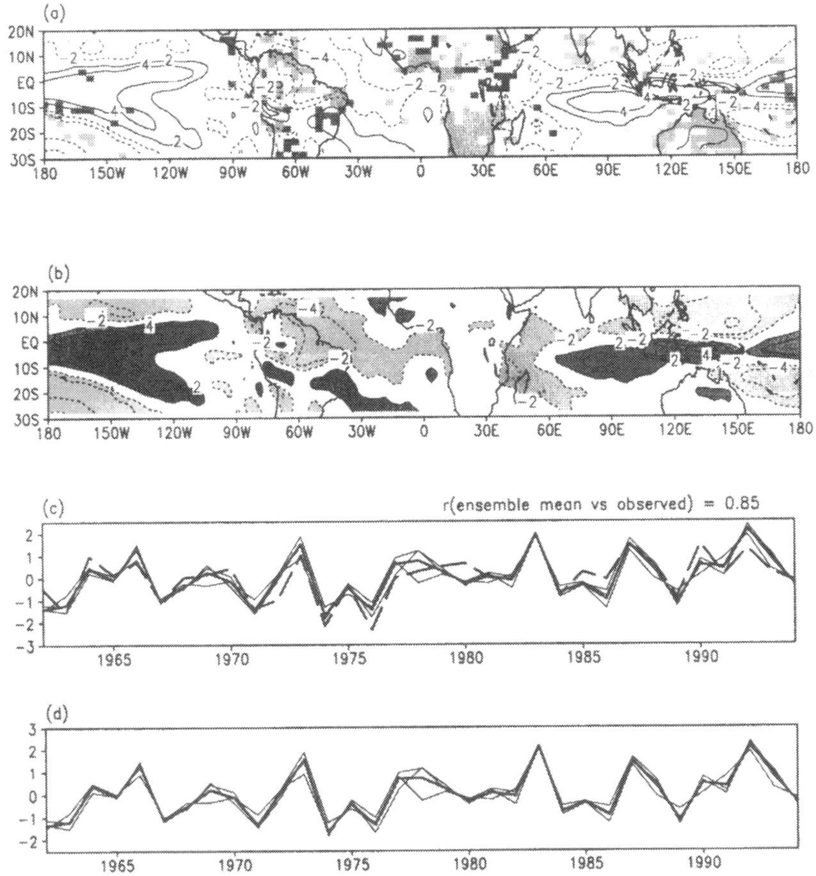

Chapter 13

Spatial Patterns: EOFs and CCA

by Hans von Storch

13.1 Introduction

Many analyses of climate data sets suffer from high dimensions of the variables representing the state of the system at any given time. Often it is advisable to split the full phase space into two subspaces. The "signal" space is spanned by few characteristic patterns and is supposed to represent the dynamics of the considered process. The "noise subspace", on the other hand, is high-dimensional and contains all processes which are purportedly irrelevant in their details for the "signal subspace".

The decision of what to call "signal" and what to call "noise" is non-trivial. The term "signal" is not a well-defined expression in this context. In experimental physics, the signal is well defined, and the noise is mostly the uncertainty of the measurement and represents merely a nuisance. In climate research, the signal is defined by the interest of the researcher and the noise is everything else unrelated to this object of interest. Only in infrequent cases is the noise due to uncertainties of the measurement, sometimes the noise comprises the errors introduced by deriving "analyses", i.e., by deriving from many irregularly distributed point observations a complete map. But in most cases the noise is made up of well-organized processes whose

Acknowledgments: I am grateful to Victor Ocaña, Gabriele Hegerl, Bob Livezey and Robert Vautard for their most useful comments which led to a significant (not statistically meant) improvement of the manuscript. Gerassimos Korres supplied me with Figures 13.1 and 13.2.

details are unimportant for the "signal". In many cases the noise is not a nuisance but its statistics are relevant for the understanding of the dynamics of the signal (see also Chapter 3). Generally the signal has longer scales in time and space than the noise, and the signal has fewer degrees of freedom than the noise.

An example is oceanic heat transport - this signal is low frequent and large in spatial scale. The extratropical storms are in this context noise, since the individual storms do not matter, but the ensemble of the storms, or the *storm track* is of utmost importance as this ensemble controls the energy exchange at the interface of atmosphere and ocean. Thus, for some oceanographers the individual storms are noise. For a synoptic meteorologist an individual storm is the object of interest, and thus the signal. But to understand an individual storm does not require the detailed knowledge of each cloud within the storm, so the clouds are noise in this context.

The purpose of this chapter is to discuss how the "signal" subspace may be represented by characteristic patterns. The specification of such characteristic patterns can be done in various ways, ranging from purely subjectively defined patterns, patterns with favorable geometric properties like a powerful representation of prescribed spatial scales (such as spherical harmonics) to patterns which are defined to optimize statistical parameters. Empirical Orthogonal Functions (EOFs) are optimal in representing variance; Canonical Correlation Patterns (CCPs) maximize the correlation between two simultaneously observed fields); others such as PIPs and POPs (see Chapter 15) satisfy certain dynamical constraints.

In this contribution we first represent the general idea of projecting large fields on a few "guess patterns" (Section 13.2). Then EOFs are defined as those patterns which are most powerful in explaining variance of a *random field* $\vec{\mathbf{X}}$ (Section 13.3). In Section 13.4 the *Canonical Correlation Analysis* of two simultaneously observed random fields $(\vec{\mathbf{X}}, \vec{\mathbf{Y}})$ is introduced. In Section 13.5 two methods of determining patterns optimized to represent maxima of variance are sketched, namely *Empirical Orthogonal Teleconnections* and *Redundancy Analysis*.

13.2 Expansion into a Few Guess Patterns

13.2.1 Guess Patterns, Expansion Coefficients and Explained Variance

The aforementioned separation of the full phase space into a "signal" subspace, spanned by a few patterns $\vec{p}^{\,k}$ and a "noise" subspace may be formally written as

$$\vec{\mathbf{X}}_t = \sum_{k=1}^{K} \alpha_k(t)\vec{p}^{\,k} + \vec{n}_t \tag{13.1}$$

with t representing in most cases time. The K "guess patterns" $\vec{p}^{\,k}$ and time coefficients $\alpha_k(t)$ are supposed to describe the dynamics in the signal subspace, and the vector $\vec{n}_t$ represents the "noise subspace". When dealing with the expressions "signal" and "noise" one has to keep in mind that "noise subspace" is implicitly defined as that space which does not contain the "signal". Also, since the noise prevails everywhere in the phase space, a complete separation between "signal" and "noise" is impossible. Indeed, the "signal subspace" contains an often considerable amount of noise.

The truncated vector of state $\vec{\mathbf{X}}_t^S = \sum_{k=1}^K \alpha_k(t)\vec{p}^{\,k}$ is the projection of the full vector of state on the signal subspace. The residual vector $\vec{n}_t = \vec{\mathbf{X}}_t - \vec{\mathbf{X}}_t^S$ represents the contribution from the noise subspace.

The vector $\vec{\mathbf{X}}$ is conveniently interpreted as a *random vector* with expectation $\mathrm{E}\left(\vec{\mathbf{X}}\right) = \vec{\mu}$, covariance matrix $\mathbf{\Sigma} = \mathrm{E}\left((\vec{\mathbf{X}} - \vec{\mu})(\vec{\mathbf{X}} - \vec{\mu})^T\right)$ and the variance $\mathrm{VAR}\left(\vec{\mathbf{X}}\right) = \sum_i \mathrm{E}\left((\mathbf{X}_i - \mu_i)^2\right)$. Then, the expansion coefficients α_k are univariate random variables whereas the patterns are constant vectors. In the case of EOFs, CCA and similar techniques the patterns are derived from $\vec{\mathbf{X}}$ so that they represent ***parameters*** of the random vector $\vec{\mathbf{X}}$.

The expansion coefficients[1] $\vec{\alpha} = (\alpha_1, \alpha_2 \ldots \alpha_K)^T$ are determined as those numbers which minimize

$$\epsilon(\vec{\alpha}) = \langle \vec{\mathbf{X}} - \textstyle\sum_k \alpha_k \vec{p}^{\,k}, \vec{\mathbf{X}} - \sum_k \alpha_k \vec{p}^{\,k} \rangle \tag{13.2}$$

with the "dot product' $\langle \vec{a}, \vec{b} \rangle = \sum_j a_j b_j$. The optimal vector of expansion coefficients is obtained as a zero of the first derivative of ϵ:

$$\sum_{i=1}^{K} \vec{p}^{\,k^T} \vec{p}^{\,i} \alpha_i = \vec{p}^{\,k^T} \vec{\mathbf{X}} \tag{13.3}$$

After introduction of the notation $\mathcal{A} = \left(\vec{a}|\vec{b}\cdots\right)$ for a matrix $\mathcal{A}$ with the first column given by the vector $\vec{a}$ and the second column by a vector $\vec{b}$, (13.3) may be rewritten as

$$\mathcal{P}\vec{\alpha} = \left(\vec{p}^{\,1}|\cdots|\vec{p}^{\,K}\right)^T \vec{\mathbf{X}} \tag{13.4}$$

with the symmetric $K \times K$-matrix $\mathcal{P} = \left(\vec{p}^{\,k^T} \vec{p}^{\,i}\right)$. In all but pathological cases the matrix $\mathcal{P}$ will be invertible such that a unique solution of (13.3) exists:

$$\vec{\alpha} = \mathcal{P}^{-1} \left(\vec{p}^{\,1}|\cdots|\vec{p}^{\,K}\right)^T \vec{\mathbf{X}} \tag{13.5}$$

Finally, if we define K vectors $\vec{p}_A^{\,1} \ldots \vec{p}_A^{\,K}$ so that

$$\left(\vec{p}_A^{\,1}|\cdots|\vec{p}_A^{\,K}\right) = \left(\vec{p}^{\,1}|\cdots|\vec{p}^{\,K}\right) \mathcal{P}^{-1} \tag{13.6}$$

[1] The expansion coefficients may be seen as the transformed coordinates after introducing the guess patterns as new basis of the phase space.

Then the k-th expansion coefficient α_k is given as the dot product of the vector of state $\vec{\mathbf{X}}$ and the "adjoint pattern" $\vec{p}_A^{\,k}$:

$$\alpha_k = \langle \vec{p}_A^{\,k}, \vec{\mathbf{X}} \rangle \tag{13.7}$$

In some cases, and in particular in case of EOFs, the patterns $\vec{p}^{\,k}$ are *orthogonal* such that $\mathcal{P}$ is the identity matrix and $\vec{p}^{\,k} = \vec{p}_A^{\,k}$. In this case (13.7) reads

$$\alpha_k = \langle \vec{p}^{\,k}, \vec{\mathbf{X}} \rangle \tag{13.8}$$

A convenient measure to quantify the relative importance of one pattern or of a set of patterns $\{\vec{p}^{\,k}\}$ is the "amount of explained variance", or, more precisely, the "proportion of variance accounted for by the $\{\vec{p}^{\,k}\}$" [formally similar to the "Brier-based score" β introduced in (10.6)]:

$$\eta = \frac{\mathrm{VAR}\left(\vec{\mathbf{X}}\right) - \mathrm{VAR}\left(\vec{\mathbf{X}} - \sum_k \alpha_k \vec{p}^{\,k}\right)}{\mathrm{VAR}\left(\vec{\mathbf{X}}\right)} \tag{13.9}$$

where we have assumed that the random vector $\vec{\mathbf{X}}$ would have zero mean $(\mathrm{E}\left(\vec{\mathbf{X}}\right) = 0)$. If the data are not centered then one may replace the variance-operator in (13.9) by the sum of second moments $\mathrm{E}\left(\vec{\mathbf{X}}^T \vec{\mathbf{X}}\right)$ and refer to the explained second moment.

The numerical value of the explained variance is bounded by $-\infty < \eta \leq 1$.[2] If $\eta = 0$ then $\mathrm{VAR}\left(\vec{\mathbf{X}} - \sum_k \alpha_k \vec{p}^{\,k}\right) = \mathrm{VAR}\left(\vec{\mathbf{X}}\right)$ and the representation of $\vec{\mathbf{X}}$ by the patterns is useless since the same result, in terms of explained variance, would have been obtained by arbitrary patterns and $\alpha_k = 0$. On the other end of the scale we have $\eta = 1$ which implies $\mathrm{VAR}\left(\vec{\mathbf{X}} - \sum_k \alpha_k \vec{p}^{\,k}\right) = 0$ and thus a perfect representation of $\vec{\mathbf{X}}$ by the guess patterns $\vec{p}^{\,k}$.

The amount of explained variance, or, sloppily formulated, "the explained variance", can also be defined locally for each component j:

$$\eta(j) = 1 - \frac{\mathrm{VAR}(\mathbf{X}_j - \sum_k \alpha_k p_j^k)}{\mathrm{VAR}(\mathbf{X}_j)} \tag{13.10}$$

If the considered random vector $\vec{\mathbf{X}}$ can be displayed as a map then also the amount of explained variance η can be visualized as a map.

[2]The number η can indeed be negative, for instance when $\sum_k \alpha_k \vec{p}^{\,k} = -\vec{\mathbf{X}}_t$. Then, $\eta = -3$.

13.2.2 Example: Temperature Distribution in the Mediterranean Sea

As an example[3], we present here an expansion (13.1) of a time-dependent 3-dimensional temperature field of the Mediterranean Sea. The output of a 9-year run of an OGCM, forced by monthly mean atmospheric conditions as analysed by the US-NMC for the years 1980 to 1988, was decomposed such that

$$\vec{\mathrm{T}}(\vec{r}, z, t) = \sum_k \alpha_k(z, t)\vec{p}^{\,k}_{\vec{r}} \tag{13.11}$$

with $\vec{r}$ representing the horizontal coordinates, z the vertical coordinate and t the time. The temperature field is given on a $(\vec{r}, z)$-grid with better resolution in the upper levels. In the representation (13.11), coefficients α_k depend on depth and time. The orthogonal patterns $\vec{p}^{\,k}$ depend only on the horizontal distribution and are independent of the depth z and of the time t. The decomposition was determined by a "Singular Value Decomposition" but the technical aspects are not relevant for the present discussion.

Prior to the analysis, the data have been processed. For each depth z the horizontal spatial mean and standard deviation have been calculated. Then, the temperature values at each depth are normalized by subtracting the (spatial) mean and dividing by the (spatial) standard deviation of the respective depth. The annual cycle is *not* subtracted so that the time series are not stationary (but cyclo-stationary) with an annual cycle of the time mean and of the temporal standard deviation.

The first two patterns are shown in Figure 13.1. The first one, which represents 57% of the total second moment of the normalized temperature, exhibits a dipole, with about half of the basin being warmer than average and the other half being cooler than average. The second mode, which represents 32% of the second moment, is relatively uniform throughout the Mediterranean Sea. The relative importance of the two modes for different layers of the ocean is described by the amount of the 2nd moment accounted for by the two modes (Figure 13.2). The second mode explains most of the 2nd moment above 100 m, whereas the first mode dominates below the top 4 layers. An inspection of the time series $\alpha_k(z,t)$ for different depths z reveals that the two modes represent different aspects of the climatology of the Mediterranean Sea. The time series $\alpha_1(z,t)$ is always positive with irregular variations superimposed. Such a behavior is indicative that the first mode describes mostly the overall mean and its interannual variability. The time series $\alpha_2(z,t)$ describe a regular annual cycle (with a negative minimum in winter so that $\alpha_2\vec{p}^{\,1}_{\vec{r}}$ is a negative distribution; and a positive maximum in

[3]This material was presented by Gerrasimos Korres in a "student paper" during the Autumn School. It will be available as a regular paper co-authored by Korres and Pinardi in 1994.

Figure 13.1: *First two characteristic horizontal distributions* $\vec{p}_{\vec{r}}^{\,k}$ *of normalized temperature in the Mediterranean Sea, as inferred from the output of a 9-year run with a numerical ocean model. (From Korres and Pinardi).*

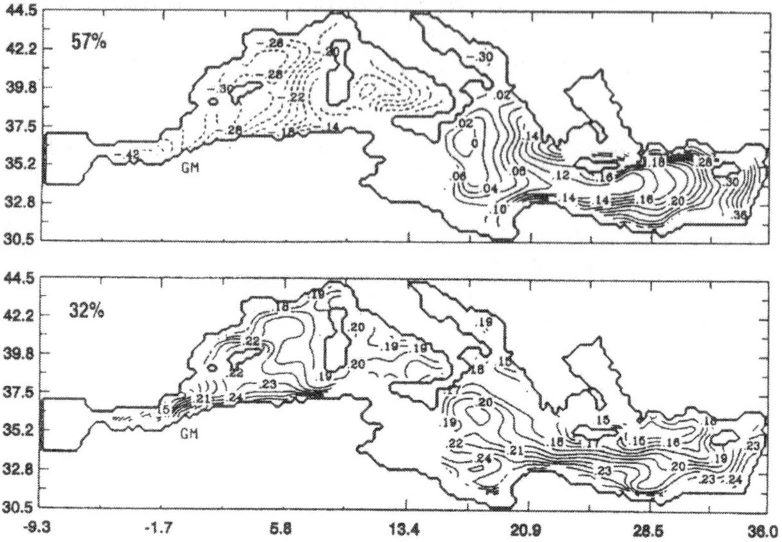

summer - indicating warmer than average conditions) plus a slight upward trend (gradual warming).

13.2.3 Specification of Guess Patterns

There are various ways to define the patterns $\vec{p}^{\,k}$:[4]

- The very general approach is Hasselmann's "Principal Interaction Patterns" formulation (PIP; Hasselmann, 1988). The patterns are implicitly defined such that their coefficients $\alpha_k(t)$ approximate certain dynamical equations, which feature unknown parameters.

- A simplified version of the PIPs are the "Principal Oscillation Patterns" (POPs, H. von Storch et al., 1988, 1993), which model linear dynamics and which have been successfully applied for the analysis of various processes (see Chapter 15).

- A standard statistical exercise in climate research aims at the identification of expected signals, such as the atmospheric response to enhanced greenhouse gas concentrations or to anomalous sea-surface temperature conditions (see Chapter 8). This identification is often facilitated by the specification of patterns determined in experiments with general circulation models (H. von Storch, 1987; Santer et al., 1993, Hegerl et al., 1996).

[4]See also Section 8.3.2.

Figure 13.2: *Vertical distribution of the percentage of the 2nd moment of the normalized temperature accounted for by the first and second patterns shown in Figure 13.1.*

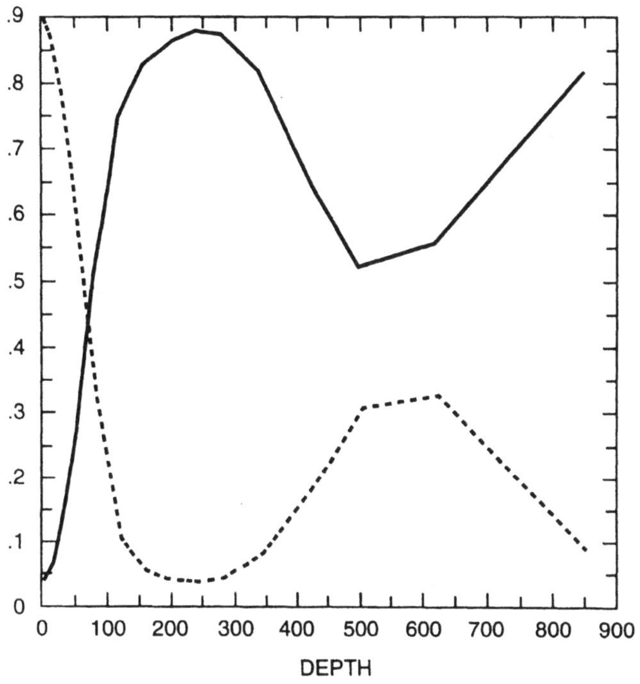

Also patterns "predicted" by simplified dynamical theory (for instance, linear barotropic equations) are in use (Hannoschöck and Frankignoul, 1985; Hense et al., 1990).

- A frequently used class of patterns are orthogonal functions such as trigonometric functions or spherical harmonics. In all "spectral" atmospheric general circulation models the horizontal fields are expanded according to (13.1) with spherical harmonics as guess patterns. In the spectral analysis of time series the trigonometric functions are used to efficiently represent fields.

- The Empirical Orthogonal Functions (EOFs) and Canonical Correlation Patterns (CCPs) are very widely used guess patterns. These choices will be discussed in some length in the next two Sections 13.3 and 13.4. Offsprings of these techniques are *Extended EOFs* (EEOFs) and *Complex EOFs* (CEOFs). In the EEOFs (Weare and Nasstrom, 1982; see also Chapter 14) the same vector at *different* times is concatenated; in the CEOF (Wallace and Dickinson, 1972; Barnett, 1983; also Section 15.3.4) the original vector real-valued time series is made complex by adding its *Hilbert transform* as imaginary component. The Hilbert transform may be seen as a kind of "momentum". Both techniques are successfully applied in climate research but we will not go into details in the present review. They are presented in more detail in H. von Storch and Zwiers (1999). A variant of Canonical Correlation Analysis is Redundancy Analysis and was proposed by Tyler (1982). It identifies pairs of patterns, so that a maximum of variance of the predictand is obtained through regression from the predictor. We will briefly introduce this technique in Section 13.5.2.

- A new approach named Empirical Orthogonal Teleconnections (EOTs), proposed by van den Dool et al. (2000), constructs stepwise orthogonal patterns by determining points with maximum skill in linearly specifying all other data points. This technique is sketched in Section 13.5.1. EOTs may be seen as a variant of conventional teleconnections patterns (see Chapter 12), which allows to split up the total variance into a sum of contributions from a limited number of patterns.

- The "wavelet" analysis is a technique which projects a given time series on a set of patterns, which are controlled by a location and a dispersion parameter. See Meyers et al. (1993) or Farge et al. (1993).

13.2.4 Rotation of Guess Patterns

For EOFs there exists a widely used variant named *Rotated EOFs* (for instance, Barnston and Livezey, 1987). The name is somewhat misleading as it indicates that the "rotation" would exploit properties special to the EOFs.

This is not the case. Instead, the general concept of "rotation" is to replace the patterns $\vec{p}^{\,k}$ in (13.1) by "nicer" patterns $\vec{p}_R^{\,k}$:

$$\sum_{k=1}^{K} \alpha_k \vec{p}^{\,k} = \sum_{k=1}^{K} \alpha_k^R \vec{p}_R^{\,k} \tag{13.12}$$

The patterns $\vec{p}_R^{\,k}$ are determined such that they minimize a certain (nonlinear) functional of "simplicity" F_R and that they span the same space as the original set of vectors $\{\vec{p}^{\,k}\}$. Constraints like unit length $({\vec{p}_R^{\,k}}^T \vec{p}_R^{\,k} = 1)$ and, sometimes, orthogonality $({\vec{p}_R^{\,k}}^T \vec{p}_R^{\,i} = 0)$ are invoked. Richman (1986) lists five vague criteria for patterns being "simple" and there are many proposals of "simplicity" functionals. If the patterns are not orthogonal the term ***oblique*** is used. The minimization of functionals such as (13.13) is in general non-trivial since the functionals are nonlinear. Numerical algorithms to approximate the solutions require the number of involved dimensions K to be not too large.

A widely used method is the "varimax", which generates a set of orthogonal patterns which minimize the joint "simplicity" measure

$$F_R(\vec{p}_R^{\,1} \cdots \vec{p}_R^{\,K}) = \sum_{k=1}^{K} f_R(\vec{p}_R^{\,k}) \tag{13.13}$$

with functions f_R such as

$$f_R(\vec{p}) = \frac{1}{m}\sum_{i=1}^{m} \left(p_i^2\right)^2 - \frac{1}{m^2}\left(\sum_{i=1}^{m} p_i^2\right)^2 \quad \text{or,} \tag{13.14}$$

$$f_R(\vec{p}) = \frac{1}{m}\sum_{i=1}^{m} \left[\left(\frac{p_i}{s_i}\right)^2\right]^2 - \frac{1}{m^2}\left[\sum_{i=1}^{m} \left(\frac{p_i}{s_i}\right)^2\right]^2 \tag{13.15}$$

The number p_i is the ith component of a m-dimensional vector $\vec{p}$, s_i is the standard deviation of the ith component of $\vec{\mathbf{X}}^S$, which is the projection of the original full random vector $\vec{\mathbf{X}}$ in the signal subspace spanned by the K vectors $\{\vec{p}^{\,1} \cdots \vec{p}^{\,K}\}$.

Both definitions (13.14,13.15) have the form of a variance: in the "raw varimax" set-up (13.14) it is the (spatial) variance of the squares of the components of the pattern $\vec{p}$ and in the "normal varimax" (13.15) it is the same variance of a normalized version $\vec{p}' = (p_i/s_i)$ with $(s_1^2 \cdots s_m^2)^T = \mathrm{VAR}(\mathbf{X}_i^S) = \sum_{k=1}^{K} \alpha_k p_i^k$. Minimizing (13.13) implies therefore finding a set of K patterns $\vec{p}_R^{\,k}$ such that their squared patterns have (absolute or relative) minimum spatial variance. The functions f_R are always positive and are zero if all $p_i = 0$ or 1 (13.14) or if all $p_i = s_i$ (13.15).

The results of a rotation exercise depend on the number K and on the choice of the measure of simplicity. The opinion in the community is divided on the subject of rotation. Part of the community advocates the use of rotation fervently as a means to define physically meaningful, statistically stable patterns whereas others are less convinced because of the hand-waving character of specifying the simplicity functions, and the implications of this specification for the interpretation of the result. The successful application of the rotation techniques needs some experience and it might be a good idea for the novice to have a look into Richman's (1986) review paper on that topic. Interesting examples are offered by, among many others, Barnston and Livezey (1987) and Chelliah and Arkin (1992). Cheng et al. (1995) found that a conventional EOF analysis yields statistically less stable patterns than a rotated EOF analysis.

In the present volume, Section 6.3.5 is dealing with a varimax-rotation (13.14) of a subset of EOFs.

13.3 Empirical Orthogonal Functions

For the sake of simplicity we assume in this Section that the expectation of the considered random vector $\vec{\mathbf{X}}$ is zero: $\vec{\mu} = 0$. Then the covariance matrix of $\vec{\mathbf{X}}$ is given by $\mathbf{\Sigma} = \mathrm{E}\left(\vec{\mathbf{X}}\vec{\mathbf{X}}^T\right)$.

The vector $\vec{\mathbf{X}}$ may represent very different sets of numbers, such as

- Observations of different parameters at one location (such as daily mean temperature, sunshine, wind speed etc.)
- Grid-point values of a continuous field which was spatially discretized on a regular grid (as is often the case for horizontal distributions) or on an irregular grid (such as the vertical discretization in GCMs).
- Observations of the same parameter (such as temperature) at irregularly distributed stations (see the example of Central European temperature in Section 13.3.4; also Briffa dealt with this case in Section 5.6.1 when he considered tree ring data from different sites).

There is some confusion with the terms, since several alternative sets of expressions are in use for the same object. What is labeled an EOF here is also named a ***principal vector*** or a ***loading***, whereas ***EOF coefficients*** are sometimes ***principal components*** (for instance in Chapter 8) or ***scores***.[5]

13.3.1 Definition of EOFs

Empirical Orthogonal Functions are defined as that set of K *orthogonal* vectors (i.e., $\vec{p}^{k^T}\vec{p}^{i} = \delta_{ki}$) which minimize the variance of the residual $\vec{n}$ in

[5]The expressions "principal vector" stems from the geometrical interpretation that these vectors are the principal vectors of an ellipsoid described by the covariance matrix. The terms "loading" and "scores" come from factor analysis, a technique widely used in social sciences. The term "EOF" seems to be in use only in meteorology and oceanography.

(13.1).[6] Because of the enforced orthogonality the coefficients $\vec{\alpha}$ are given by (13.8).

The EOFs are constructed consecutively: In a first step the pattern $\vec{p}^{\,1}$ of unit length (${\vec{p}^{\,1}}^T \vec{p}^{\,1} = 1$) is identified which minimizes

$$\mathrm{E}\Big(\big(\vec{\mathbf{X}} - \alpha_1 \vec{p}^{\,1}\big)^T \big(\vec{\mathbf{X}} - \alpha_1 \vec{p}^{\,1}\big)\Big) = \epsilon_1 \tag{13.16}$$

After the first EOF $\vec{p}^{\,1}$ is determined, the second EOF is derived as the pattern minimizing

$$\mathrm{E}\Big(\Big[\big(\vec{\mathbf{X}} - \alpha_1 \vec{p}^{\,1}\big) - \alpha_2 \vec{p}^{\,2}\Big]^T \Big[\big(\vec{\mathbf{X}} - \alpha_1 \vec{p}^{\,1}\big) - \alpha_2 \vec{p}^{\,2}\Big]\Big) = \epsilon_2 \tag{13.17}$$

with the constraints ${\vec{p}^{\,1}}^T \vec{p}^{\,2} = 0$ and ${\vec{p}^{\,2}}^T \vec{p}^{\,2} = 1$. In similar steps the remaining EOFs are determined. A K-dimensional field has in general K EOFs but we will see below that in practical situations the number of EOFs is limited by the number of samples.

We demonstrate now how to get the first EOF. The derivation of the other EOFs is more complicated but does not offer additional significant insights (for details, see H. von Storch and Hannoschöck, 1986). Because of the orthogonality we may use (13.8) and reformulate (13.16) such that

$$\begin{aligned} \epsilon_1 &= \mathrm{E}\big(\vec{\mathbf{X}}^T \vec{\mathbf{X}}\big) - 2\mathrm{E}\Big(\big(\vec{\mathbf{X}}^T \vec{p}^{\,1}\big) {\vec{p}^{\,1}}^T \vec{\mathbf{X}}\Big) + \mathrm{E}\big(\vec{\mathbf{X}}^T \vec{p}^{\,1} \vec{\mathbf{X}}^T \vec{p}^{\,1}\big) \\ &= \mathrm{VAR}\big(\vec{\mathbf{X}}\big) - {\vec{p}^{\,1}}^T \mathbf{\Sigma} \vec{p}^{\,1} \end{aligned} \tag{13.18}$$

To find the minimum, a Lagrange multiplier λ is added to enforce the constraint ${\vec{p}^{\,1}}^T \vec{p}^{\,1} = 1$. Then the expression is differentiated with respect to $\vec{p}^{\,1}$ and set to zero:

$$\mathbf{\Sigma} \vec{p}^{\,1} - \lambda \vec{p}^{\,1} = 0 \tag{13.19}$$

Thus, the first (and all further) EOF must be an eigenvector of the covariance matrix $\mathbf{\Sigma}$. Insertion of (13.19) into (13.18) gives

$$\epsilon_1 = \mathrm{VAR}\big(\vec{\mathbf{X}}\big) - \lambda \tag{13.20}$$

so that a minimum ϵ_1 is obtained for the eigenvector $\vec{p}^{\,1}$ with the largest eigenvalue λ.

More generally, we may formulate the **Theorem:**

[6] The approach of minimizing the variance of the residual has a mathematical background: the variance of the residual is a measure of the "misfit" of $\vec{\mathbf{X}}$ by $\sum_{k=1}^{K} \alpha_k(t) \vec{p}^{\,k}$. Other such measures of misfit could be chosen but this quadratic form allows for a simple mathematical solution of the minimization problem.

The first K eigenvectors $\vec{p}^k$, for any $K \leq m$, of the covariance matrix $\Sigma = \mathrm{E}\left(\vec{\mathbf{X}}\vec{\mathbf{X}}^T\right)$ of the m-variate random vector $\vec{\mathbf{X}}$ form a set of pairwise orthogonal patterns. They minimize the variance

$$\epsilon_K = \mathrm{E}\left(\left(\vec{\mathbf{X}} - \textstyle\sum_{k=1}^K \alpha_k \vec{p}^k\right)^2\right) = \mathrm{VAR}\left(\vec{\mathbf{X}}\right) - \sum_{k=1}^{K} \lambda_k \tag{13.21}$$

with $\alpha_k = \vec{\mathbf{X}}^T \vec{p}^k$ and $\vec{p}^{k^T} \vec{p}^k = 1$. The patterns are named "Empirical Orthogonal Function".

From the construction of the EOFs it becomes clear the patterns represent an optimal potential to compress data into a minimum number of patterns. *Sometimes* the first EOF of the first few EOFs represent a meaningful physical summary of relevant processes, which go with characteristic patterns. For further discussion see Section 13.3.2

Favorable aspects of the EOFs are the geometrical orthogonality of the patterns and the ***statistical independence***, or, more correctly, the zero-correlation of the "EOF coefficients" α_i:

$$\mathrm{E}(\alpha_i \alpha_k) = \vec{p}^{i^T} \mathrm{E}\left(\vec{\mathbf{X}}\vec{\mathbf{X}}^T\right)\vec{p}^k = \vec{p}^{i^T} \Sigma \vec{p}^k = \lambda_k \vec{p}^{i^T} \vec{p}^k = \lambda_k \delta_{k,i} \tag{13.22}$$

A byproduct of the calculation (13.22) is $\mathrm{VAR}(\alpha_k) = \lambda_k$.

EOFs are ***parameters*** of the random vector $\vec{\mathbf{X}}$. If $\vec{\mathbf{X}}$ is a Gaussian distributed random vector then the set of coefficients α_k form a set of univariate normally distributed independent random variables (with zero means and standard deviations given by the square root of the respective eigenvalues, if the mean of $\vec{\mathbf{X}}$ is zero.)

The relative importance of the EOFs may be measured by their capability to "explain" $\vec{\mathbf{X}}$-variance. This amount of explained variance η can be calculated for individual EOFs or for sets of EOFs, for the complete m-variate vector $\vec{\mathbf{X}}$ or for its components separately [see (13.9, 13.10)]. For the first K EOFS, we find with the help of (13.21).

$$\eta_{\{1\ldots K\}} = 1 - \frac{\epsilon_K}{\mathrm{VAR}\left(\vec{\mathbf{X}}\right)} = 1 - \frac{\sum_{k=K+1}^m \lambda_k}{\sum_{k=1}^m \lambda_k} = \frac{\sum_{k=1}^K \lambda_k}{\sum_{k=1}^M \lambda_k} \tag{13.23}$$

If η_j and η_k are the explained variances by two single EOFs $\vec{p}^k$ and $\vec{p}^j$ with indices $j > k$ such that $\lambda_j \leq \lambda_k$, then the following inequality holds: $0 < \eta_j \leq \eta_k \leq \eta_{\{j,k\}} \leq 1$, with η_{jk} representing the variance explained by both EOFs $\vec{p}^j$ and $\vec{p}^k$).

If the original vector $\vec{\mathbf{X}}$ has m components - what is an adequate truncation K in (13.1)? There is no general answer to this problem which could also be phrased "Which are the (physically) significant[7] EOFs?" A good answer will depend on the physical problem pursued. One relevant piece of information is the amount of explained variance. One might select K so that the percentage of $\vec{\mathbf{X}}$-variance explained by the first K EOFs, $\eta_{(1\ldots K)}$, passes a certain threshold. Or such that the last kept EOF accounts for a certain minimum variance:

$$\eta_{(1\ldots K)} \leq \kappa_1 < \eta_{(1\ldots K+1)} \qquad \text{or} \qquad \eta_K > \kappa_2 > \eta_{K+1} \tag{13.24}$$

Typical values for κ_1 are 80% or 90% whereas choices of $\kappa_2 = 5\%$ or 1% are often seen.

We have introduced EOFs as patterns which minimize the variance of the residual (13.21). The variance depends on the chosen geometry, and we could replace in (13.21) the square by a ***scalar product*** $\langle\cdot,\cdot\rangle$ such that

$$\epsilon_K = \mathrm{E}\Big(\langle\vec{\mathbf{X}} - \textstyle\sum_{k=1}^K \alpha_k\vec{p}^{\,k}, \vec{\mathbf{X}} - \sum_{k=1}^K \alpha_k\vec{p}^{\,k}\rangle\Big) \tag{13.25}$$

The EOF coefficients are then also given as dot products

$$\alpha_k(t) = \langle\vec{\mathbf{X}}_t, \vec{p}^{\,k}\rangle$$

Obviously the result of the analysis depends on the choice of the dot product, which is to some extend arbitrary.

13.3.2 What EOFs are *Not* Designed for ...

There are some words of caution required when dealing with EOFs. These patterns are constructed to represent in an optimal manner *variance* and *covariance* (in the sense of ***joint variance***), not ***physical connections*** or *maximum correlation* (see Chen and Harr, 1993). Therefore they are excellent tools to *compress* data into a few variance-wise significant components. Sometimes people expect more from EOFs, for instance a description of the "coherent structures" (as, for instance, teleconnections). This goal can be achieved only when the data are normalized to variance one, i.e., if the correlation matrix instead of the covariance matrix is considered (see Wallace and Gutzler, 1981). Another expectation is that EOFs would tell us something about the structure of an underlying continuous field from which the data vector $\vec{\mathbf{X}}$ is sampled. Also often EOFs are thought to represent modes of "natural" or "forced" variability. We will discuss these expectations in the following.

---

[7] Note that the word "significant" used here has nothing to do with "statistical significance" as in the context of testing null hypotheses (see Chapters 8 and 9). Instead, the word "significance" is used in a colloquial manner. We will return to the buzz-word "significance" in Section 13.3.3.

- To demonstrate the limits of an EOF analysis to identify *coherent structures* let us consider the following example with a two-dimensional random vector $\vec{\mathbf{X}} = (\mathbf{X}_1, \mathbf{X}_2)^T$. The covariance matrix $\mathbf{\Sigma}$ and the correlation matrix $\mathbf{\Sigma}'$ of $\vec{\mathbf{X}}$ is assumed to be

$$\mathbf{\Sigma} = \begin{pmatrix} 1 & \rho a \\ \rho a & a^2 \end{pmatrix} \quad \text{and} \quad \mathbf{\Sigma}' = \begin{pmatrix} 1 & \rho \\ \rho & 1 \end{pmatrix} \tag{13.26}$$

The correlation matrix $\mathbf{\Sigma}'$ is the covariance matrix of the normalized random vector $\vec{\mathbf{X}}' = \mathcal{A}\vec{\mathbf{X}}$ with the diagonal matrix $\mathcal{A} = \begin{pmatrix} 1 & 0 \\ 0 & 1/a \end{pmatrix}$.

Obviously both random vectors, $\vec{\mathbf{X}}$ and $\vec{\mathbf{X}}'$, represent the same correlation structure. The relative distribution of variances in the two components $\mathbf{X}_1$ and $\mathbf{X}_2$ depends on the choice of a. Also the eigen-structures of $\mathbf{\Sigma}$ and $\mathbf{\Sigma}'$ differ from each other since the transformation matrix $\mathcal{A}$ is not orthogonal, i.e., it does not satisfy $\mathcal{A}^T = \mathcal{A}^{-1}$.

We will now calculate these eigen-structures for two different standard deviations a of $\mathbf{X}_2$. The eigenvalues of $\mathbf{\Sigma}$ are given by

$$\lambda_{1,2} = \frac{1}{2}\left[1 + a^2 \pm \sqrt{1 - 2a^2 + a^4 + 4(\rho a)^2}\right] \tag{13.27}$$

and the eigenvectors are, apart from proper normalization, given by

$$\vec{p}^{\,1} = \begin{pmatrix} 1 \\ \frac{\lambda_1 - 1}{\rho a} \end{pmatrix} \quad \text{and} \quad \vec{p}^{\,2} = \begin{pmatrix} \frac{1-\lambda_1}{\rho a} \\ 1 \end{pmatrix} \tag{13.28}$$

because of the orthogonality constraint.

In the case of $a^2 \ll 1$ we find

$$\begin{aligned} \lambda &\approx \frac{1}{2}\left[1 + a^2 \pm \sqrt{1 - 2a^2 + 4(\rho a)^2}\right] \\ &\approx \frac{1}{2}\left[1 + a^2 \pm \left(1 - \frac{a^2 - 2(\rho a)^2}{2}\right)\right] = \left\{ \begin{array}{l} 1 + (\rho a)^2 \\ a^2(1 - \rho^2) \end{array} \right. \end{aligned} \tag{13.29}$$

If the two components $\mathbf{X}_1$ and $\mathbf{X}_2$ are perfectly correlated with $\rho = 1$ then the first EOF represents the full variance $\mathrm{VAR}\left(\vec{\mathbf{X}}\right) = \mathrm{VAR}(\mathbf{X}_1) + \mathrm{VAR}(\mathbf{X}_2) = 1 + a^2 = \lambda_1$ and the second EOF represents no variance ($\lambda_2 = 0$). If, on the other hand, the two components are independent, then $\lambda_1 = \mathrm{VAR}(\mathbf{X}_1) = 1$ and $\lambda_2 = \mathrm{VAR}(\mathbf{X}_2) = a^2$.

The first EOF is $\vec{p}^{\,1} \approx \begin{pmatrix} 1 \\ \rho a \end{pmatrix} \approx \begin{pmatrix} 1 \\ 0 \end{pmatrix}$, which is reasonable since the first component represents almost all variance in the case of $a^2 \ll 1$. Because of the orthogonality constraint the second EOF is $\vec{p}^{\,2} \approx$

$\begin{pmatrix} -\rho a \\ 1 \end{pmatrix} \approx \begin{pmatrix} 0 \\ 1 \end{pmatrix}$. Thus in the case $a \ll 1$ the EOFs are the unit vectors *independently of the size of* ρ.

If we deal with the correlation matrix $\mathbf{\Sigma}'$ the difference of relative importance of the two components is erased. The eigenvalues are given by (13.27) with $a = 1$:

$$\lambda' = \frac{1}{2}\left[2 \pm \sqrt{4\rho^2}\right] = 1 \pm \rho \tag{13.30}$$

The eigenvectors are given by (13.28): If $\rho = 0$ then the eigenvalue (13.30) is double and no unique eigenvectors can be determined. If $\rho > 0$, then the non-normalized EOFs are

$$\vec{p}^{\,1} = \begin{pmatrix} 1 \\ 1 \end{pmatrix} \quad \text{and} \quad \vec{p}^{\,2} = \begin{pmatrix} 1 \\ -1 \end{pmatrix} \tag{13.31}$$

Because of the positive correlation, the first EOF describes in-phase variations of $\mathbf{X}_1$ and $\mathbf{X}_2$. The orthogonality constraint leaves the second pattern with the representation of the out-of-phase variations.

The "patterns" (13.31) are markedly different from the eigenvectors of the $a \ll 1$-covariance matrix calculated above. Thus, the result of the EOF analyses of two random vectors with the same correlation structure depends strongly on the allocation of the variance within the vector $\vec{\mathbf{X}}$.

This example demonstrates also the impact of using vectors $\vec{\mathbf{X}}$ which carry numbers subjective to different units. If air pressure from midlatitudes is put together with pressure from low latitudes then the EOFs will favor the high-variance midlatitude areas. If a vector is made up of temperatures in units of K and of precipitation in m/sec[8], then patterns of the EOFs will concentrate on the temperature entries.

- An EOF analysis deals with a *vector* of observations $\vec{\mathbf{X}}$. This vector may entertain physically very different entries, as outlined at the beginning of this Section. The EOF analysis does not *know* what type of vector it is analyzing. Instead all *components* of $\vec{\mathbf{X}}$ are considered as equally relevant, independently if they represent a small or a large grid box in case of a longitude × latitude grid, or a thin or a thick layer (in case of ocean general circulation model output). If we study Scandinavian temperature as given by 10 stations in Denmark and one station in the other Scandinavian states, then the first EOFs will invariably concentrate on Denmark.

  If we deal with a relatively uniform distribution of variance, and if we know that the *characteristic spatial scale* of the considered variable, such

[8]The unit mm/sec is, admittedly, not widely used for precipitation. But precipitation is a rate, often given in mm/day - which is in standard units expressible as m/sec.

as temperature, is comparable to the considered area, then the first EOF will in *most* cases be a pattern with the same sign at all points - simply because of the system's tendency to create anomalies with the same sign in the entire domain. The need to be orthogonal to the first EOF then creates a second EOF with a dipole pattern (which is the largest-scale pattern orthogonal to the uniform-sign first EOF). Our 2-dimensional ($a = 1, \rho > 0$)-case, discussed above, mimics this situation. If, however, the characteristic spatial scale is smaller than the analysis domain then often the first EOF is not a monopole [see, for instance, the SST analysis of Zorita et al. (1992)].

- Do EOFs represent *modes* or *processes* of the physical system from which the data are sampled? In many cases the first EOF may be identified with such a mode or process. For the second and higher indexed EOFs, however, such an association is possible only under very special circumstances (see North, 1984). A severe limitation to this end is the imposed spatial orthogonality of the patterns and the resulting temporal independence of the coefficients (13.22). Thus EOFs can represent only such physical modes which operate independently, and with orthogonal patterns. In most real-world cases, however, processes are interrelated.

13.3.3 Estimating EOFs

The EOFs are parameters of the covariance matrix $\mathbf{\Sigma}$ of a random variable $\vec{\mathbf{X}}$. In practical situations, this covariance matrix $\mathbf{\Sigma}$ is unknown. Therefore, the EOFs have to be estimated from a finite sample $\{\vec{\mathbf{x}}(1)\dots\vec{\mathbf{x}}(n)\}$. In the following, estimations are denoted by $\hat{}$. We assume that the observations represent anomalies, i.e., deviations from the true mean or from the sample mean. However, the analysis can be done in the same way also with data without prior subtraction of the mean.

To *estimate* EOFs from the finite sample $\{\vec{\mathbf{x}}(1)\dots\vec{\mathbf{x}}(n)\}$ two different strategies may be pursued. One strategy considers the finite sample as a finite random variable and calculates orthogonal patterns $\hat{\vec{p}}^k$ which minimize

$$\sum_{l=1}^{n}\left[\vec{\mathbf{x}}(l) - \sum_{k=1}^{K}\hat{\alpha}_k(l)\hat{\vec{p}}^k\right]^2 = \hat{\epsilon}_K \tag{13.32}$$

with coefficients $\hat{\alpha}_k(l) = \sum_{j=1}^{n}\vec{\mathbf{x}}(l)_j\hat{p}^k{}_j$ given by (13.8).

An alternative approach is via the Theorem of Section 13.2, namely to use the eigenvectors $\hat{\vec{p}}^k$ of the *estimated* covariance matrix

$$\hat{\mathbf{\Sigma}} = \frac{1}{n}\sum_{l=1}^{n}\vec{\mathbf{x}}(l)\vec{\mathbf{x}}(l)^T = \frac{1}{n}\left[\sum_{l=1}^{n}\mathbf{x}_i(l)\mathbf{x}_j(l)\right]_{i,j} \tag{13.33}$$

as *estimators* of the true EOFs $\vec{p}^k$. Interestingly, both approaches result in the same patterns (H. von Storch and Hannoschöck, 1986).

For the actual computation the following comments might be helpful:

- The samples, which determine the estimated EOFs, enter the procedure only in (13.33) - and in this equation the ordering of the samples is obviously irrelevant. The estimated covariance matrix $\hat{\Sigma}$ and thus the estimated EOFs, are invariant to the order of the samples.

- When m is the dimension of the analyzed vector $\vec{\mathbf{X}}$ the true covariance matrix Σ as well as the estimated covariance matrix $\hat{\Sigma}$ have dimension $m \times m$. Therefore the numerical task of *calculating* the eigenvectors and eigenvalues of a sometimes huge $m \times m$ matrix is difficult or even impossible. A wholesale alternative is based on the following little algebraic trick (H. von Storch and Hannoschöck, 1984): *If $\mathcal{Y}$ is a $n \times m$ matrix, then $\mathcal{A} = \mathcal{Y}\mathcal{Y}^T$ and $\mathcal{A}^T = \mathcal{Y}^T\mathcal{Y}$ are $n \times n$- and $m \times m$ matrices which share the same nonzero eigenvalues. If $\mathcal{Y}\vec{q}$ (or $\vec{r}$) is an eigenvector of $\mathcal{A}$ to the eigenvalue $\lambda \neq 0$ then $\vec{q}$ (or $\mathcal{Y}^T\vec{r}$) is an eigenvector of $\mathcal{A}^T$ to the same eigenvalue λ.*

  The estimated covariance matrix $\hat{\Sigma}$ may be written as $\hat{\Sigma} = \frac{1}{n}\mathcal{X}\mathcal{X}^T$ with the *data matrix*

$$\mathcal{X} = \begin{pmatrix} x_1(1) & x_1(2) & \dots & x_1(n) \\ x_2(1) & x_2(2) & \dots & x_2(n) \\ \vdots & \vdots & \ddots & \vdots \\ x_m(1) & x_m(2) & \dots & x_m(n) \end{pmatrix} = (\vec{\mathbf{x}}(1)|\dots|\vec{\mathbf{x}}(n)) \tag{13.34}$$

  The n columns of the $n \times m$ data matrix $\mathcal{X}$ are the sample vectors $\vec{\mathbf{x}}(j), j = 1, \dots n$; the rows mark the m coordinates in the original space. The matrix product $\mathcal{X}\mathcal{X}^T$ is a quadratic $m \times m$ matrix even if $\mathcal{X}$ itself is not quadratic. The product $\mathcal{X}^T\mathcal{X}$, on the other hand, is a $n \times n$-matrix. The above mentioned trick tells us that one should calculate the eigenvalues and eigenvectors of the smaller of the two matrices $\mathcal{X}^T\mathcal{X}$ and $\mathcal{X}\mathcal{X}^T$. In practical situations we often have the number of samples n being much smaller than the number of components m.

  A byproduct of the "trick" is the finding that we can estimate only the first n EOFs (or $n - 1$ if we have subtracted the overall mean to get anomalies) of the m EOFs of the m-variate random variable.

- Numerically, the EOF analysis of a finite set of observed vectors may by done by a *Singular Value Decomposition* (SVD, see Chapter 14).

$$\mathcal{X}^T = \begin{pmatrix} \tilde{\alpha}_1(1) & \tilde{\alpha}_2(1) & \dots & \tilde{\alpha}_n(1) \\ \tilde{\alpha}_1(2) & \tilde{\alpha}_2(2) & \dots & \tilde{\alpha}_n(2) \\ \vdots & \vdots & & \vdots \\ \tilde{\alpha}_1(n) & \tilde{\alpha}_2(n) & \dots & \tilde{\alpha}_n(n) \end{pmatrix} \mathcal{D}\left(\hat{\vec{p}}^1|\dots|\hat{\vec{p}}^m\right)^T \tag{13.35}$$

with a rectangular $n \times m$ matrix $\mathcal{D}$ with zero elements outside the diagonal and positive elements on the diagonal: $d_{ij} = s_i\delta_{ij} \geq 0$. The quadratic $n \times n$ and $m \times m$ matrices to the right and left of $\mathcal{D}$ are orthogonal.

The eigenvalues of the estimated covariance matrix are $\hat{\lambda}_i = s_i^2$. The coefficients of the estimated EOFs are given by $\hat{\alpha}_i = s_i\tilde{\alpha}_i$. Again, there is a maximum of $min(n,m)$ nonzero s_i-values so that at most $min(n,m)$ useful EOFs can be determined.

- The choice of the numerical algorithm is irrelevant for the mathematical character of the product - EOFs are the eigenvectors of the estimated covariance matrix independently if the number crunching has been done via the eigenvector problem or via SVD.

As always, when estimating parameters of a random variable from a finite sample of observations, one may ask how accurate the estimation probably is:

- *Biases*
  If $\hat{\lambda}_k$ is an estimate of the true eigenvalue λ_k and $\hat{\alpha}_k$ the EOF coefficient of the kth estimated EOF the equality of eigenvalues and variance of EOF coefficients is biased (cf. H. von Storch and Hannoschöck, 1986):

  – For the largest eigenvalues λ_k:

  $$\mathrm{E}\left(\hat{\lambda}_k\right) > \lambda_k = \mathrm{VAR}(\alpha_k) > \mathrm{E}(\mathrm{VAR}(\hat{\alpha}_k)) \tag{13.36}$$

  – for the smallest eigenvalues λ_k:

  $$\mathrm{E}\left(\hat{\lambda}_k\right) < \lambda_k = \mathrm{VAR}(\alpha_k) < \mathrm{E}(\mathrm{VAR}(\hat{\alpha}_k)) \tag{13.37}$$

  The relation (13.36,13.37) means that the large (small) eigenvalues are systematically over-(under)estimated, and that the variance of the random variable $\hat{\alpha}_k = \langle \vec{\mathbf{X}}, \hat{\vec{p}}^k \rangle$ which are expansion coefficients when projecting $\vec{\mathbf{X}}$ on the random variable "estimated EOFs" is systematically over- or underestimated by the sample variance $\widehat{\mathrm{VAR}(\hat{\alpha}_k)} = \hat{\lambda}_k$ derived from the sample $\{\vec{\mathbf{x}}(1)\ldots\vec{\mathbf{x}}(n)\}$. Similarly, $\mathrm{COV}(\hat{\alpha}_k,\hat{\alpha}_j) \neq \widehat{\mathrm{COV}(\hat{\alpha}_k,\hat{\alpha}_j)} = 0$.

- *"Selection Rules"*
  So-called ***selection rules*** have been proposed. One often used is named "Rule N" (Preisendorfer and Overland, 1982), which is supposed to determine the physically "significant" EOFs. The basic concept is that the full phase space is the sum of a subset in which all variations are purely noise and of a subset whose variability is given by dynamical processes. The signal-subspace is spanned by well-defined EOFs whereas in the noise-subspace no preferred directions exist. For the eigenvalue-spectrum this

assumption implies that the eigenvalues of the EOFs spanning the signal-subspace are unequal and that the eigenvalues in the noise-subspace are all identical.

The selection rules compare the distributions of sample eigenvalue-spectra, representative for the situation that all or the $m - K$ smallest true eigenvalues (K being specified a-priori or determined recursively) are all alike, with the actually observed sample eigenvalue spectrum. All those estimated eigenvalues which are larger than the, say, 95%-percentile of the (marginal) distribution of the reference "noise spectra", are selected as ***significant*** at the 5%-level.

The problem with this approach is that this selection rule is claimed to be a *statistical test* which supposedly is capable of accepting, with a given risk, the alternative hypothesis that all EOFs with an index smaller than some number $m-K$ represent "signals" of the analyzed data field. The null hypothesis tested would be "all eigenvalues are equal", and the rejection of this null hypothesis would be the acceptance of the alternative "not all eigenvalues are equal". The connection between this alternative and the determination of a "signal subspace" is vague. Also the above sketched approach does not consider the quality of the estimation of the patterns; instead the selection rules are concerned with the eigenvalues only.

I recommend forgetting about the identification of "significant" EOFs by means of selection rules and resorting to more honest approaches like *North's rule-of-thumb* outlined in the next paragraph.

- *North's Rule-of-Thumb*
  Using a scale argument North et al. (1982) found as the "typical errors"

$$\Delta\lambda_k \approx \sqrt{\frac{2}{n}}\lambda_k \tag{13.38}$$

$$\Delta\hat{\vec{p}}^{\,k} \sim \frac{\Delta\lambda_k}{\lambda_j - \lambda_k}\vec{p}^{\,j} \tag{13.39}$$

with λ_j being the eigenvalue closest to λ_i and n being the number of *independent* samples. Approximation (13.39) compares patterns, and not the lengths of vectors, since we are dealing with normalized vectors.

- The first order error $\Delta\hat{\vec{p}}^{\,k}$ is of the order of $\sqrt{\frac{1}{n}}$. The convergence to zero is slow.
- The first order error $\Delta\hat{\vec{p}}^{\,k}$ is orthogonal to the true EOF $\vec{p}^{\,k}$.
- The estimation of the EOF $\vec{p}^{\,k}$ is most contaminated by the patterns of those other EOFs $\vec{p}^{\,j}$ which belong to eigenvalues λ_j closest to λ_k. The contamination will be the more severe the smaller the difference $\lambda_j - \lambda_k$ is.

North et al. (1982) finally formulated the following "rule-of-thumb":

> "If the sampling error of a particular eigenvalue $\Delta\lambda$ is comparable or larger than the spacing between λ and a neighboring eigenvalue, then the sampling error $\Delta\vec{p}$ of the EOF will be comparable to the size of the neighboring EOF."

When using this rule-of-thumb one should be careful not to oversee the condition of independent samples - in most geophysical data this assumption is not valid.

13.3.4 Example: Central European Temperature

As an example we consider the covariance structure of the winter mean temperature anomalies (i.e., deviations from the overall winter mean) at eleven Central European stations (Werner and H. von Storch, 1993). Thus $m = 11$. Relatively homogeneous time series were available for eighty winters from 1901 to 1980. For both of the 40-year interval before and after 1940 an EOF analysis was performed. The results of the analysis are very similar in the two intervals - as a demonstration we show in Figure 13.3 the first two EOFs for both periods. The representation of the patterns deviates from the definition introduced above: The contribution of the k-th EOF to the full signal is given by $\alpha_k(t)\vec{p}^{\,k}$. According to our definitions the variance of the coefficient is given by the k-th eigenvalue λ_k and the vector has unit length. For a better display of the results sometimes a different normalization is convenient, namely $\alpha_k\vec{p}^{\,k} = (\alpha_k/\sqrt{\lambda_k}) \times (\vec{p}^{\,k}\sqrt{\lambda_k}) = \alpha'_k\vec{p}^{\,\prime k}$. In this normalization the coefficient time series has variance one for all indices k and the relative strength of the signal is in the patterns $\vec{p}^{\,\prime}$. A typical coefficient is $\alpha' = 1$ so that the typical reconstructed signal is $\vec{p}^{\,\prime}$. In this format the first two EOFs and their time coefficients obtained in the analysis of Central European winter temperature are shown in Figures 13.3 and 13.4.

In both time periods the first EOF has a positive sign at all locations, represents about 90% of the total variance and exhibits "typical anomalies" of the order of $1 - 2K$. The second EOF represents a northeast-southwest gradient, with typical anomalies of $\pm 0.5K$, and accounts for 6% and 7% of the variance in the two time periods. The remaining 9 EOFs are left to represent together the variance of mere 5%.

In Figure 13.4 the EOF coefficients are shown. As mentioned above, they are normalized to variance one. The first coefficient $\alpha_1(t)$ varies most of the time between ± 1 but exhibits a number of spiky excursions to large negative values < -2. Together with the information provided by the patterns (Figure 13.3) such large negative coefficients represent extremely cold winters, such as 1940 and 1941, with mean negative anomalies of the order of $< -4K$. The distribution of the first EOF coefficient is markedly skewed (Figure 13.5) whereas the distribution of the second coefficient is fairly symmetric. The

Figure 13.3: *First two EOFs of January-February mean temperature at 11 Central European stations derived from the winters 1901-40 (left) and from the winters 1941-80 (right). The percentage of explained variance is given in the upper left corner of the diagrams. The contour lines are added to help the reader to understand the distribution of numbers. Units:* 10^{-2} ^oC. *(From Werner and H. von Storch, 1993)*

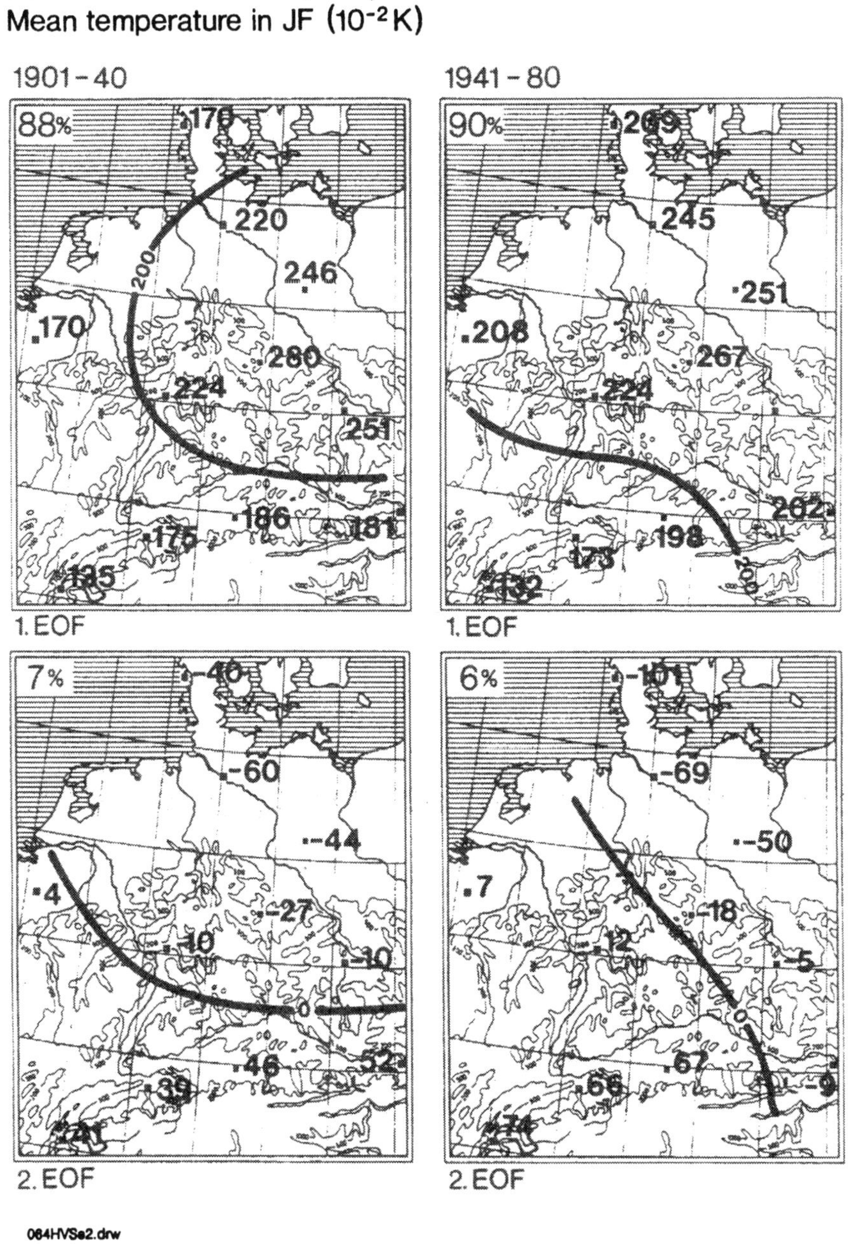

Figure 13.4: *EOF coefficient time series* $\alpha_1(t)$ *and* $\alpha_2(t)$ *of the first two EOFs of winter mean temperature at 11 Central European stations. Note that the time series have been normalized to one so that the information about the strength of the variation is carried by the patterns. (From Werner and H. von Storch, 1993).*

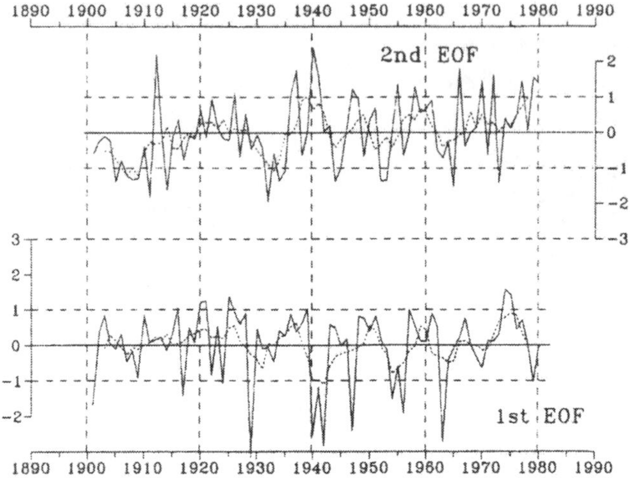

time series of the 2nd coefficient, depicted in Figure 13.4 shows no dramatic outliers but, interestingly, an upward trend translates at the stations with a slow warming of the Alpine region ($\approx 0.005K/yr$) and a gradual cooling ($\approx 0.01K/yr$) in the lowlands.

This is about all that the EOFs can tell us about the evolution of winter mean temperature in Central Europe in the years 1901-80. We will come back to this example in Section 13.4.4.

13.4 Canonical Correlation Analysis

We assume for this section again that the expectations of the considered random vectors $\vec{\mathbf{X}}$ and $\vec{\mathbf{Y}}$ vanish: $\vec{\mu}_X = \vec{\mu}_Y = 0$.

13.4.1 Definition of Canonical Correlation Patterns

In the Canonical Correlation Analysis [CCA, proposed by Hotelling (1936) and introduced into climate research by, among others, Barnett and Preisendorfer (1987)] not one random vector $\vec{\mathbf{X}}$ is expanded into a finite set of

Figure 13.5: *Frequency distribution of the EOF coefficient time series shown in Figure 13.4. (From Werner and H. von Storch, 1993)*

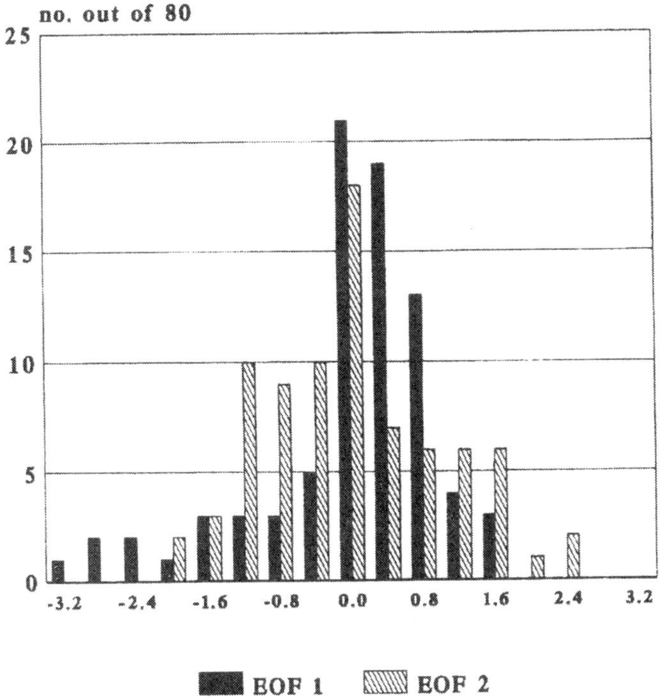

vectors but a pair of two simultaneously observed vectors $\vec{\mathbf{X}}$ and $\vec{\mathbf{Y}}$:

$$\vec{\mathbf{X}}_t = \sum_{k=1}^{K} \alpha_k^X(t)\vec{p}_X^{\,k} \quad \text{and} \quad \vec{\mathbf{Y}}_t = \sum_{k=1}^{K} \alpha_k^Y(t)\vec{p}_Y^{\,k} \tag{13.40}$$

with the same number K. The dimensions m_X and m_Y of the vectors $\vec{p}_X^{\,k}$ and $\vec{p}_Y^{\,k}$ will in general be different. The expansion is done in such a manner that

1. The coefficients $\alpha_k^X(t)$ and $\alpha_k^Y(t)$ in (13.40) are *optimal* in a least square sense [i.e., for given patterns $\vec{p}_X^{\,k}$ and $\vec{p}_Y^{\,k}$ the norms $\| \vec{\mathbf{X}}_t - \sum_{k=1}^{K} \alpha_k^X(t)\vec{p}_X^{\,k} \|$ and $\| \vec{\mathbf{Y}}_t - \sum_{k=1}^{K} \alpha_k^Y(t)\vec{p}_Y^{\,k} \|$ are minimized, as in (13.2)]. This condition implies (see Section 13.2.1) that

$$\alpha_k^X = \langle (\vec{p}_X^{\,k})_A, \vec{\mathbf{X}} \rangle \quad \text{and} \quad \alpha_k^Y = \langle (\vec{p}_Y^{\,k})_A, \vec{\mathbf{Y}} \rangle \tag{13.41}$$

   with certain ***adjoint patterns*** $(\vec{p}_X^{\,k})_A$ and $(\vec{p}_Y^{\,k})_A$ given by (13.6).

2. The correlations

   - between α_k^X and α_l^X
   - between α_k^Y and α_l^Y
   - between α_k^X and α_l^Y

   are zero for all $k \neq l$.

3. The correlation between α_1^X and α_1^Y is maximum.

4. The correlation between α_2^X and α_2^Y is the maximum under the constraints of 2) and 3). The correlations for the higher indexed pairs of coefficients satisfy similar constraints (namely of being maximum while being independent with all previously determined coefficients.)

It can be shown (see, for instance, Zorita et al., 1992) that the adjoint patterns are the eigenvectors of somewhat complicated looking matrices, namely:

$$\mathcal{A}_X = \mathbf{\Sigma}_X^{-1}\mathbf{\Sigma}_{XY}\mathbf{\Sigma}_Y^{-1}\mathbf{\Sigma}_{XY}^T \quad \text{and} \quad \mathcal{A}_Y = \mathbf{\Sigma}_Y^{-1}\mathbf{\Sigma}_{XY}^T\mathbf{\Sigma}_X^{-1}\mathbf{\Sigma}_{XY} \tag{13.42}$$

Here $\mathbf{\Sigma}_X$ and $\mathbf{\Sigma}_Y$ are the covariance matrices of $\vec{\mathbf{X}}$ and $\vec{\mathbf{Y}}$. $\mathbf{\Sigma}_{XY}$ is the cross-covariance matrix of $\vec{\mathbf{X}}$ and $\vec{\mathbf{Y}}$, i.e., $\mathbf{\Sigma}_{XY} = \mathrm{E}\left(\vec{\mathbf{X}}\vec{\mathbf{Y}}^T\right)$ if $\mathrm{E}\left(\vec{\mathbf{X}}\right) = \mathrm{E}\left(\vec{\mathbf{Y}}\right) = 0$. The matrix $\mathcal{A}_X$ is a $m_X \times m_X$ matrix and $\mathcal{A}_Y$ is a $m_Y \times m_Y$ matrix. The two matrices $\mathcal{A}_X$ and $\mathcal{A}_Y$ may be written as products $\mathcal{B}_1\mathcal{B}_2$ and $\mathcal{B}_2\mathcal{B}_1$ with two matrices $\mathcal{B}_1$ and $\mathcal{B}_2$. Therefore the two matrices share the same nonzero eigenvalues, and if $\vec{p}_X^{\,k}$ is an eigenvector of $\mathcal{A}_X$ with an eigenvalue $\lambda \neq 0$ then $\mathbf{\Sigma}_Y^{-1}\mathbf{\Sigma}_{XY}^T\vec{p}_X^{\,k}$ is an eigenvector of $\mathcal{A}_Y$ with the same eigenvalue.

Note that for univariate random variables $\vec{\mathbf{X}} = \mathbf{X}$ and $\vec{\mathbf{Y}} = \mathbf{Y}$ the two matrices $\mathcal{A}_X$ and $\mathcal{A}_Y$ in (13.42) reduce to the squared correlations between $\mathbf{X}$ and $\mathbf{Y}$.

The k-adjoint pattern is given by the eigenvector with the k-largest eigenvalue of $\mathcal{A}$. The correlation between α_k^X and α_k^Y is given by the k-th largest nonzero eigenvalue of $\mathcal{A}_X$ or $\mathcal{A}_Y$.

The covariance between the "Canonical Correlation Coefficients" α_k^X and the original vector $\vec{\mathbf{X}}$ is given by

$$\mathrm{E}\left(\alpha_k^X \vec{\mathbf{X}}\right) = \mathrm{E}(\alpha_k^X \textstyle\sum_i \alpha_i^X \vec{p}_X^{\,i}) = \vec{p}_X^{\,k} \tag{13.43}$$

so that, because of $\alpha_k^X = \vec{\mathbf{X}}^T (\vec{p}_X^{\,k})_A$:

$$\vec{p}_X^{\,k} = \mathbf{\Sigma}_X (\vec{p}_X^{\,k})_A \quad \text{and} \qquad \vec{p}_Y^{\,k} = \mathbf{\Sigma}_Y (\vec{p}_Y^{\,k})_A \tag{13.44}$$

Thus, to determine the "Canonical Correlation Patterns" (CCP) and the canonical correlation coefficients one has first to calculate the covariance matrices and cross covariance matrices. From products of these matrices (13.42) the adjoint patterns are derived as eigenvectors. With the adjoint pattern the CCPs are calculated via (13.44) and the coefficients through (13.41). Because of the specific form of the matrices, it is advisable to solve the eigenvector problem for the smaller one of the two matrices.

13.4.2 CCA in EOF Coordinates

A simplification of the mathematics may be obtained by first transforming the random vectors $\vec{\mathbf{X}}$ and $\vec{\mathbf{Y}}$ into a low-dimensional EOF-space, i.e., by expanding

$$\vec{\mathbf{X}} \approx \vec{\mathbf{X}}^S = \sum_{i=1}^{K} (\beta_i^X)(\sqrt{\nu_i^X} \vec{e}_X^{\,i}) \quad \text{and} \quad \vec{\mathbf{Y}} \approx \vec{\mathbf{Y}}^S = \sum_{i=1}^{K} (\beta_i^Y)(\sqrt{\nu_i^Y} \vec{e}_Y^{\,i}) \tag{13.45}$$

with EOFs $\vec{e}_X^{\,i}$ of $\vec{\mathbf{X}}$ and $\vec{e}_Y^{\,i}$ of $\vec{\mathbf{Y}}$. The numbers ν_i^X and ν_i^Y, which are the eigenvalues associated with the EOFs, are introduced to enforce $\mathrm{VAR}(\beta_i^X) = \mathrm{VAR}(\beta_i^Y) = 1$. Equations (13.45) may be written more compactly with the help of matrices $\mathcal{E} = (\vec{e}^{\,1}|\dots|\vec{e}^{\,K})$ with the EOFs in their columns (so that $\mathcal{E}\mathcal{E}^T = 1$ and $\mathcal{E}^T\mathcal{E} = 1$) and diagonal matrices $\mathcal{S} = (diag\sqrt{\nu_i})$:

$$\vec{\mathbf{X}}^S = \mathcal{E}_X \mathcal{S}_X \vec{\beta}^X \quad \text{and} \quad \vec{\mathbf{Y}}^S = \mathcal{E}_Y \mathcal{S}_Y \vec{\beta}^Y \tag{13.46}$$

When we operate with objects in the EOF coordinates we add a tilde $\tilde{}$. In these coordinates we have $\tilde{\mathbf{\Sigma}}_X = 1$ and $\tilde{\mathbf{\Sigma}}_Y = 1$ and the CCA matrices (13.42) are of the simpler and symmetric form

$$\tilde{\mathcal{A}}_X = \widetilde{\mathbf{\Sigma}_{XY}}\,\widetilde{\mathbf{\Sigma}_{XY}}^T \quad \text{and} \quad \tilde{\mathcal{A}}_Y = \widetilde{\mathbf{\Sigma}_{XY}}^T\,\widetilde{\mathbf{\Sigma}_{XY}} \tag{13.47}$$

In the EOF coordinates the CCA patterns are orthogonal, so that *in these coordinates* $\vec{p}_X^k = (\widetilde{\vec{p}_X^k})_A$. The procedure to get the CC patterns and the adjoints in the original Euclidean-space is the same for $\vec{\mathbf{X}}$ and $\vec{\mathbf{Y}}$, so that we consider only the $\vec{\mathbf{X}}$-case and drop the index "X" as well as the index "i" in the following for convenience. Also we identify the full representation $\vec{\mathbf{X}}$ with the truncated presentation $\vec{\mathbf{X}}^S$.

- The CCA coefficients α at any given time should be independent of the coordinates. Thus, if $\vec{\beta}(t)$ is the state of $\vec{\mathbf{X}}$ in the EOF coordinates (13.45) and $\vec{x}(t)$ in the original Euclidean coordinates, then the CCA coefficients shall be given as the dot product of this vector of state with adjoint patterns $\vec{p}_A$ and $\widetilde{\vec{p}_A}$:

$$\alpha(t) = \widetilde{\vec{p}_A}^T \vec{\beta} = \vec{p}_A{}^T \vec{x}(t) \tag{13.48}$$

- The initial transformation (13.45), $\vec{x} = \mathcal{E}\mathcal{S}\vec{\beta}$, describes the transformation of the CC patterns from the EOF coordinates to the Euclidean coordinates:

$$\vec{p}^{\,k} = \mathcal{E}\mathcal{S}\widetilde{\vec{p}^{\,k}} \tag{13.49}$$

- To get the back transformation of the adjoints we insert (13.45) into (13.48) and get $\vec{p}_A{}^T\vec{x} = \widetilde{\vec{p}_A}^T\vec{\beta} = \widetilde{\vec{p}_A}^T\mathcal{S}^{-1}\mathcal{E}^T\vec{x}$ and

$$\vec{p}_A^{\,k} = \mathcal{E}\mathcal{S}^{-1}\widetilde{\vec{p}_A^{\,k}} \tag{13.50}$$

In general we have $\mathcal{S} \neq \mathcal{S}^{-1}$ so that neither the property "self adjoint", i.e, $\widetilde{\vec{p}_A^{\,k}} = \widetilde{\vec{p}^{\,k}}$, nor the property "orthogonal", i.e. $\widetilde{\vec{p}^{\,k}}^T\widetilde{\vec{p}^{\,l}} = 0$ if $k \neq l$, are valid after the back transformation into the Euclidean space.

In the EOF coordinates we can establish a connection to the EOF calculus (Vautard; pers. communication). For convenience we drop now the ˜ marking objects given in the EOF coordinates. First we concatenate the two vectors $\vec{\mathbf{X}}$ and $\vec{\mathbf{Y}}$ to one vector $\vec{\mathbf{Z}} = (\vec{\mathbf{X}}, \vec{\mathbf{Y}})$ and calculate the EOFs $\vec{e}^{\,i}$ of this new random vector. These EOFs are the eigenvectors of the joint covariance matrix

$$\mathbf{\Sigma}_Z = \begin{pmatrix} \mathbf{\Sigma}_X & \mathbf{\Sigma}_{XY} \\ \mathbf{\Sigma}_{XY}^T & \mathbf{\Sigma}_Y \end{pmatrix} = \begin{pmatrix} 1 & \mathbf{\Sigma}_{XY} \\ \mathbf{\Sigma}_{XY}^T & 1 \end{pmatrix} \tag{13.51}$$

A vector $\vec{p} = (\vec{p}_X, \vec{p}_Y)$ is an eigenvector of $\mathbf{\Sigma}_Z$ if

$$\frac{1}{\lambda - 1}\mathbf{\Sigma}_{XY}\vec{p}_Y = \vec{p}_X \quad \text{and} \quad \frac{1}{\lambda - 1}\mathbf{\Sigma}_{XY}^T\vec{p}_X = \vec{p}_Y \tag{13.52}$$

so that $\vec{p}_X$ and $\vec{p}_Y$ have to satisfy

$$\mathbf{\Sigma}_{XY}\mathbf{\Sigma}_{XY}^T\vec{p}_X = (\lambda - 1)^2\vec{p}_X \quad \text{and} \quad \mathbf{\Sigma}_{XY}^T\mathbf{\Sigma}_{XY}\vec{p}_Y = (\lambda - 1)^2\vec{p}_Y \tag{13.53}$$

Thus the two components $\vec{p}_X$ and $\vec{p}_Y$ of the joint "extended" EOF of $\vec{\mathbf{X}}$ and $\vec{\mathbf{Y}}$ form a pair of canonical correlation patterns of $\vec{\mathbf{X}}$ and $\vec{\mathbf{Y}}$. Note that this statement depends crucially on the nontrivial assumption $\mathbf{\Sigma}_X = \mathbf{\Sigma}_Y = 1$.

13.4.3 Estimation: CCA of Finite Samples

The estimation of CC patterns, adjoints and CC coefficients is made in a straightforward manner by estimating the required matrices $\mathbf{\Sigma}_X$, $\mathbf{\Sigma}_Y$ and $\mathbf{\Sigma}_{XY}$ in the conventional way [as in (13.33)], and multiply the matrices to get estimates $\widehat{\mathcal{A}_X}$ and $\widehat{\mathcal{A}_Y}$ of $\mathcal{A}_X$ and $\mathcal{A}_Y$. The calculation is simplified if the data are first transformed (13.45) into EOF coordinates.

In the case of EOFs we had seen that the first eigenvalues of the estimated covariance matrix *over*estimate the variance which is accounted for by the first EOFs. This overestimation makes sense if one considers the fact that the EOFs must represent a certain amount of variance of the full (*infinite*) random variable $\vec{\mathbf{X}}$, whereas the estimated EOF represents a fraction of variance in the *finite* sub-space given by the samples. In the case of the CCA we have a similar problem: The correlations are *over*estimated - since they are fitted to describe similar behavior only in a finite subspace, given by the samples, of the infinite space of possible random realizations of $\vec{\mathbf{X}}$ and $\vec{\mathbf{Y}}$. This overestimation decreases with increasing sample size and increases with the number of EOFs used in the a-priori compression (13.45).

Zwiers (1993) discusses the problem of estimating the correlations. Following Glynn and Muirhad (1978) he proposes to improve the straightforward estimator $\hat{\rho}_k$ by using the formula

$$\hat{\theta}_k = Z_k - \frac{1}{2n\hat{\rho}_k}\left[m_X + m_y - 2 + \hat{\rho}_k^2 + 2(1-\hat{\rho}_k^2)\sum_{j\neq k}\frac{\hat{\rho}_j^2}{\hat{\rho}_k^2 - \hat{\rho}_j^2}\right] \tag{13.54}$$

with $Z_k = tanh^{-1}(\hat{\rho}_k)$ and $\theta = tanh^{-1}(\rho_k)$. Glynn and Muirhead show that $\hat{\theta}_k$ is an unbiased estimator of θ_k up to $\mathcal{O}(n^{-2})$ and that $\mathrm{VAR}\left(\hat{\theta}\right) = \frac{1}{n}$ up to order $\mathcal{O}(n^{-2})$. An approximate 95% confidence interval for the $k-$th canonical correlation is then given by $tanh(\hat{\theta}_k) \pm \frac{2}{\sqrt{n}}$. However, Zwiers (1993) found in Monte Carlo experiments that the correction (13.54) represents an improvement over the straightforward approach but that substantial bias remains when the sample size is small.

13.4.4 Example: Central European Temperature

We return now to the Central European temperature in winter, with which we have dealt already in Section 13.3.4. Werner and H. von Storch (1993) analysed simultaneously the large-scale circulation, as given by the seasonal mean sea-level air-pressure (SLP) anomaly over the North Atlantic and Europe, and the Central European temperature given at $m_T = 11$ locations.

The objective of this exercise was to determine to what extent the regional temperature is controlled by large-scale circulation anomalies.[9]

With first 40 years of the full 1901-1980 data set the CCA was done. The data were first projected on EOFs as given by (13.45). The number of EOFs retained was determined in such a way that an increase by one EOF would change the canonical correlations only little. The first pair of patterns goes with a correlation of 70%. The patterns (Figure 13.6), which are plotted here as a "typical simultaneously appearing pair of pattern" (by normalizing the canonical correlation coefficients to variance one) indicate a simple physically plausible mechanism: In winters with a persistent anomalous southwesterly flow, anomalous warm maritime air is advected into Europe so that the winter mean temperatures are everywhere in central Europe above normal, with typical anomalies of the $1^o - 2^oC$. If, however, an anomalous high pressure system is placed south of Iceland, then the climatological transport of maritime air into Central Europe is less efficient, and temperature drops by one or two degrees.

With the full data set, from 1901-80 the correlation of the coefficients α_1^T and α_1^{SLP} is recalculated and found to be only 0.64 compared to 0.70 derived from the "fitting" interval 1901-40. Also the percentage η of temperature variance at the eleven Central European locations accounted for by the temperature-pattern $\vec{p}_T^1$ is computed from the full data set. The results, shown as lower numbers in the top panel of Figure 13.6, vary between 22% (at the northernmost station Fanø) to 58% (at the northern slope of the Alps, at station Hohenpeissenberg).

With two CCA pairs a "downscaling model" was designed to estimate the temperature variations only from the SLP variations without any local information. The result of this exercise is shown for the station Hamburg in Figure 13.7: The year-to-year variations are nicely reproduced - and more than 60% of the 80-year variance is represented by the statistical model - but the long-term (downward) trend is captured only partly by the CCA model: the SLP variations allude to a decrease by -2.6^oC whereas the real decrease has only been -1.0^oC.

Finally, the output of a climate GCM has been analysed whether it reproduces the connection represented by the CCA-pairs. The regional temperature from a GCM output is given at grid points (the proper interpretation of what these grid points represent is not clear). Therefore the 11 Central European stations are replaced by 6 grid points. The first pair of CC patterns, derived from 100 years of simulated data, is shown in Figure 13.8. The patterns are similar to the patterns derived from observed data, with an anomalous southwesterly flow being associated with an overall warming

[9]In any kind of correlation-based analysis a positive result (i.e., the identification of a high correlation) is no proof that the two considered time series are connected through a cause-effect relationship. It can very well be that a third (unknown) is controlling both analysed time series. In the present case the physical concept that the large-scale circulation affects regional temperatures is invoked to allow for the interpretation "control".

Figure 13.6: *First pair of canonical correlation patterns $\vec{p}_T^1$ of Central European temperature (top panel, upper numbers; in 10^{-2} ^oC) and $\vec{p}_{SLP}^1$ of North Atlantic/European sea-level air-pressure (in hPa) in winter. The patterns are normalized such that the coefficients α_1^T and α_1^{SLP} have variance one. Therefore the patterns represent a pair of typical simultaneously appearing anomalies.*
The (lower) percentage numbers in the top panel are the amount η of temperature variance accounted for by the $\vec{p}_T^1$-pattern. (From Werner and H. von Storch, 1993)

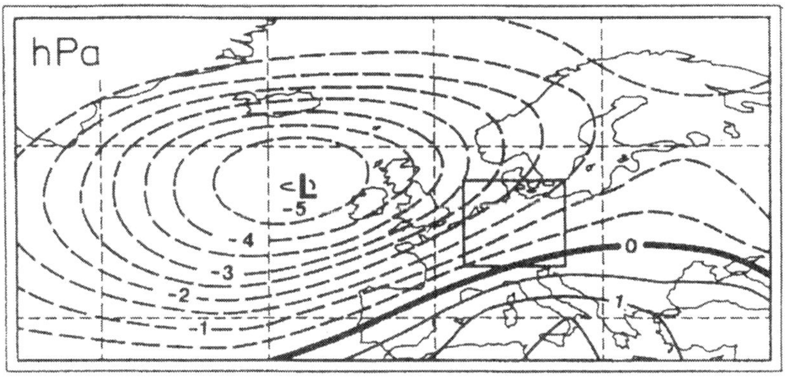

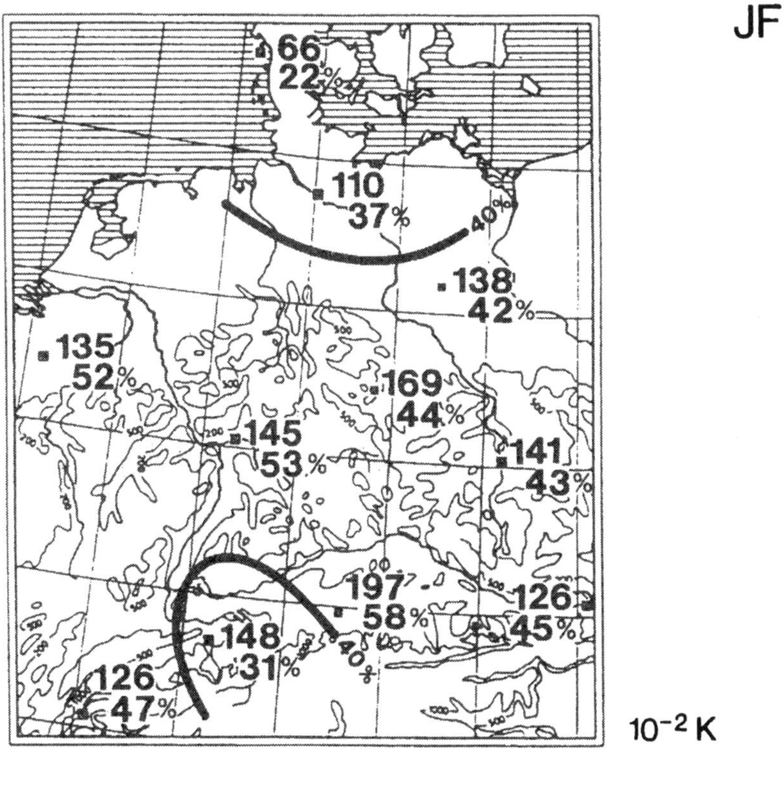

Figure 13.7: *Time series of winter mean temperature in Hamburg (deviations from the 1901-40 mean) as derived from in-situ observations (dotted) and as derived indirectly from large-scale SLP patterns by means of CC patterns (solid line). (From Werner and H. von Storch, 1993).*

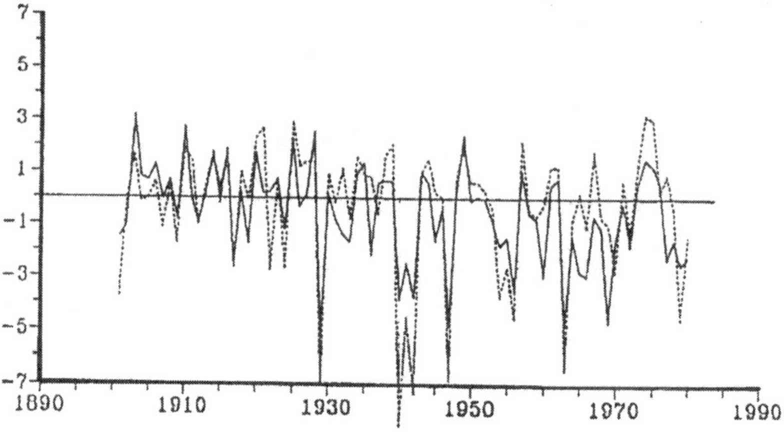

Figure 13.8: *First pair of CC patterns derived from the output of a climate model. Left: SLP pattern; right: regional temperature at 6 grid points located in central Europe. (From Werner and H. von Storch, 1993).*

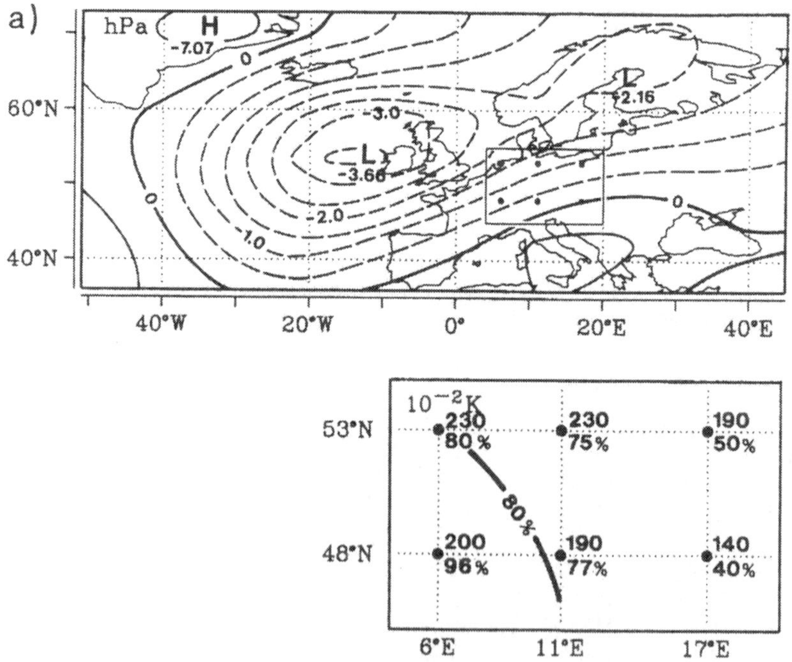

of the order of one to two degrees. The details, however, do not fit. First, the correlation is only 0.53 compared to 0.64 in the real world. Second, the structure within Central Europe is not reproduced: Maximum temperature anomalies are at the westernmost grid points and minimum values at the easternmost. The local explained variances η are much higher for the GCM output (with a maximum of 96% and a minimum of 50% compared to 58% and 22% in the real world).

13.5 Optimal Regression Patterns

In the past years, two new techniques for the derivation of spatial patterns have been introduced. Both techniques are based on the idea of optimizing linear regression equations. In the following we sketch both approaches.

13.5.1 Empirical Orthogonal Teleconnections

In the this technique, proposed by van den Dool et al. (2000), a random vector $\vec{\mathbf{X}}_t$ is expanded into a series (13.1): $\vec{\mathbf{X}}_t = \sum_{k=1}^K \alpha_k(t)\vec{p}^{\,k} + \vec{n}_t$. The patterns, named *Orthogonal Empirical Teleconnections* (EOTs), are determined sequentially.

To do so, we label the random vector $\vec{\mathbf{X}}_t$ as $\vec{\mathbf{X}}_t^{(0)}$ and its components as as $\mathbf{X}_{i;t}^{(0)}$. Then regression equations are derived, mapping the time series $\mathbf{X}_{i;t}^{(0)}$ at a component i on all other components. If the regression from component i on component j is called $r_{i;j}^{(0)}$ then the total proportion of variance explained by these regressions, given by

$$V_i^{(0)} = \sum_j \text{VAR}\left(r_{i;j}^{(0)}\mathbf{X}_{i;t}^{(0)} - \mathbf{X}_{j;t}^{(0)}\right) \tag{13.55}$$

measures how representative the point i is for the full vector. As first EOT is chosen the vector of regression coefficients $r_{i;j}^{(0)}$ with the index i_1 with maximum V_{i_1}. That is $\vec{p}^{\,1} = (r_{i_1;j}^{(0)})_j$, and $\alpha_1(t) = \mathbf{X}_{i_1;t}^{(0)}$.

In a second step, and in all further steps, the "already explained" part is subtracted, i.e.,

$$\vec{\mathbf{X}}_t^{(1)} = \vec{\mathbf{X}}_t^{(0)} - \alpha_1(t)\vec{p}^{\,1} \tag{13.56}$$

is formed, and again regression equations for all points calculated. The second pattern $\vec{p}^{\,2}$ is then the vector of regression coefficients $r_{i;j}^{(1)}$ with maximum $V_i^{(1)}$. Of course, $r_{i_2;i_1}^{(1)} = 0$ and $r_{i_2;i_2}^{(1)} = 1$. The time series $\alpha_2(t)$ is the reduced component time series i_2, i.e., $\alpha_2(t) = \mathbf{X}_{i_2;t}^{(1)}$. Because of the construction with regression equations and the subtraction (13.56), the coefficient time series are uncorrelated: $\text{COV}(\alpha_1, \alpha_2) = 0$.

In this way, a series of pattern $\vec{p}$ are constructed, with uncorrelated coefficient time series and specified skill in specifying the random vector as a whole. Since the coefficients are uncorrelated, the total variance of $\vec{\mathbf{X}}$ is the sum of the variances of the α's. When the data $\vec{\mathbf{X}}$ can be displayed as a map, then also the EOTs may be represented as maps.

In a number of cases, presented by van den Dool et al. (2000), the EOTs were almost as good in specifying variance as EOFs were. Van den Dool et al. (2000) note a number of advantages of their approach, among others that the patters are linked to specific geographical points (when $\vec{\mathbf{X}}$ is made up of spatially distributed variables). In that sense, EOTs are automatically "regionalized" and may therefore serve as an alternative to Rotated EOFs. They may help to identify "important" sampling locations. For instance, when analysing January mean temperature of the contiguous US, they found as first two EOTs a location in West Kentucky, representing 38% of the total variance, and at another location in East Wyoming, representing 31%.

An interesting variant of this method concerns the choice of the indices i_1 etc. There is no need to maximize $V_i^{(0)}$; instead a component could be chosen independently of the variance for dynamical reasons, for instead a center of action of a teleconnection pattern.

Another variant is to reverse the role of space and time; then the EOTs are regression patterns relating the same point at some time with itself for all other times. The series α is then a spatial distribution at a certain fixed time. In the example mentioned above, van den Dool et al. (2000) identified Januaries 1977 and 1937 as such "characteristic" times.

13.5.2 Redundancy Analysis

Redundancy Analysis (RDA) is a variant of Canonical Correlation Analysis. CCA determines in a pair of random vectors patters whose coefficients share a maximum of correlation,; thus for CCA the two random vectors may be swapped without changing the result. In contrast, in RDA the two random vectors $\vec{\mathbf{X}}$ and $\vec{\mathbf{Y}}$ play different roles, with one, say $\vec{\mathbf{X}}$, being a predictor, and the other, say $\vec{\mathbf{Y}}$, being the predictand.[10] Patterns are determined so that a maximum proportion of variance of $\vec{\mathbf{Y}}$ is accounted for by regressing "the $\vec{\mathbf{X}}$-patterns" on $\vec{\mathbf{Y}}$.

The concept was invented by Tyler (1982), and worked out for climate applications by H. von Storch and Zwiers (1999). Since the mathematics are a bit involved, we limit ourselves here to a short heuristic introduction.

We begin with the same formalism as in case of CCA, i.e., with expansion (13.40). For a given number K RDA determines a set of $\vec{\mathbf{X}}$-patterns $\vec{p}_X^{\,j}$, $j = 1 \ldots k$, so that the regression of the time coefficients α_k^X on $\vec{\mathbf{Y}}$ is maximum.

[10] The use of the conventional expressions "predictor" and "predictand" does not imply that forecasts in time are made; indeed the terms "specificator" and "specificand" would be more precise.

This regression maps the pattern $\vec{p}_X^j$ on the $\vec{\mathbf{Y}}$-pattern $\vec{p}_Y^j$. These pattern can be chosen so that the $\vec{\mathbf{Y}}$-patterns form a set of orthonormal patterns, whereas the $\vec{\mathbf{X}}$-patterns are linearly independent.

The $\vec{\mathbf{X}}$-patterns are invariant under non-singular transformations; this is meaningful as the regression should not depend on the units of the predictor; however, since the $\vec{p}_Y^j$-patterns are constructed for an optimal representation of $\vec{\mathbf{Y}}$-variance, these patterns are invariant only under orthonormal transformation - since the variance would otherwise be changed.

The patterns are given as solutions of eigen-problems, namely

$$\left[\boldsymbol{\Sigma}_{xx}^{-1}\boldsymbol{\Sigma}_{xy}\boldsymbol{\Sigma}_{yx}\right]\vec{p}_X^j = \lambda_j\vec{p}_X^j \tag{13.57}$$

$$\left[\boldsymbol{\Sigma}_{yx}\boldsymbol{\Sigma}_{xx}^{-1}\boldsymbol{\Sigma}_{xy}\right]\vec{p}_Y^j = \lambda_j\vec{p}_Y^j \tag{13.58}$$

The eigenvalues λ_j represent the proportion of $\vec{\mathbf{Y}}$-variance accounted for through the regression from $\vec{p}_X^j$ on $\vec{p}_Y^j$.

The technique has been used for the purpose of empirical downscaling, for instance for estimating wave height statistics from monthly mean air pressure distributions (WASA, 1998).

Chapter 14

Patterns in Time: SSA and MSSA

by Robert Vautard

14.1 Introduction

Singular Spectrum Analysis (SSA) is a particular application of the EOF expansion. In classical EOF analysis, the random field $\vec{\mathbf{X}}$ to be studied, called also the *state vector*, contains values measured or estimated at a given time, that is, the coordinates of $\vec{\mathbf{X}}$ represent different locations in space at the same time. By diagonalising the covariance matrix of $\vec{\mathbf{X}}$, one tries therefore to capture the dominant spatial patterns. The SSA expansion (Vautard et al., 1992) is an EOF expansion, but the state vector $\vec{\mathbf{X}}$ now contains values *at the same location* but at *different* lags. The leading eigenelements of the corresponding covariance matrix represent thus the leading *time patterns* of the random field. SSA is a *time series* analysis, in the sense that a single signal is analysed.

When both space and lag vary in the state vector $\vec{\mathbf{X}}$, the analysis is called Multichannel Singular Spectrum Analysis (MSSA; Plaut and Vautard, 1994), or Extended EOF analysis (EEOF; Weare and Nasstrom,1982). These latter two techniques are mathematically equivalent, but differ in their practical domain of application: In general, MSSA deals with more temporal degrees

---

Acknowledgments: The NMC data were kindly provided by Kingtse Mo at the Climate Analysis Center. Most of the results shown here were carried out by Guy Plaut at the Institut Non-Linaire de Nice. The long-range forecasting scheme has been developed recently by Carlos Pires in his Ph.D. work.

of freedom than spatial ones, allowing the development of interesting spectral properties, whereas in EEOFs, the state vector contains only a few lags and a large number of spatial points. In both cases, one seeks the dominant space-time patterns of the analysed signals, taking into account both their spatial and temporal correlations.

By contrast with EOF analysis, time or space-time patterns provide *dynamical* information about the underlying physical system, that is, about the evolution of the state vector. SSA was originally devised by Broomhead and King (1986) as a method of dynamical reconstruction; without any knowledge of the evolution equations of a dynamical system, one can reconstruct its complete dynamics with the sole knowledge of a large set of observations from only one variable, under certain hypothesis. In Section 14.2, we recall the main motivations for the development of SSA within the framework of dynamical systems and we define the delay-coordinate space which is the space where the application of EOF expansion is called SSA. SSA and MSSA turn out to be powerful statistical tools. Their characteristics and properties are developed in Section 14.3. Two applications of MSSA are presented in Section 14.4.

14.2 Reconstruction and Approximation of Attractors

14.2.1 The Embedding Problem

The climate's time evolution is quantified by a set of partial differential equations in a manner similar to other physical systems. Schematically speaking, these equations can be written in the form of a *continuous dynamical system*:

$$\frac{\partial \vec{\mathbf{X}}_t}{\partial t} = F(\vec{\mathbf{X}}_t) \tag{14.1}$$

where $\vec{\mathbf{X}}$ is the vector describing the state of the climate at a given instant. Such a system has the fundamental *deterministic* property that, given the state vector $\vec{\mathbf{X}}_t$ at time t, only one state vector $\vec{\mathbf{X}}_{t+s}$ is possible at a future time $t+s$. When computers are to be used to approximate solutions of (14.1), both time and space discretisation are applied, so that the system (14.1) is transformed into a discrete dynamical system,

$$\vec{\mathbf{X}}_{t+\Delta} = G(\vec{\mathbf{X}}_t) \tag{14.2}$$

where Δ is the integration time step. Spatial discretisation enforces $\vec{\mathbf{X}}_t$ to be a vector of finite dimension m. In the case of the climatic system, the function G is a nonlinear function that contains forcing and dissipation. In that case, the dynamical system (14.2) is generally chaotic, that is, has an evolution that does not tend, as t goes to infinity, to be stationary or periodic, or, equivalently, initially close-by states diverge exponentially with time. For such systems, the limit set of possible states after a long time of integration, called the attractor, has a complex topological structure. A complete picture of the attractor, which is possible only for low-order systems such as the Lorenz (1963) system, provides considerable insight into the system's dynamics. In order to compute this attractor, one is left with two possibilities. Either one uses approximate equations, which is the modeler's job, or we use observations, which is the analyser's job.

There has been recently a considerable interest in the second possibility, that is, the reconstruction of the nonlinear dynamics from time series of observations. This interest has been motivated by the Takens (1981) theory, based on the less recent Whitney embedding theorem: If one disposes of an infinite series of successive values of a *single* coordinate of $\vec{\mathbf{X}}_t$, say $\mathbf{X}_{i,t}$, it is possible to define a state vector $\vec{\mathbf{Y}}_t$ that has the same topological properties as $\vec{\mathbf{X}}_t$. That is to say, only one variable suffices to reconstruct the attractor. In fact, the theorem is more general: For *almost every* measurement function, i.e., function of the variables only, the vector $\vec{\mathbf{Y}}_t$ can be constructed, and has the same topology as $\vec{\mathbf{X}}_t$. Naturally, if one is unlucky by measuring a quantity that is not coupled to the full system, the theorem fails. This situation has a zero probability to occur, however. Therefore, in principle, the global climate could be completely reconstructed by an infinite series of measurements at a given location. This new state vector is defined by embedding the time series into the delay-coordinate space:

$$\vec{\mathbf{Y}}_t = (\mathbf{X}_{i,t+\Delta}, \mathbf{X}_{i,t+2\Delta}, \ldots \mathbf{X}_{i,t+m^*\Delta}) \tag{14.3}$$

where Δ is the sampling rate of the observations. The Takens theorem says that, provided that the *embedding dimension* m^* is large enough, $m^* \geq 2m+1$, there is an embedding[1] Φ from R^m to R^{m^*} that is such that at each time t, $\vec{\mathbf{Y}}_t = \Phi(\vec{\mathbf{X}}_t)$.

A simple example of embedding can be given by the dynamical system defining an oscillator:

$$\begin{aligned} \frac{dx}{dt} &= -\Omega y \\ \frac{dy}{dt} &= \Omega x \end{aligned}$$

It is easy to verify that with only successive values of x, lagged by a time step t, the following relations, defining Φ, form an embedding of dimension 2 satisfying the above requirements:

$$\begin{aligned} x(t) &= x(t) \\ x(t+\Delta) &= cos(\Omega\Delta)x(t) - sin(\Omega\Delta)y(t) \end{aligned}$$

[1]An embedding is a differentiable, one-to-one function, with an one-to-one derivative.

The reconstruction of nonlinear attractors by embedding a single time series has been demonstrated by Broomhead and King (1986), who compared phase portraits of the original Lorenz attractor with the reconstructed one. The result of Takens emphasized the need of analysing the structure of the data embedded in the delay coordinates. SSA is precisely an analysis of the second-order moments in this new coordinate system.

14.2.2 Dimension and Noise

The reconstruction problem, in practice, suffers from two major problems:

- One does not have any knowledge of the underlying dimension m of the system, so that the condition on the embedding dimension is hard to verify;
- Data are always polluted by noise due to round-off, measurement, estimation errors.

In the absence of noise, and with large sets of data, it is still possible to estimate the required value of m^*, by increasing it and calculating the dimension of the attractor, using the Grassberger-Procaccia algorithm, for instance; this dimension increases with m^* up to a saturation value reached when the attractor is fully resolved in the delay-coordinate space.

Unfortunately, noise and the shortness of records mean that, generally, no reliable estimate of the proper dimension is possible. In fact, due to these problems, there is a scale down to which it is impossible to describe the detailed structure of the attractor (Vautard and Ghil, 1989). With a finite, noisy data series, only a *macroscopic* dimension can be estimated. We show now how SSA may help to reconstruct an *approximate* attractor: one gives up the ambition of reconstructing the very fine structure of the attractor including complex foldings and fractal patterns, but one would like still to conserve its macroscopic aspect.

14.2.3 The Macroscopic Approximation

A product of the EOF analysis is, as explained in Chapter 13, the distinction between "signal" and "noise" components. Noisy components are usually reflected by the relatively flat behaviour of the tail of the eigenvalue spectrum. The truncation of the EOF expansion (13.1) to the significant components constitutes therefore an approximation of the cloud of data points, by projecting it onto a finite, restricted number of directions, spanned by the significant "guess patterns". In a sense, the attractor is approximated by a flat subspace. This approximation depends of course both on the noise variance and the sample length, and is not characteristic of the physical system.

The EOF expansion in the delay-coordinate space (SSA) has the same property; the state vector $\vec{\mathbf{Y}}_t$ can be decomposed, as in (13.1) as the sum of a "signal" vector, reflected by a few components, and a "noise" vector reflected by the flat tail of the eigenvalue spectrum. The truncation to the significant components provides an approximation of the attractor as a flat subspace of the delay-coordinate space. In the original-coordinate space, this flat subspace is distorted by the nonlinearity of the embedding Φ, leading to a more general approximation of the attractor by a smooth manifold of the same dimension as the number of significant directions.

This "macroscopic" approximation has, however, a dynamical implication that is not shared by classical EOF analysis. Since each principal component is a linear combination of the value of $\mathbf{X}_{t,i}$ taken at different lags, cancelling one component automatically leads to the assumption that the time series comes from a linear *autoregressive* model (of order $m^* - 1$) of the form

$$\mathbf{X}_{t,i} = a_1 \mathbf{X}_{t-\Delta,i} + a_2 \mathbf{X}_{t-2\Delta,i} + \ldots + a_{m^*-1} \mathbf{X}_{t-(m^*-1)\Delta,i} \qquad (14.4)$$

Such a linear system has solutions under the form of complex exponentials. In fact, it is possible to show (Vautard et al., 1992) that the macroscopic approximation leads to the approximation of the dynamics by a quasi-periodic system, that is, a system that is the sum of a few uncoupled oscillators. Conversely, a system containing only a finite number q of oscillators has only $2q$ non-vanishing components in the delay-coordinate space. In that case, the attractor is approximated by a *torus*. Therefore, in the delay-coordinate space, the truncation to non-noisy components retains the most significant oscillators of the system. Intuitively, SSA will be particularly suited for studying oscillatory behaviour.

14.3 Singular Spectrum Analysis

This section is devoted to properties of SSA and MSSA as statistical analysis techniques, and we leave for the moment the dynamical aspects developed above. Most of the technical details can be found in Vautard et al. (1992) for SSA and Plaut and Vautard (1994) for MSSA.

14.3.1 Time EOFs

Let us start now with the analysis of a scalar time series $\mathbf{X}_t$. In the delay-coordinate space, the associated state vector is defined by lagging m^* times the scalar values like in (14.3). This vector is then considered as a random vector and analysed by diagonalising its covariance matrix. In the case of the delay-coordinate space, the covariance matrix is called the Toeplitz matrix of the signal, since it has constant diagonals. Its principal diagonal contains the variance of $\mathbf{X}$. The second diagonal contains the lag-1 covariance coefficient $\mathrm{E}(\mathbf{X}_t \mathbf{X}_{t+\Delta})$, the third diagonal contains the lag-2 covariance, and so on. Note that in the SSA case, EOFs are the same if one uses the correlation matrix instead of the covariance matrix. In that case, the first diagonal is constant equal to 1, and there is no such arbitrariness in the choice of the dot product as mentioned in Chapter 13.

The k-th EOF $\vec{p}^k$ is the k-th eigenvector of the Toeplitz matrix (sorted into decreasing order of the eigenvalues). It is is a lag sequence of length m^*. Thus, the time patterns associated to these EOFs are called time-EOFs or T-EOFs. The SSA expansion in terms of the T-EOFs writes

$$(\mathbf{X}_{t+\Delta}, \mathbf{X}_{t+2\Delta}, \ldots \mathbf{X}_{t+m^*\Delta}) = \sum_{k=1}^{m^*} \alpha_k(t) \vec{p}^k \tag{14.5}$$

where again the projection coefficients $\alpha_k(t)$, that we shall call the *time-principal components* (T-PCs), are obtained by calculating the dot product of the state vector with the T-EOFs,

$$\alpha_k(t) = \sum_{j=1}^{m^*} \mathbf{X}_{t+j\Delta} p_j^k \tag{14.6}$$

Under this form, the T-PCs can be interpreted as moving averages of the original signal, the averages being weighted by the coordinates of the T-EOFs. T-PCs are therefore filtered versions of the time series. This has important spectral consequences (see Section 14.3.4).

14.3.2 Space-Time EOFs

MSSA is designed to analyse multichannel time series. The original signal to be analysed consists of a random vector $\vec{\mathbf{X}}_t$ of dimension L. In general L is smaller than m^*, the total number of variables. Embedding this series into the delay-coordinate space is obtained in the same way as for SSA. The (big) new state vector $\vec{\mathbf{Y}}_t$ writes

$$\begin{aligned} \vec{\mathbf{Y}}_t \;=\; & (\mathbf{X}_{t+\Delta,1}, \mathbf{X}_{t+\Delta,2} \ldots \mathbf{X}_{t+\Delta t,L}; \\ & \mathbf{X}_{t+2\Delta,1} \ldots \mathbf{X}_{t+2\Delta,L}; \mathbf{X}_{t+m^*\Delta,1} \ldots \mathbf{X}_{t+m^*\Delta,L}) \end{aligned} \tag{14.7}$$

This state vector contains both space and time information. It consists, as in EEOF analysis, of sequences of consecutive maps. The big covariance matrix contains the covariance between each pair of coordinates of $\vec{\mathbf{X}}$ at lags smaller than m^*. The EOFs are now sequences of patterns and are called ST-EOFs. The MSSA expansion takes a form similar to (14.4) and (14.5).

From a dynamical point of view, the embedding theorem also applies with multichannel time series provided that $m^* L \geq 2m + 1$. In the same way as SSA does, the "macroscopic approximation" consists in the approximation of the dynamics by a linear system having oscillatory components.

14.3.3 Oscillatory Pairs

Let us first take the following example: consider a progressive wave in the finite interval $-Y \leq y \leq Y$

$$W(y,t) = Acos(\mu y - \omega t) \tag{14.8}$$

where A is the amplitude (constant), μ the wavenumber and ω the frequency. Assume that the signal is discretized in time by $t_i = i\Delta$, and in space by $y_l = l\Delta y$. The MSSA expansion is easy to determine, if one notices that for any real number ϕ,

$$W(y,t+s) = Acos(\omega t+\phi)cos(\mu y-\omega s+\phi)+Asin(\omega t+\phi)sin(\mu y-\omega s+\phi) \tag{14.9}$$

Indeed, in discrete form, the MSSA expansion is achieved by taking (14.5)

$$\begin{aligned}
X_{l,i} &= W(y_l, t_i) \\
p^1_{l,j} &= F_1 cos(\mu\Delta y - \omega j\Delta + \phi) \\
p^2_{l,j} &= F_2 sin(\mu\Delta y - \omega j\Delta + \phi) \\
a_1(t_i) &= \frac{A}{F_1} cos(\omega i\Delta + \phi) \\
a_2(t_i) &= \frac{A}{F_2} sin(\omega i\Delta + \phi) \\
\phi &= \frac{1}{2}(\mu(L+1)\Delta y - \omega(m^* + 1)\Delta)
\end{aligned}$$

In the above equations, F_1 and F_2 are normalization constants depending on the parameters $(m^*, \Delta y, \Delta)$, calculated by imposing that the ST-EOFs $\vec{p}^{\,1}$ and $\vec{p}^{\,2}$ be of unit norm. The phase ϕ is governed by the orthogonality condition. Under these definitions, the two ST-EOFs are orthogonal and in *phase quadrature*. The expansion therefore contains only two non-vanishing terms. The covariance, at lag zero, of the two PCs vanishes clearly, and they are also in phase quadrature. The non-vanishing part of the eigenvalue spectrum is restricted to the first two eigenvalues $\lambda_1 = (A/2F_1)^2$ and $\lambda_2 = (A/2F_2)^2$. These eigenvalues, although not equal, are equivalent as m^* becomes large: the ratio $\frac{\lambda_1 - \lambda_2}{\lambda_1 + \lambda_2}$ goes to zero as m^* goes to infinity.

More generally, an oscillation present in the signal stands out as a pair of almost-degenerate eigenvalues $(\lambda_k, \lambda_{k+1})$, with T-EOFs or ST-EOFs $(\vec{p}^{\,k}, \vec{p}^{\,k+1})$ in phase quadrature. The associated T-PCs $a_k(t), a_{k+1}(t+1)$ or ST-PCs are also in phase quadrature. SSA and MSSA can therefore be used to isolate oscillatory components in chaotic signals. Several objective criteria have been developed in Vautard et al. (1992) in order to extract these oscillatory pairs. Moreover, the associated PCs serve as an index of phase and amplitude: their quadrature relationship allows one to represent the state of the oscillation as a complex number from which phase $\Theta(t)$ and amplitude $\rho(t)$ can be determined, $z_k(t) = a_k(t) + ia_{k+1}(t) = \rho(t)e^{i\Theta(t)}$.

In the study of oscillatory behaviour, SSA acts as a simplified wavelet analysis, since it allows the amplitude to vary with time. It has been shown to be very efficient in detecting strongly intermittent oscillations. The essential difference with classical spectral analysis is the relaxation of the constraint on the basis functions which are local in time and determined adaptively, instead of being prescribed constant sine and cosine functions.

14.3.4 Spectral Properties

Let us consider, for the sake of simplicity, the SSA case. A review of MSSA spectral properties can be found in Plaut and Vautard (1994). As we have seen above, T-EOFs correspond to data-adaptive moving-average filters. A T-PC is in fact equal to the convolution of the original signal by the corresponding T-EOF. The power spectrum $\Gamma_k(f)$ of the k-th T-PC α_k, at frequency f is given by

$$\Gamma_k(f) = \Gamma_x(f)|\tilde{p}^k(f)|^2, \tag{14.10}$$

where $\Gamma_x(f)$ is the power spectrum of the signal $\mathbf{X}_t$, and

$$\tilde{p}^k(f) = \sum_{j=1}^{m^*} p_j^k e^{2\pi i j f} \tag{14.11}$$

is the Fourier transform of the k-th T-EOF $\vec{p}^{\,k}$. Its square modulus is therefore the *gain function* of the linear filter. The orthogonality constraints of the problem give the identity

$$1 = \frac{1}{m^*}\sum_{k=1}^{m^*} |\tilde{p}^k(f)|^2 \tag{14.12}$$

Thus, for any frequency f, by summing (14.11) for all k, one obtains that the sum of the spectra of the T-PCs is identical to the power spectrum of $\mathbf{X}_t$, i.e.,

$$\Gamma_x(f) = \frac{1}{m^*}\sum_{k=1}^{m^*} \Gamma_k(f) \tag{14.13}$$

This identity permits the examination of the fraction of explained variance by each component at *each frequency* f, by dividing the power spectrum of the T-PC by the total spectrum of the series. It is also interesting to build *stack spectra* by piling up the contributions of the various components.

14.3.5 Choice of the Embedding Dimension

The choice of the embedding dimension m^* (or window length) is crucial. It should be made according to the frequency range under study. In the asymptotic limit $m^* \to \infty$, with fixed sampling interval, i.e., the window length goes to infinity, the eigenvectors tend to pairs of sines and cosines and the associated eigenvalues tend to the corresponding ***spectral density*** values (Devijver and Kittler, 1982). For finite values of m^*, all eigenvalues fall between the maximum and the minimum of the spectral density.

A key problem in SSA is the proper choice of m^*. It can be shown that SSA does not resolve periods longer than the window length. Hence, if one wishes to reconstruct a strange attractor, whose spectrum includes periods of arbitrary length, the larger m^* the better, as long as statistical errors do not dominate the last values of the autocovariance function. To prevent this, one should not exceed in practice $m^* = \frac{n}{3}$.

In many physical and engineering applications, however, one wishes to concentrate on oscillatory phenomena in a certain frequency band, which may be associated with the least-unstable periodic orbits embedded in a strange attractor. Such periodic orbits typically generate oscillations of strongly-varying amplitude: the system's trajectory approaches and follows them for a certain time, comparable to or longer than the period in question, only to wander off into other parts of phase space. When the ratio of m^* to the life time of such an intermittent oscillation, the typical time interval of sustained high amplitudes spells, is large, the oscillatory eigenvector pair suffers from the same Gibbs effect as classical spectral analysis. Spells of the oscillation will be smoothed out. The following arguments should clarify the difficulty and help in making the correct choice of m^*.

If m^* is too small, the coarse spectral resolution may mix together several neighboring peaks in the spectrum of $\mathbf{X}_t$. When there is an intermittent oscillation, reflected by a broad spectral peak, on the contrary, large m^* values (high resolution) will split the peak into several components with neighboring frequencies. In fact, it can be shown that, given a peak in the power spectrum $\Gamma_x(f)$ of $\mathbf{X}_t$, with maximal spectral density at f_o and width $2\Delta f$, SSA will isolate correctly the intermittent oscillation if

$$\frac{1}{f_o} \leq m^* \leq \frac{1}{2\Delta f} \tag{14.14}$$

In other words, the embedding dimension has to be chosen between the period of the oscillation and the average life time of its spells. In practice, this latter quantity cannot be estimated a priori, but SSA is typically successful at analyzing periods in the range $(\frac{m^*}{5}, m^*)$.

14.3.6 Estimating Time and Space-Time Patterns

The estimation of SSA elements does not really differ from the estimation of EOFs, since the calculations are the same, except that they are transposed in the delay-coordinate phase space. However, one has to be careful since generally the SSA eigenvalue spectra are generally flatter than EOFs eigenvalue spectra. Hence eigenvalues are more "degenerate" and the associated patterns tend to be mixed from one sample to another. Typically, for short samples, the order of a "significant" component is statistically unstable, whereas its spectral properties are not. Also, pairs of oscillatory components are almost degenerate (see the example of Section 14.3.4). Thus, from one sample to another, one does not in general expect to recover the right phase in the same oscillatory T-EOF or ST-EOF.

The estimation of power spectra associated with SSA elements can be done in a very neat way by combining Maximum entropy estimation with SSA (Vautard et al., 1992). For SSA, this implies an autoregressive model of order $m^* - 1$, and for MSSA, this involves building a *multichannel* autoregressive model of the same order. This latter model looks like the POP model described in Chapter 15, but with a higher order than 1. The estimation of the coefficients of these models is tricky. A very nice description of the problems encountered in this estimation can be found in Ulrych and Bishop (1975).

14.4 Climatic Applications of SSA

Along the line of statistical climate analysis, SSA and MSSA have been applied to climatic data, on various time scales. In the analysis of oceanic sediment cores (Vautard and Ghil, 1989), the dominant cycles (100 ka, 40 ka and 20 ka)[2] of the paleoclimate have been extracted as pairs of oscillatory components in the oxygen isotopes ratio series, and SSA allowed to exhibit strong modulations of the amplitude of these cycles along the quaternary epoch. On a shorter time scale, SSA has been applied to global surface temperature records (Ghil and Vautard, 1991), like the ones presented in Chapter 4, and allowed to distinguish an interdecadal oscillation, albeit poorly significant, owing to the shortness of the record. The analysis of the Southern oscillation index (SOI) has been carried out by Rasmusson et al. (1990), who demonstrated the existence of two dominant periods in the ENSO signal. Finally, the intraseasonal variability has been investigated by the analysis of geopotential heights (Ghil and Mo, 1991; Plaut and Vautard, 1994), and the atmospheric angular momentum (Penland et al., 1991). Strongly intermittent intraseasonal oscillations have been found in the mid-latitudes. We shall, as an example, focus on the intraseasonal variability, by showing two applications of MSSA.

14.4.1 The Analysis of Intraseasonal Oscillations

For details about the contents of this section, the reader is referred to Plaut and Vautard (1994). The series to be analyzed here is the NMC final analysis of 700 hPa geopotential heights, covering a period of 32 years (1954-1986).

[2] 1 ka represents 1000 years, i.e., a "kiloyear".

The analysis has been carried out over the Atlantic domain, the Pacific domain, and the global Northern hemisphere, but for the sake of conciseness, we shall present here results for the Atlantic domain only. The NMC grid on which data was provided is the diamond grid, including 113 points over the mid-latitude Atlantic (80^oW – 40^oE; 30° – 70^oN). Since we are interested in periods in the range of 10-100 days, the remarks of Section 14.3 about the window length lead us to consider a window of 200 days. In order to avoid having enormous covariance matrices to diagonalize, a prior EOF expansion is performed on the data, and the first 10 principal components, the "channels", are retained for the analysis. This reduction has the advantage also of filtering synoptic motions which we are not interested in anyway. With these $L = 10$ "channels", a window length m^* of 200 days and a sampling rate Δ of 5 days, the size of the covariance matrix is 400.

Since no time filter is applied beforehand, the first two ST-EOFs sticking out represent the seasonal cycle. They explain about 50% of the total variance. Next, in the eigenvalue spectrum, one finds an oscillatory pair (satisfying the phase quadrature relationships) for eigenmodes 7 and 8, with a period of about 70 days. Figure 14.1 displays ST-EOFs 7 and 8 in each of the 10 channels. Each pair of curves represents the value of the ST-EOFs as a function of the lag, each channel being associated to one spatial EOF. First of all, one remarks that the amplitude varies a lot from a channel to another. The channel where the amplitude is the largest is the second, meaning that the associated oscillation projects mostly on the second EOF, which is the famous NAO pattern of Wallace and Gutzler (1981). If its components on other channels were exactly zero, then the oscillation would be a standing oscillation of the second EOF. This is not the case, and we see particularly that the phase differs from a channel to another. Therefore we expect a propagative space-time pattern.

In Figure 14.1 we displayed the ST-EOFs 7 and 8 in the lag space, i.e., for each channel, as a function of the lag. ST-EOFs can also be represented in physical space as a sequence of maps. In order to be slightly more general, we display next on Figure 14.2 the composite half-cycle of the oscillation as a function of its phase. One could have used ST-PCs in the complex plane as explained in Section 13.3.2. In fact, the phase used to key the composites of Figure 14.2 are based on a slightly more accurate index, the reconstruction filter, calculated also from the two ST-PCs 7 and 8 [see Plaut and Vautard (1994) for more details]. From this figure one sees that the oscillation has a poleward-propagating component in the region of the Atlantic jet, with a standing component over Siberia. Its amplitude is maximal, as expected, when it is in phase with the NAO (Figure 14.2d).

As explained above, SSA or MSSA allow the study of the explained variance in the frequency domain. Figure 14.3 shows the Maximum Entropy spectra of several ST-PCs, divided by the total variance. Each curve represents therefore the fraction of variance explained by one ST-PC as a function

Figure 14.1: *ST-EOFs 7 and 8 in each channel as a function of lag (in days), for the 700 hPa geopotential height over the Atlantic area, using a sampling rate of 5 days and a window length of 200 days. The 10 channels represent the coefficient of projection of the ST-EOFs on the first 10 spatial EOFs. ST-EOFs are dimensionless and normalised in such a way that the sum of squares of their coefficients is 1. (After Plaut and Vautard, 1994).*

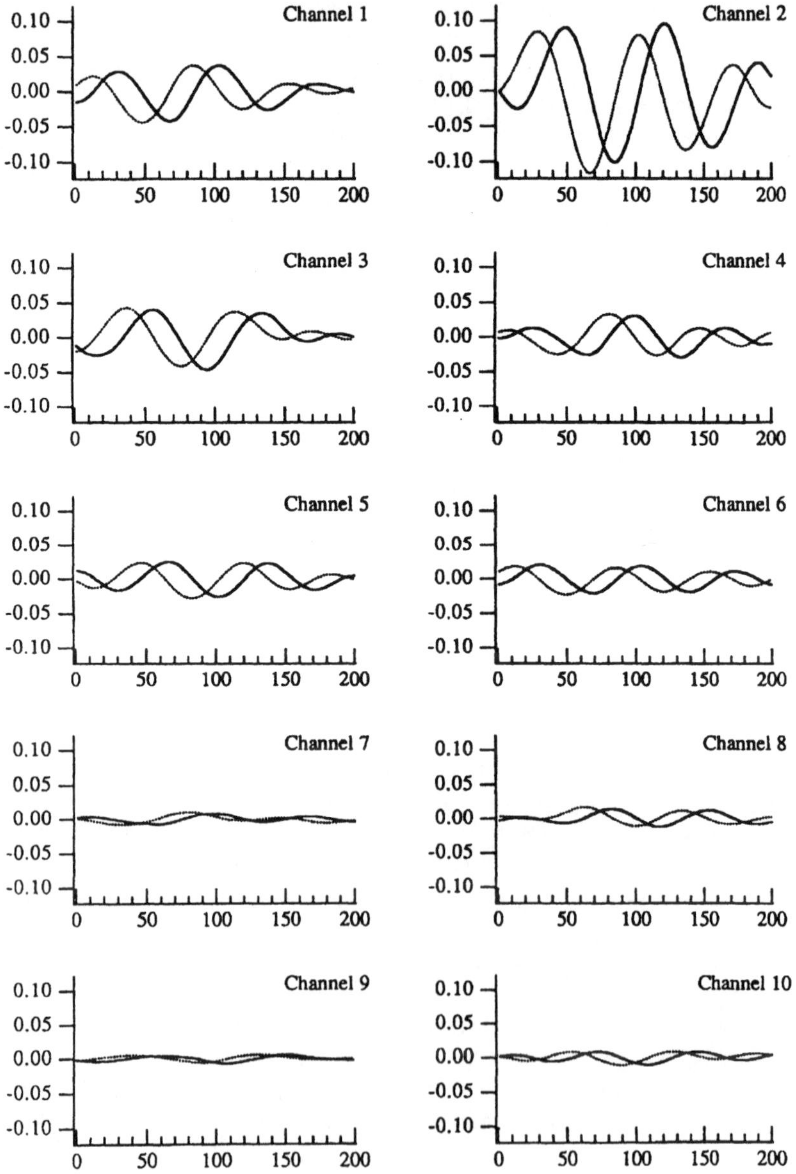

Figure 14.2: *Composite of the anomaly of the 700 hPa keyed to half-a-cycle of the 70-day oscillation. Panels a) to d) show the composite evolution of the anomaly averaged over 4 among 8 equally-populated categories of the cycle. (After Plaut and Vautard, 1994).*

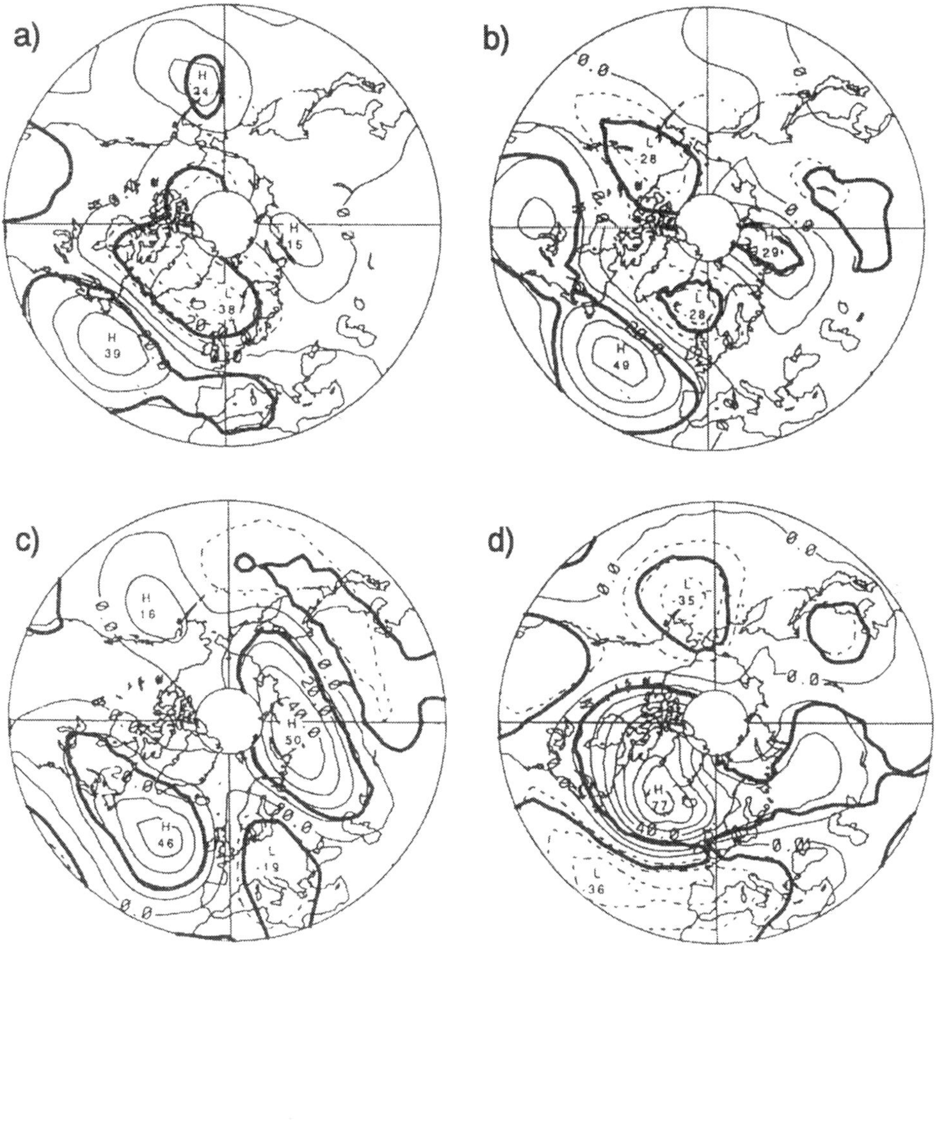

Figure 14.3: *Fraction of explained variance, in percent, of the first 16 ST-PCs, as a function of the frequency. This variance fraction is calculated as the ratio between the power spectrum of the ST-PCs and the total power spectrum of the multichannel time series. Pairs of eigenelements are emphasized (see legend). (After Plaut and Vautard, 1994).*

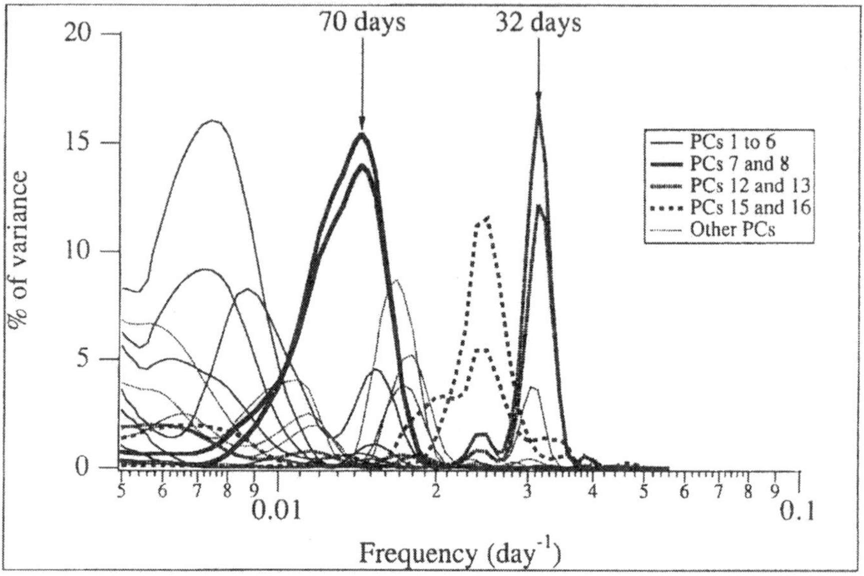

of frequency. In particular, we recover the two ST-EOFs 7 and 8 peaking near 70 days, explaining about 30% of the variance at this period. Another remarkable oscillatory pair is the pair 12 and 13 peaking at a period of 32 days. ST-PCs 1 and 2 peak out of the graph exactly at a period of 1 year. ST-PCs 3 to 6 explain most of the variance between 60 days and 1 year.

Another interesting feature is the fact that the dominant oscillations have periods of, roughly speaking, 70 and 35 days. The study of the phase index of the above two oscillations shows in fact that these two oscillations are phase-locked. That is, the 35-day oscillation is in fact a harmonic of the 70-day oscillation. This has important dynamical consequences since the two phenomena cannot be studied separately. This also emphasizes the nonlinear nature of these oscillations. Indeed, should the underlying dynamical system be linear, the solutions would result in two periodic orbits that would be symmetric and ellipsoidal in phase-space, whereas here they are more likely to result in one periodic orbit of period 70 days distorted in such a way that it generates a non-vanishing harmonic.

The time-local nature of MSSA allows one also to distinguish periods of high amplitude, called ***oscillation spells***. Such is the case, for instance, during winters 61-62 and 77-78, for the 70-day oscillation. ST-EOFs 7 and 8 gathered explain globally about 3% of the total variance (annual cycle removed). However, the ratio of the sum of squares of the two ST-PCs involved to the total sum of squares of all ST-PCs represent the explained *local variance* (in time), and can be as high as 20% during high-amplitude spells. Thus, MSSA allows the quantification of the variance explained in the time domain, and to extract high-amplitude spells.

In order to illustrate the fact that high-amplitude spells result in important changes in the flow circulation, Figure 14.4 shows the composite average of the 700 hPa heights over days selected in (i) the 20 highest-amplitude spells of the 70 day oscillation, and (ii) found simultaneously in particular phases of the two oscillations. This composite is particularly interesting since it exhibits a pronounced European blocking structure. Thus, one possible origin of blocking may simply be the interference pattern of two oscillating phenomena, which would contradict the strongly nonlinear character of blocking. In fact, this interference situation cannot explain blocking in general, since only a small fraction of blocking events turn out to be present in the composite of Figure 14.4.

A more systematic study of the correlation between weather regimes (Molteni et al., 1990; Vautard, 1990; Cheng and Wallace, 1993; Kimoto and Ghil; 1993), such as blocking, and these low-frequency oscillations would show that the occurrence of the former is influenced by the phase of the latter (Plaut and Vautard, 1994). In other words, one can think of the oscillations as creating favorable environments for the onset of particular weather regimes. Such a phenomenon can also be observed with the environment created by El Niño on the mid-latitude flow (see Chapter 6).

This last point has important consequences for climate modeling and long-range forecasting. First, it implies that a good climate model must simulate correctly the intraseasonal oscillations in order to simulate also correctly the weather regimes. Second, the occurrence of weather regimes is not random, since they are influenced by these oscillations. As the oscillations are intrinsically predictable phenomena, they are a source of long-range predictability in the mid-latitudes. We shall test this idea now by constructing an empirical model based on MSSA predictors.

14.4.2 Empirical Long-Range Forecasts Using MSSA Predictors

The expansion of a signal into T-PCs or ST-PCs provides a decomposition of the flow in time patterns that are generally band-limited in frequency. The first ST-EOFs are usually associated with large-scale slow or oscillatory motions, that are presumably more predictable than synoptic eddies (Shukla, 1984). In a certain manner, MSSA not only compresses the space-time information, but also gathers the most predictable components into a few time series, the first ST-PCs. Hence, using these ST-PCs as empirical predictors

Figure 14.4: *Composite of the raw 700 hPa geopotential height keyed to simultaneous occurrences of one particular phase of the 70-day oscillation and one particular phase of the 35-day oscillation, during high-amplitude spells. Contour interval is 20m. (After Plaut and Vautard, 1994)*

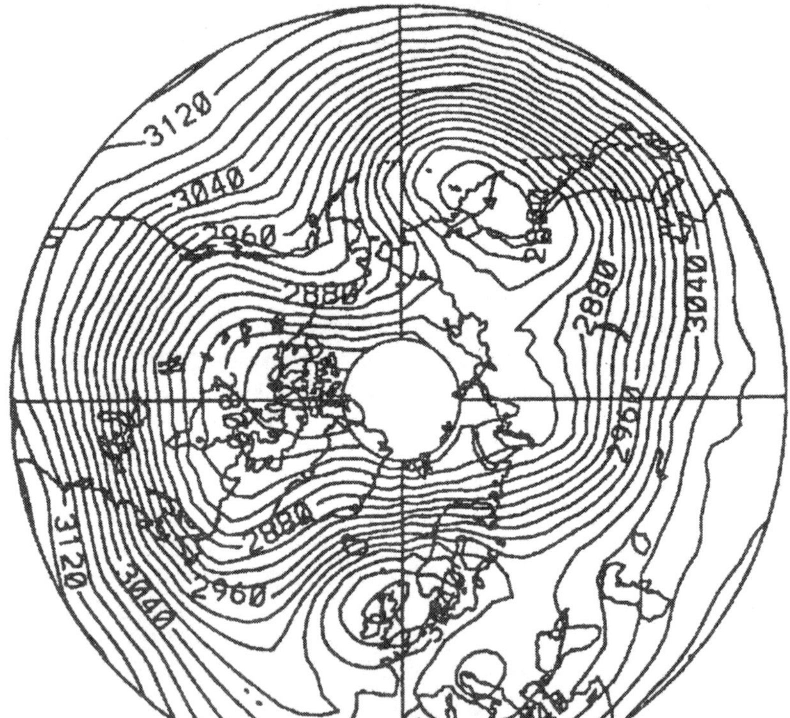

should provide an improved basis for long-range forecasting. In order to illustrate this point, we describe here a hindcast experiment in which we attempt to forecast the average of the 700 hPa geopotential height (Z700) over the forthcoming month, and over the Atlantic area. For technical details, the reader is referred to Vautard et al. (1994).

The dataset, consisting of NMC final analysis up to December 1992, and ECMWF data during 1993, covers the 42-year period from July 1951 to July 1993. Data are gathered in pentads (5-day averages). Here, MSSA is used with a shorter window of 3 months, with data where the annual cycle is removed beforehand, and only the first 6 spatial EOFs are used. Thus, $L = 6$ channels and $m^* = 18$ lags are used. The results, however, are very stable to changes in the MSSA parameters. The sampling rate Δ is also 5 days.

Assume that today is time $t = 0$, and we want to forecast the forthcoming monthly mean of Z700 over the Atlantic domain. Since it would be far too ambitious to forecast the exact monthly mean values, our goal is to forecast, at each Atlantic grid point, the *tercile* in which this value falls (high, near-average, or low). Moreover, it would also be too ambitious to give deterministic forecasts, hence only ***probabilities*** of falling within each tercile are estimated. We proceed in three steps:

1. Analysis: An empirical model is built from a learning period that does not contain the forecast verification period. From this period only, MSSA is performed and ST-PCs are calculated. A linear regression of the ST-PCs is performed 6 pentads ahead. The regression coefficients are estimated from the learning period, as well as the climatologies.

2. Forecasting: The ST-PCs at time $t = 0$ are calculated and forecast at lead time $t = 30$, using the regression model. The regressors are the first 100 ST-PCs at time 0, and the forecast values are the first 15 ST-PCs. We emphasize that the filtered values given in the ST-PCs are calculated, each forecast day, using only past data covering the latest 18 pentad period. However, the MSSA analysis, i.e. the calculation of the ST-EOFs is performed over the whole learning period.

3. Specification: The first 15 ST-PCs are used to estimate conditional probabilities of occurrence within each tercile. This is done by seeking in the past for the 200 best analog ST-PCs of the forecast ST-PCs and counting the number of analogs associated with each tercile. For the validation of the experiment, we choose as a deterministic forecast the tercile that has the highest conditional probability.

This procedure is a poor-man's version of the long-range empirical forecast scheme used at the U.K. Meteorological office (Folland et al., 1986b), where first a weather type is forecast (among 6 clusters), and then the specification procedure in order to determine categories is performed using discriminant analysis and other more subtle techniques.

Figure 14.5: *Cross-validations scores, averaged over the Atlantic area, of a hindcast experiment using MSSA predictors (solid), of persistence forecasts (shaded bars). The forecast quality is the tercile of the forthcoming 30-day average of the 700 hPha. Bars show the LEPS categorical score of each 3-winter period of the cross-validation experiment. Triplets of winters are annotated at the bottom of the figure. LEPS scores are normalized by the scores of perfect forecasts. Winter is defined as 1 December to 31 March. See the text for a description of the hindcast experiment.*

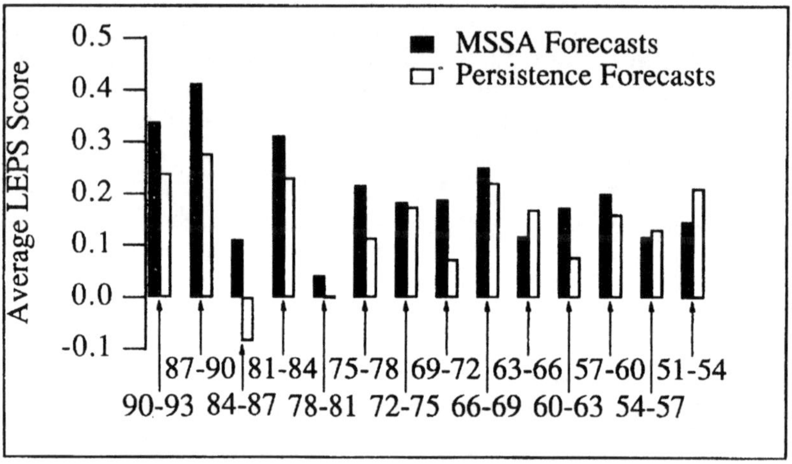

The forecasting scheme presented here is validated using a cross-validation technique described in Chapter 10. Verification data are gathered in 14 groups of 3 years, each being removed from the total dataset in order to form a learning period from which data can be used to construct the models. Given a 3-year verification period, the coefficients of the forecasting model are calculated from data in the associated learning period, and the forecast first 15 ST-PCs are compared with all the observed first 15 ST-PCs occurring at each pentad of the learning set, using the cosine of the angle between the two (pattern correlation) as a similarity measure. The 200 best learning analogs are extracted. At each Atlantic grid point, their associated monthly mean tercile is calculated. Thus, the ***conditional probability*** of occurrence of a tercile is the percentage of these 200 analogs falling within the tercile. Next, for validation purposes, a deterministic forecast is built by choosing the tercile that has the highest conditional probability. Verification scores are stored in a 3×3 contingency table, where the diagonal counts the number of successful forecasts, at each grid point. This table is used to calculate the LEPS categorical score (see Chapter 10) of the hindcast experiment.

Figure 14.5 shows the LEPS score counted over all Atlantic grid points for the 14 periods of the cross-validation experiment. The scores of "persistence"

Figure 14.6: *Spatial distribution of the time-average wintertime LEPS score over the full 42-year cross-validation period, for the hindcast experiment presented in the text.*

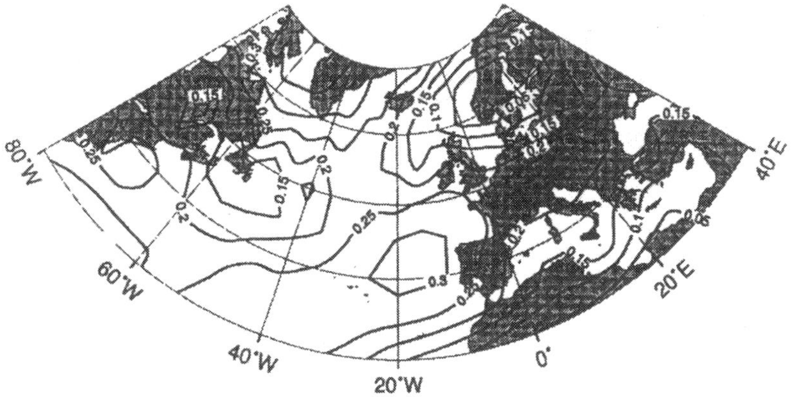

forecasts are also represented. Persistence forecasts consist in forecasting, for the forthcoming month, the same tercile as the one of the past month. The MSSA scores are generally better than persistence scores which are themselves better than random forecast scores, except in the earliest periods. The scores lie, in general, around 0.2, except for the latest periods, which seem to have exceptionally good scores, possibly due to the improvements in the data themselves, leading to better initial conditions of the forecast.

Figure 14.6 shows the spatial distribution of the average score over the whole cross-validation period. It is noteworthy that this distribution is not homogeneous at all, with two areas of good scores near Greenland and off the European western coasts. This reflects the fact that one pattern is particularly well forecast: The NAO pattern. Half-way between the two centers of action of this pattern, there is a band of relatively poor forecasts. NAO is the most persistent feature of the Atlantic area, and is also strongly involved in the 70-day oscillation. Yet, notice that the score is everywhere positive.

Recently, Vautard et al. (1999) used ST-PCs of the SST over the world ocean in order to hindcast the seasonal anomaly of the surface temperature over North America. One of their results is shown in Figure 14.7. In general, they found results quite in good agreement with the experiments performed by Barnston (1994), who used Canonical Correlation Analysis (CCA) with the same predictors. The method gave actually a slightly better score than CCA, although the differences are minor. Statistically significant forecast skill is found, for relatively long lead times (such as 6-9 months), for the prediction of the late winter and mid-summer periods. As shown by Figure 14.7, the skill is not uniform. Canada seem to benefit from a uniformly significant skill during winter, while no skill at all is found over the Rocky

Figure 14.7: *Geographical distribution of the correlation between hindcasts of seasonal-mean surface temperatures and their observed values (the verification stations are marked by dots), using 10 ST-PCs of the world-ocean SSTs (originating from the COADS data set) as predictors, and the same model as for the prediction of geopotential heights (see above). The skill distribution is shown for the 4 seasons: winter (JFM), spring (AMJ), summer (JAS) and fall (OND). After Vautard et al., 1999, to which the reader is referred for further details.*

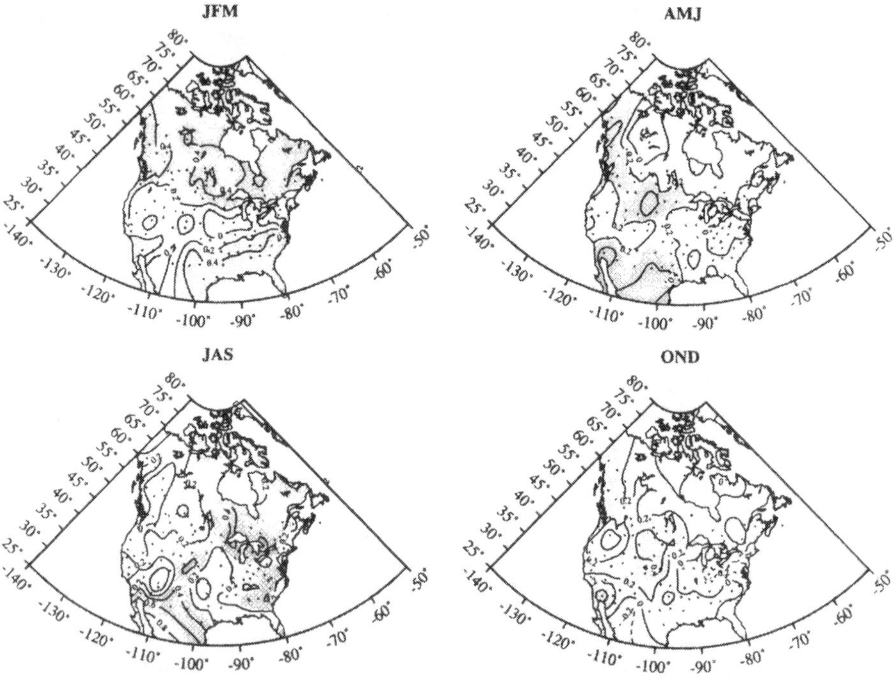

mountains in the United States. Note also the relatively good skill of seasonal forecasts over the South-East U.S. The reason for this skill distribution of skill remains unclear. However, the SST ST-PC predictors were identified as components of the El-Niño Southern Oscillation, as also confirmed by Moron et al. (1998).

14.5 Conclusions

The analysis of time patterns provides a dynamical insight to the climate. SSA and MSSA are statistical analysis tools that describe the structure of second-order moments of time sequences or space-time sequences, just as well as EOF analysis does with second-order moments of spatial distributions. There is however a difference in the interpretation of spatial patterns and time patterns: as explained in Chapter 13, spatial EOFs depend strongly on the dot product used, and this arbitrariness leads to a difficulty in assessing the dynamical significance of the patterns. They cannot be considered as "dynamical modes" of variability. They can still be used as "guess patterns", since a few of them approximate generally quite well the spatial structure of instantaneous flows. This arbitrariness is still present with the MSSA for the spatial directions, but it is removed in the delay-coordinate space, i.e., in the lag domain. Space-time patterns are generally band limited. Spatial structures are tied to a time scale. The natural oscillators of the physical system under study will come out as pairs of time patterns in phase quadrature.

When a projection is made onto the first few time patterns, one makes the implicit assumption that the system is the sum of a finite number of oscillators. Thus SSA and MSSA are particularly suited to study oscillatory behaviour. At first sight SSA might seem inadequate for the study of strongly nonlinear systems such as the climate. Such systems may still possess intermittent oscillations due to the presence of unstable periodic orbits. The fact that the basis functions of SSA are given on a lag window of finite length allows the localisation of these intermittent spells, unlike classical spectral analysis which does not allow amplitude modulation.

Finally, we showed two applications of MSSA:

- the description of intraseasonal mid-latitude variability and its dominant oscillations. This application demonstrates that it is possible to detect oscillatory phenomena without any prior guess of the underlying physics. MSSA provides an objective index of phase and amplitude of an oscillation that can serve as a basis for composite studies. We showed in particular that weather regimes and intraseasonal oscillations are related. This emphasizes the need for GCMs to reproduce this oscillatory behaviour in order to be able to produce valuable long-range forecasts.

- MSSA time-patterns can be used as a basis of predictors for long-range forecasting. These patterns are optimal in the sense that they achieve a good compromise between compression of information and extraction of slowly-varying oscillatory components. We have presented here a possible probabilistic forecasting scheme that is an improved-persistence scheme, and that can be used, at least for testing long-range forecast scores of dynamical models.

Chapter 15

Multivariate Statistical Modeling: POP-Model as a First Order Approximation

by Jin-Song von Storch

15.1 Introduction

The study of a time series is a standard exercise in statistical analysis. A time series, which is an ordered set of random variables, and its associated probability distribution are called a stochastic process. This mathematical construct can be applied to time series of climate variables. Strictly speaking, a climate variable is generated by deterministic processes. However since a myriad of processes contribute to the behavior of a climate variable, a climate time series behaves like one generated by a stochastic process. More detailed discussion of this problem is given by H. von Storch and Zwiers (1999).

The most important assumption made when modeling a time series is that the corresponding stochastic process is stationary, and that a stationary stochastic process may be adequately described by the lower moments of its probability distribution. The lower moments include the mean, variance, covariance function and the Fourier transform of the covariance function, the power spectrum. For multivariate time series, one has additionally cross-covariance functions between components of the multivariate time series and the Fourier transform of the cross-covariance functions, the cross-spectra. Given a stationary time series, the goal of statistical modeling is to describe the stochastic process by means of a model containing a few parameters which provide information about the low order moments of the stochastic process.

For univariate time series, the statistical models normally used are autoregressive (AR) or moving average (MA) processes. The covariance functions and the power spectra of univariate time series generated by AR or MA processes are functions of the process parameters. Since low order processes are described by a few parameters, the second moment information can be easily derived from a univariate time series. In the example of a time series generated by a first order autoregressive [AR(1)] process with parameter α, the covariance function and the shape of the spectrum is determined by α.

The situation becomes more complicated for a multivariate time series. Climate time series are usually multivariate with high dimensions. In the case of observations, the dimension is of the order of 10^2 to 10^3, whereas the dimension of General Circulation Model outputs is normally larger than 10^3. The second moment information of such a multivariate time series is not only difficult to derive, but also difficult to interpret. For an m-dimensional time series generated by an AR(1) process, the process parameter is no longer just one number, as discussed before, but an $m \times m$ matrix. Even if one could interpret cross-covariance function at lag Δ which now is a matrix, it does not guarantee that one could also interpret the cross-covariance function at lag $\Delta + 1$. It will be shown later that because of the nonlinearity of matrix multiplication, there is a strong limitation in deriving second moment information from parameters of multivariate AR or MA processes.

Besides the random characters, a climate time series displays also deterministic features. Among a myriad of processes, frequently only a few stand out. They represent the dominant dynamics of the system and are responsible for the time evolution of our time series. This fact suggests that a climate time series can be considered to be generated separately by a deterministic process which represents *the* process and a stochastic one which represents all other processes. It is therefore reasonable to build problem-specific models which are, to our knowledge, suitable for the dynamical processes. This kind of statistical modeling was first proposed by Hasselmann (1988).

The assumption made for Hasselmann's statistical models is that the climate phase space can be split into two subspaces called "signal" and "noise" subspaces. This notion has been introduced in Chapter 13. In our context, it is assumed that the deterministic processes operate in the "signal" subspace whereas the other remaining ones operate in the "noise" subspace. Furthermore, the "signal" subspace can be spanned by a few characteristic spatial patterns. The coefficients of these patterns are described by a set of equations which represent the dynamics of the processes and control the temporal evolution of the "signal" subspace. Thus, depending on the problem considered, the model can take any (linear or non-linear) form. A linear (non-linear) model is called, according to Hasselmann (1988), a POP (PIP) model , and the patterns which spanned the "signal" subspace are called Principal Oscillation Patterns (POPs) for a linear model and Principal Interaction Patterns (PIPs) for a nonlinear model. The contribution from the "noise" subspace is introduced in a PIP or POP model by a stochastic forcing term.

The consideration of a phase space consisting of "signal" and "noise" subspaces is also suitable for an AR or MA process. In the case of the AR(1) process, the signal is represented by the fact that the time series at time t has a fixed relation to time series at time $t-1$. The noise part is represented by a white noise forcing. In this respect, both AR/MA model and PIP/POP model represent the same category of models. The difference lies in the way the models are constructed. An AR/MA model is designed for inferring statistics of *a* process whereas a POP/PIP model for inferring statistics of *the* process. Thus, an AR/MA model provides information about the low-order moments, whereas a POP/PIP model provides information about the dominant dynamics.

In the case that there is no a-priori knowledge about the process of our interest, it is reasonable to consider the first order approximations of the process. A linearization of a process can be considered as a first order approximation. A POP model which is the discrete form of such a linearization is designed to study the eigenstructure of the process. If $\vec{\mathbf{X}}_t$ represents a time series in the multivariate "signal" subspace, the model states

$$\vec{\mathbf{X}}_t = \mathcal{A}\vec{\mathbf{X}}_{t-1} + \vec{\mathbf{N}}_{t-1} \tag{15.1}$$

where $\mathcal{A}$ is a $m \times m$ real matrix, called hereafter the system matrix of $\vec{\mathbf{X}}$. $\mathcal{A}$ described the linear dynamics of the "signal" subspace, $\vec{\mathbf{N}}_t$ represents all other remaining processes. It is assumed that $\vec{\mathbf{N}}_t$ is independent of all $\vec{\mathbf{X}}_{t-\Delta}$ $(\Delta \geq 1)$.

Equation (15.1) has also the form of a multivariate AR(1) process which is the simplest AR process. Thus, a consideration of (15.1) would provide information both of the first order approximation of the second moments and of the linear dynamics. In this Chapter, some examples are shown to emphasize the ability of the model (15.1) in describing second moments and linear dynamics. The first application of the POP model was presented by H. von Storch et al. (1988). A more detailed summary of the POP analysis with various examples on the POPs as a diagnostic and predictive tool is given by H. von Storch et al. (1995).

Section 15.2 gives the general definition of cross-covariance functions and cross-spectra of a multivariate time series. It will be shown in Section 15.3 that the system matrix or process parameter, $\mathcal{A}$, provides the full second moment information of a multivariate time series generated by an AR(1) process. The difference in identifying the second moment information by considering an AR(1) process or by considering one cross-spectrum matrix defined over a frequency band is discussed in Section 15.3.4. The latter is normally called a complex EOF analysis. In terms of one example, Section 15.4 considers the question of to what extent the POPs can be interpreted as normal modes of the considered system.

15.2 The Cross-Covariance Matrix and the Cross-Spectrum Matrix

For a stationary real (complex) m-dimensional discrete time series $\vec{\mathbf{X}}_t$, the cross-covariance matrix at lag Δ, $\mathbf{\Sigma}(\Delta)$, takes the form of an $m \times m$ real (complex) matrix and is defined by:

$$\mathbf{\Sigma}(\Delta) = \mathrm{E}\Big((\vec{\mathbf{X}}_t - \vec{\mu})(\vec{\mathbf{X}}_{t-\Delta} - \vec{\mu})^{*T}\Big) \qquad (15.2)$$

where ^{T} denotes matrix transposition and * complex conjugation [1]. E( ) is expectation. For a complex time series, one has:

$$\mathbf{\Sigma}(-\Delta) = \mathbf{\Sigma}(\Delta)^{*T} \qquad (15.3)$$

The cross-spectrum matrix of $\vec{\mathbf{X}}$ at frequency f is a complex $m \times m$ matrix and is defined by:

$$\begin{aligned} \mathbf{\Gamma}(f) &= \sum_{\Delta=-\infty}^{\infty} \mathbf{\Sigma}(\Delta) e^{-i2\pi\Delta f} \qquad (15.4) \\ &= \sum_{\Delta=0}^{\infty} \left(\mathbf{\Sigma}(\Delta) e^{-i2\pi\Delta f} + \mathbf{\Sigma}(\Delta)^{T*} e^{i2\pi\Delta f}\right) - \mathbf{\Sigma}(0) \end{aligned}$$

It holds:

$$\mathbf{\Gamma}^{*T}(f) = \mathbf{\Gamma}(f) \qquad (15.5)$$

thus, $\mathbf{\Gamma}(f)$ is a Hermitian matrix. $\mathbf{\Sigma}(\Delta)$ is the Fourier transform of $\mathbf{\Gamma}(f)$ and $\mathbf{\Gamma}(f)$ is the Fourier transform of $\mathbf{\Sigma}(\Delta)$. The j-th diagonal component of $\mathbf{\Gamma}(f)$ is real and corresponds to the autospectrum of $\mathbf{X}_j$ at frequency f, whereas its ij-th off-diagonal component is complex with the real part being the co-spectrum and the imaginary part the quadrature spectrum between $\mathbf{X}_i$ and $\mathbf{X}_j$ at frequency f.

The structures of $\mathbf{\Sigma}$ and $\mathbf{\Gamma}$ are in general complicated. They are functions of time lag Δ and frequency f. In principle, in order to describe the second moments of a multivariate time series of infinite length, one needs infinite cross-covariance matrices at lags $\Delta = -\infty$ to $\Delta = +\infty$, or infinite cross-spectrum matrices at frequency $f = \frac{1}{2}$ to $f = 0$. This causes immense difficulties in describing the full temporal characteristics of a multivariate time series. A simplification of the problem can be achieved by considering the approximation (15.1).

[1] In order to be consistent with (15.2), the scalar product of $\vec{a}$ and $\vec{b}$ is defined in this chapter by $\vec{a}^{*T}\vec{b}$.

15.3 Multivariate AR(1) Process and its Cross-Covariance and Cross-Spectrum Matrices

For a real time series $\vec{\mathbf{X}}$ with $\vec{\mu} = 0$ generated by an AR(1)-process of the form (15.1):

$$\begin{aligned}
\mathbf{\Sigma}(1) &= \mathrm{E}\left(\vec{\mathbf{X}}_t \vec{\mathbf{X}}_{t-1}^T\right) && (15.6)\\
&= \mathrm{E}\left((\mathcal{A}\vec{\mathbf{X}}_{t-1} + \vec{\mathbf{N}}_{t-1})\vec{\mathbf{X}}_{t-1}^T\right)\\
&= \mathcal{A}\mathbf{\Sigma}(0)
\end{aligned}$$

since the "noise" $\vec{\mathbf{N}}_{t-1}$ is independent from $\vec{\mathbf{X}}_{t-1}$, that is $\mathrm{E}\left(\vec{\mathbf{N}}_{t-1}\vec{\mathbf{X}}_{t-1}^T\right) = 0$. The system matrix $\mathcal{A}$ in (15.1) is then given by:

$$\mathcal{A} = \mathbf{\Sigma}(1)\mathbf{\Sigma}(0)^{-1} \quad (15.7)$$

It can be shown that the cross-covariance matrix and the cross-spectrum matrix of an AR(1) process are related to $\mathcal{A}$:

$$\begin{aligned}
\mathbf{\Sigma}(\Delta) &= \mathcal{A}^{\Delta}\mathbf{\Sigma}(0) && (15.8)\\
\mathbf{\Gamma}(f) &= \sum_{\Delta=-\infty}^{\infty} \mathbf{\Sigma}(\Delta) e^{-i2\pi\Delta f} && (15.9)\\
&= \sum_{\Delta=0}^{\infty} \left[\mathcal{A}^{\Delta}\mathbf{\Sigma}(0)e^{-i2\pi\Delta f} + \left(\mathcal{A}^{\Delta}\mathbf{\Sigma}(0)\right)^{T*} e^{+i2\pi\Delta f}\right] - \mathbf{\Sigma}(0)
\end{aligned}$$

Thus the spectral characteristics of a multivariate time series generated by an AR(1) process are described entirely by two matrices, the system matrix $\mathcal{A}$ and the lag-0 cross-covariance matrix $\mathbf{\Sigma}(0)$. The following section shows how to calculate the summation in (15.9) and to obtain a cross-spectrum matrix which is a function of f. Such calculation is only possible for an AR(1) process. For higher order AR process, because of the non-linearity of matrix multiplication, there is no simple way to express cross-covariance function or cross-spectrum matrix in terms of process parameters.

15.3.1 The System Matrix $\mathcal{A}$ and its POPs

The product of the matrices $\mathcal{A}^{\Delta}$ can be simplified by decomposing $\mathcal{A}$ into its eigenvector matrix $\mathcal{P}$ and its eigenvalue matrix $\mathbf{\Lambda}$, which are defined below.

- The eigenvectors, $\vec{p}^{\,j}$, of the system matrix $\mathcal{A}$ are the Principal Oscillation Patterns (POPs)

- The POP coefficient $\mathbf{Z}_j$ is expansion coefficient with $\mathbf{Z}_j = \vec{p}_A^{j\,*T}\vec{\mathbf{X}}$, where $\vec{p}_A^{\,j}$ is the adjoint vector of the eigenvector $\vec{p}^{\,j}$. [2]
- If $\vec{\mathbf{X}}$ is generated by (15.1), the POP coefficient is then generated by a univariate AR(1)-process:

$$\mathbf{Z}_{j,t} = \lambda_j \mathbf{Z}_{j,t-1} + \mathbf{R}_{j,t-1} \tag{15.10}$$

with λ_j being the j-th eigenvalue of $\mathcal{A}$. $\mathbf{R}_j = \vec{p}_A^{j\,*T}\vec{\mathbf{N}}$ is independent of $\mathbf{Z}_{j,t}$.

Because $\mathcal{A}$ is not symmetric, some or all of its eigenvalues λ_j and eigenvectors $\vec{p}^{\,j}$ can be complex. Since $\mathcal{A}$ is a real matrix, if $\vec{p}^{\,j}$ is a complex eigenvector of $\mathcal{A}$ with complex eigenvalue λ_j, the complex conjugate $\vec{p}^{\,j*}$ is also an eigenvector of $\mathcal{A}$ with eigenvalue λ_j^*.

If $\mathcal{A}$ has m different eigenvalues, which is normally the case for a climate time series, $\vec{\mathbf{X}}$ may be uniquely expressed in terms of its eigenvectors. The transformation from the real vector $\vec{\mathbf{X}}$ to the generally complex vector $\vec{\mathbf{Z}}$ is defined by:

$$\begin{aligned} \vec{\mathbf{X}} &= \mathcal{P}\vec{\mathbf{Z}} \quad or \\ \vec{\mathbf{Z}} &= \mathcal{P}_A^{*T}\vec{\mathbf{X}} \end{aligned} \tag{15.11}$$

where $\mathcal{P}$ and $\mathcal{P}_A$ are complex $m \times m$ matrices. $\mathcal{P}$ is the eigenvector matrix, whose columns are the eigenvectors of $\mathcal{A}$. $\mathcal{P}_A$ is the adjoint matrix, whose columns are the corresponding adjoint vectors. One has $\mathcal{P}^{*T}\mathcal{P}_A = \mathcal{P}_A^{*T}\mathcal{P} = \mathcal{I}$.

15.3.2 Cross-Spectrum Matrix in POP-Basis: Its Matrix Formulation

The eigenvalue matrix $\mathbf{\Lambda}$ is a diagonal one

$$\mathbf{\Lambda} = \begin{pmatrix} \lambda_1 & \dots & 0 \\ \vdots & \ddots & \vdots \\ 0 & \dots & \lambda_m \end{pmatrix} \tag{15.12}$$

with the eigenvalues of $\mathcal{A}$ in the main diagonal. $\mathcal{A}$ can then be decomposed into

$$\mathcal{A} = \mathcal{P}\mathbf{\Lambda}\mathcal{P}_A^{*T} \tag{15.13}$$

which leads to the desired simplification

$$\mathcal{A}^{\Delta} = \mathcal{P}\mathbf{\Lambda}^{\Delta}\mathcal{P}_A^{*T} \tag{15.14}$$

[2] The eigenvector $\vec{p}^{\,j}$ and its adjoint vector $\vec{p}_A^{\,j}$ satisfy $(\vec{p}_A^{\,i})^{*T}\vec{p}^{\,j} = \delta_{i,j}$.

with

$$\Lambda^{\Delta} = \begin{pmatrix} \lambda_1^{\Delta} & \dots & 0 \\ \vdots & \ddots & \vdots \\ 0 & \dots & \lambda_m^{\Delta} \end{pmatrix} \tag{15.15}$$

The cross-spectrum matrix of $\vec{\mathbf{X}}_t$ can be further simplified by transforming $\vec{\mathbf{X}}_t$ to $\vec{\mathbf{Z}}_t$.

Using the transformation of (15.11), the expression of cross-covariance matrix of $\vec{\mathbf{X}}$ (15.8), the eigenvector decomposition of $\mathcal{A}$ (15.14), the cross-covariance matrix of $\vec{\mathbf{Z}}$ is given by

$$\Sigma_Z(+\Delta) = \mathcal{P}_A^{*T}\Sigma_X(+\Delta)\mathcal{P}_A = \mathcal{P}_A^{*T}\mathcal{A}^{\Delta}\Sigma_X(0)\mathcal{P}_A = \Lambda^{\Delta}\mathcal{P}_A^{*T}\Sigma_X(0)\mathcal{P}_A$$

$$\Sigma_Z(-\Delta) = \left(\Lambda^{\Delta}\mathcal{P}_A^{*T}\Sigma_X(0)\mathcal{P}_A\right)^{T*}$$

The cross-spectrum matrix in the POP-basis can be then calculated from (15.4):

$$\begin{aligned} \Gamma_Z(f) &= \sum_{\Delta=-\infty}^{\infty} \Sigma_Z(\Delta)e^{-i2\pi f\Delta} \\ &= \sum_{\Delta=0}^{\infty} \left[\Sigma_Z(+\Delta)e^{-i2\pi f\Delta} + \Sigma_Z(-\Delta)e^{i2\pi f\Delta}\right] - \Sigma_Z(0) \\ &= \left(\sum_{\Delta=0}^{\infty} \Lambda^{\Delta}e^{-i2\pi f\Delta}\right)\mathcal{P}_A^{*T}\Sigma(0)\mathcal{P}_A + \\ &\quad \mathcal{P}_A^{*T}\Sigma(0)\mathcal{P}_A\left(\sum_{\Delta=0}^{\infty} \Lambda^{*\Delta}e^{i2\pi f\Delta}\right) - \mathcal{P}_A^{*T}\Sigma(0)\mathcal{P}_A \end{aligned} \tag{15.16}$$

In (15.16), the expressions within the big bracket are summations over diagonal matrices, which are the only terms depending on frequency f. It is now possible to carry out the summation for each diagonal element and to obtain the full expression of the cross-spectrum matrix as a function of f. Without the eigenvector decomposition (15.14), it is not possible to simplify the expression of cross-spectrum matrix in (15.9). In this respect, (15.1) is powerful in approximating the second moments of a multivariate time series. The diagonal components of $\Gamma_Z(f)$ are the autospectra of the components of $\vec{\mathbf{Z}}$. The POP-basis is represented by the directions in the "signal" subspace, and the spectral characteristics of each direction are given by the diagonal components of $\Gamma_Z(f)$.

Since $\mathcal{P}_A^{*T}\Sigma(0)\mathcal{P}_A$ is in general not a diagonal matrix, the off-diagonal components of $\Gamma_Z(f)$ which are the co- and quadrature spectrum between two different POP coefficients are normally not zero. This fact implies that spectral relationships might exist between any two POP modes.

The transformation from (15.9) to (15.16) holds also if $\vec{\mathbf{X}}$ is transformed into another basis of the "signal" subspace. The time series in the new basis is then $\vec{\mathbf{Y}} = \mathcal{L}\vec{\mathbf{X}}$ with $\mathcal{L}$ being any invertible matrix. The eigenvalues of system matrix $\mathcal{A}_Y$ are the same of those of $\mathcal{A}$, but the eigenvectors are transformed into $\mathcal{P}_Y = \mathcal{L}\mathcal{P}$. It can be shown that the diagonal elements of the cross-spectrum matrix of the POP coefficients of $\vec{\mathbf{Y}}$ takes exactly the same form as those in (15.16). A linear transformation does not change the main temporal characteristics of the considered time series.

15.3.3 Cross-Spectrum Matrix in POP-Basis: Its Diagonal Components

The jth diagonal component of (15.16) is:

$$\Gamma_j(f) = \left[\vec{p}_A^{\,j\,*T}\Sigma(0)\vec{p}_A^{\,j}\right]\left(\sum_{\Delta=0}^{\infty}[\lambda_j^{\Delta}e^{-i2\pi f\Delta} + \lambda_j^{*\Delta}e^{i2\pi f\Delta}] - 1\right) \tag{15.17}$$

It can be shown:

$$\vec{p}_A^{\,j\,*T}\Sigma(0)\vec{p}_A^{\,j} = \mathrm{VAR}(\mathbf{Z}_j) = \frac{\mathrm{VAR}(\mathbf{R}_j)}{1-\lambda_j\lambda_j^*} \tag{15.18}$$

The summation in (15.17) yields for a real eigenvalue

$$\Gamma_j(f) = \frac{\mathrm{VAR}(\mathbf{R}_j)}{1-2\lambda_j\cos 2\pi f + \lambda_j^2} \tag{15.19}$$

and for a complex eigenvalue

$$\Gamma_j(f) = \frac{\mathrm{VAR}(\mathbf{R}_j)}{(e^{i2\pi f}-\lambda_j)(e^{i2\pi f}-\lambda_j)^*} \tag{15.20}$$

where $|\lambda| < 1$ for a stationary time series is used to carry out the summation in (15.17).

The autospectra of the components of $\vec{\mathbf{Z}}$ are determined by the corresponding eigenvalues. If λ_j is real, $\Gamma_j(f)$ is the spectrum of a univariate AR(1)-process and therefore has no spectral peak. However, if $\lambda_j = |\lambda_j|e^{i\phi_j}$ is complex, $\Gamma_j(f)$ describes a spectrum with a spectral peak at $f_j = \phi_j/(2\pi)$. The width of the peak is determined by $|\lambda|$. For $|\lambda| = 1$, a resonance peak appears with zero width. The smaller the $|\lambda|$, the broader the peak. Thus, $\Gamma_j(f)$ of a complex $\mathbf{Z}_j$ can be considered as the spectrum of a univariate AR(2)-process.

In general, the autospectrum of a univariate AR(m)-process may be written as the sum of spectra of AR(1)- and AR(2)-processes. A univariate AR(m)-process can be reformulated into an m-dimensional AR(1)-process. Thus, we can expect that a multivariate AR(1) process is able to describe many spectral features and the POPs separate multivariate cross-spectral features such that each POP-direction is described by an AR(1) or AR(2)-process.

The equations (15.19) and (15.20) can also be derived by transforming (15.10) into the frequency domain as suggested by Hasselmann (1988). The consideration of (15.10) suggests also the coherent and 90°-out-of-phase relationship between the real and imaginary parts of a complex POP coefficient at the frequency $f_j = \phi_j/2\pi$. The corresponding POP direction defines two patterns in the "signal" subspace, one represented by the real part and the other by the imaginary part of the POP. Both oscillate in a coherent way with the real part following the imaginary part within one quarter of the oscillation period $T_j = \frac{1}{f_j}$. The real and imaginary patterns describe together spatially propagating structures (more detailed description see H. von Storch et al., 1995). Since they are normally not orthogonal to each other but may have any spatial phase relationship, the POPs are not restricted to describing sinusoidal propagating waves.

This points to a limitation of a POP model in describing a pure standing oscillation. One extreme of the spatial phase relationship between two patterns is the zero-phase relationship. In this limit, the two patterns coincide and we have a pure standing oscillation with well defined time period and fixed spatial structure. The amplitude of the spatial structure can become larger or smaller following an oscillation, but the form of the spatial structure is always the same. Such a standing oscillation can only be identified by one pattern, in a POP model by a *real* POP. Since the spectrum of a real POP is one which has no spectral peak within the frequency interior (15.19), the oscillation frequency cannot be detected by a POP model. The solution of this problem is suggested by Bürger (1993).

15.3.4 Eigenstructure of Cross-Spectrum Matrix at a Frequency Interval: Complex EOFs

The cross spectrum matrix of a multivariate time series can also be written as the square of the Fourier transform of the time series:

$$\Gamma_X(f) = \vec{\mathbf{F}}(f)\vec{\mathbf{F}}^{*T}(f) \tag{15.21}$$

where the m-dimensional complex vector $\vec{\mathbf{F}}(f)$ is the Fourier transform of $\vec{\mathbf{X}}$ at frequency f. One has

$$\vec{\mathbf{X}}_t = \int_{-\frac{1}{2}}^{+\frac{1}{2}} \vec{\mathbf{F}}(f)e^{i2\pi ft}df \tag{15.22}$$

Defining the random vector $\vec{\mathbf{Y}}_t$ by:

$$\vec{\mathbf{Y}}_t = \frac{1}{2}\left[\vec{\mathbf{X}}_t + i\vec{\mathbf{X}}_t^h\right] \tag{15.23}$$

where $\vec{\mathbf{X}}_t^h = \int_0^{\frac{1}{2}}(-i)\vec{\mathbf{F}}(f)e^{i2\pi ft}df + \int_{-\frac{1}{2}}^0 i\vec{\mathbf{F}}(f)e^{i2\pi ft}df$ is the Hilbert transform of $\vec{\mathbf{X}}_t$, it can be shown

$$\vec{\mathbf{Y}}_t = \int_0^{\frac{1}{2}} \vec{\mathbf{F}}(f)e^{i2\pi ft}df \tag{15.24}$$

Thus, (15.21) for positive frequencies is also the cross-spectrum matrix of the complex random vector $\vec{\mathbf{Y}}_t$.

Equation (15.21) is formally equivalent to (15.4) and satisfies also (15.5). As discussed in Section 15.2, it is difficult to investigate the cross-spectrum matrix as a function of f in this general formulation. In order to make progress, Wallace and Dickinson (1972) pointed out the possibility of studying the cross-spectrum matrix integrated over a frequency interval centered at f_o. We denote such a matrix by $\overline{\mathbf{\Gamma}}^{f_0}$, with $^{-f_0}$ indicating the integral over the frequency interval centered at f_0. They suggested to study the eigenstructure of $\overline{\mathbf{\Gamma}}^{f_0}$ in analogy to an EOF analysis. The complex eigenvectors of this Hermitian matrix are called Complex EOFs. In analogy to 1-dimensional waves which can be described by a complex function, one can derive the amplitude and spatial phase relationship of an m-dimensional wave from the complex EOFs. Another way to interpret a complex EOF is to use the notion that the cross-spectrum matrix of $\vec{\mathbf{X}}$ at a positive frequency is also the cross-spectrum matrix of $\vec{\mathbf{Y}}$. The eigenvectors of the latter can be interpreted according to the definition of $\vec{\mathbf{Y}}$ (15.23) which shows that the real and imaginary part of $\vec{\mathbf{Y}}$ are coherent with a fixed temporal phase shift of 90° at each frequency. This suggests that if the real part of the complex EOF represents spatial structure at frequency f, the imaginary part represents then the spatial structure at the same frequency but with a time lag of one quarter of the oscillation period $T = 1/f$. The Wallace and Dickinson approach is denoted hereafter as *complex EOF analysis in the frequency domain.*

The filtered time series of $\vec{\mathbf{Y}}$ in which all spectral components outside the frequency interval $[f_o - \epsilon, f_o + \epsilon]$ are removed is denoted by $\vec{\mathbf{Y}}'$. The cross-spectrum matrix of $\vec{\mathbf{Y}}$ averaged over the frequency interval $[f_o - \epsilon, f_o + \epsilon]$ is also the cross-spectrum matrix of $\vec{\mathbf{Y}}'$. Furthermore, since the cross-spectrum matrix $\mathbf{\Gamma}_{Y'}$ is the Fourier transform of cross-covariance matrix $\mathbf{\Sigma}_{Y'}$ at frequency f, one has

$$\mathbf{\Sigma}_{Y'}(\Delta) = \int_{-\frac{1}{2}}^{\frac{1}{2}} \mathbf{\Gamma}_{Y'}(f)e^{i2\pi f\Delta}df = \int_{f_0-\epsilon}^{f_0+\epsilon} \mathbf{\Gamma}_{Y'}(f)e^{i2\pi f\Delta}df \tag{15.25}$$

For $\Delta = 0$, one has

$$\overline{\mathbf{\Gamma}}_{Y'}^{f_0} = \int_{f_o-\epsilon}^{f_o+\epsilon} \mathbf{\Gamma}_{Y'}(f)df = \mathbf{\Sigma}_{Y'}(0) \tag{15.26}$$

Equation (15.26) suggests that for a filtered time series which obtains only spectral component at frequency interval, a EOF analysis of the cross-spectrum matrix integrated over $[f_o-\epsilon, f_o+\epsilon]$ and a EOF analysis of the lag-0 cross-covariance matrix of the filtered time series provide exactly the same results. The latter, which is named *Complex EOF analysis in time domain*, was suggested by Barnett (1983, 1985). The difference of the two complex EOF analyses lies in the consideration of the formally different matrices, which contain the same information about the second moment statistics at a frequency band. In Wallace and Dickinson's approach, the matrix is obtained in the frequency domain, whereas in Barnett's approach the matrix is obtained in the time domain.

In practice, it is a delicate task to choose the width 2ϵ of the frequency interval. On the one hand, one wishes to get a stable estimation of the cross-spectrum matrix. Therefore, the frequency interval should be not too narrow. The resulting matrix describes the second moment information averaged over the interval. On the other hand, one wishes to focus on a specific frequency so that the frequency interval should be as small as possible. In the limit of $\epsilon \to 0$, the contributions of the time series at frequency f_o are obtained. In the limit of $\epsilon \to \infty$, (15.26) reduces to cross-covariance matrix of the unfiltered time series. In this case, there is no separation between contributions from different frequencies.

Several "nice" mathematical features can be derived for complex EOFs. Since the matrices in (15.26) are Hermitian, the complex EOFs form an orthonormal basis and the eigenvalues are real. With $\mathcal{P}$ being the eigenvector matrix of $\overline{\mathbf{\Gamma}}^{f_0}$ or $\mathbf{\Sigma}_{Y'}(0)$, it holds $\mathcal{P}^{*T}\mathcal{P} = \mathcal{I}$. The orthogonality implies that differently to a complex POP, a complex EOF describes only sinusoidal spatial structures. In analogy to the conventional EOFs (Section 13.3.1), the complex EOFs maximize the variance of the filtered time series $\vec{\mathbf{Y}}'$ with the first eigenvector explaining the maximum amount of total variance, the second one explaining the maximum amount of the residual variance and so forth. If the original time series $\vec{\mathbf{Y}}'$ is transformed into an eigenvector basis via $\vec{\mathbf{Y}}' = \mathcal{P}\vec{\mathbf{Z}}$ with $\vec{\mathbf{Z}}' = \mathcal{P}^{*T}\vec{\mathbf{Y}}'$ being the complex EOF coefficients, it can be shown that the cross-spectrum matrix of $\vec{\mathbf{Z}}'$ satisfies:

$$\overline{\mathbf{\Gamma}}^{f_0}_{Z'} = \mathbf{\Sigma}_{Z'}(0) = \mathcal{P}^{*T}\mathbf{\Sigma}_{Y'}(0)\mathcal{P} = \mathbf{\Lambda} \tag{15.27}$$

where $\mathbf{\Sigma}_{Y'}(0) = \mathcal{P}\mathbf{\Lambda}\mathcal{P}^{*T}$ is used with $\mathbf{\Lambda}$ being the corresponding eigenvalue matrix.

The cross-spectrum matrix of complex EOF coefficient in a frequency interval around f_o is a real diagonal matrix. The j-th diagonal component indicate variances contributed by the j-th mode in this frequency interval. The zero off-diagonal components indicate that cross-spectrum between any two different complex EOF coefficients is zero in this frequency interval.

Both the frequency and the time domain approach can be used to identify dominant modes. However, the modes describe only spectral features in the chosen frequency interval. Furthermore, they must satisfy all features required for a Hermitian matrix. In practice, it is a-priori not clear why two modes should explain different amounts of total variance, and be orthogonal in space and independent in time with zero co- and quadrature spectrum between their coefficients. Different from complex EOF analysis, the POP analysis identifies modes, which vary on the entire frequency domain. Since the system matrix $\mathcal{A}$ is not Hermitian, the above mentioned constraints for the complex EOF modes do not exist for the POP modes. It is in principle possible to study the relationship between two modes using POP analysis. An example is given in the following section.

15.3.5 Example

In the tropics, there are two phenomena with similar time scales. One is the *El Niño Southern Oscillation* (ENSO) which is the dominant interannual signal in the tropical atmosphere and ocean. The other is the *Quasi-Biennial Oscillation* (QBO) in the lower stratosphere. Theoretically, it is shown (an overview is given by Holton, 1983) that the QBO is driven by the interaction between the stratospheric mean flow and vertically propagating equatorial waves generated in the troposphere. It is commonly assumed that equatorial convective disturbances which display large changes during an ENSO event might excite these waves in the upper troposphere. The fact that the wave sources probably have substantial interannual variability related to the ENSO suggests a possible link between the QBO and the SO. Such a link would be of importance for the ENSO forecast.

In order to investigate a possible link, (15.1) was fitted to time series which contained both surface data [gridded equatorial zonal wind at surface and sea surface temperature (SST)] and stratospheric data (zonal wind at 6 levels between 70 hPa to 15 hPa) (Xu, 1992). Surface wind, SST and stratospheric wind were normalized to have the same variance in the combined time series. The first 10 EOFs of the combined times series were used to form the "signal" subspace. The system matrix $\mathcal{A}$ and its eigenvectors (POPs) were estimated (see Section 15.4) from this combined time series.

Figure 15.1 shows real (thin lines) and imaginary parts (thick line) of two POPs of the estimated $\mathcal{A}$. POP1 (Figure 15.1a) has large amplitude in stratospheric winds whereas POP2 (Figure 15.1b) has large amplitude in surface winds and SST. As discussed in Sections 15.3.2 and 15.3.3, a complex POP picks up two patterns in the "signal" subspace which oscillate coherently around frequency $f = \frac{\phi}{2\pi}$ given by the corresponding eigenvalue $\lambda = |\lambda| e^{\phi}$. The estimated frequency for POP1 is $f_1 \approx 1/(28\ months)$ and for POP2 $f_2 \approx 1/(45\ months)$. Since the real and imaginary part of each complex POP are 90° out of phase around f_j, the stratospheric wind patterns in Figure 15.1a describe propagation of easterly anomalies from about 20 hPa (imaginary part) to about 40 hPa (real part) $T_1/4 = 1/(4f_1) = 7\ months$ later. A propagation is also seen in Figure 15.1b for the surface winds with

Figure 15.1: *Two complex POPs (in relative units) derived from a combined dataset including anomalies of the surface wind and SST and the stratospheric zonal wind. The imaginary (real) part of a POP is indicated by the thick (thin) line. (From Xu, 1993).*

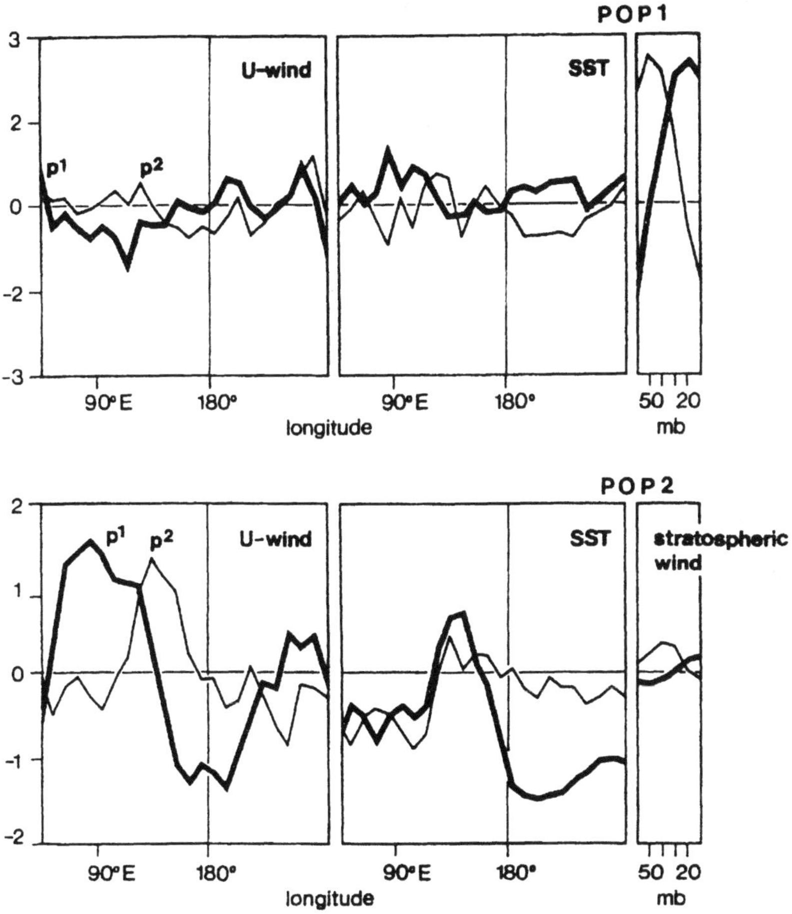

Figure 15.2: a) *Autospectra of two complex POP coefficients whose corresponding patterns are shown in Figure 15.1.*
b) *Squared coherence spectrum between these two POP coefficients.*
(For all spectra, the POP rotation, that is from imaginary part to real part, is defined by a negative frequency.)

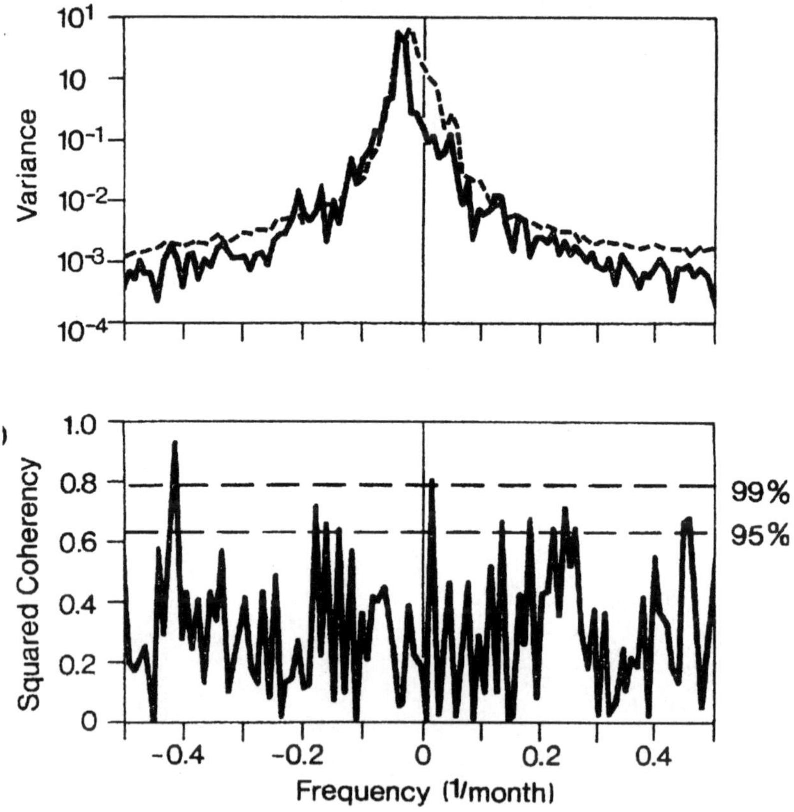

an eastward propagation of westerly anomalies from Indian Ocean to Indonesian regions within about $T_2/4 = 1/(4f_2) = 11$ *months*. The extreme values in SST of POP2 are in nearly the same locations in the real and imaginary part of POP2 (Figure 15.1b). This behavior suggests that the SST signal is characterized by a strengthening of standing anomalies, especially over the central and eastern Pacific, rather than by a propagation of the anomalies.

The autospectrum and cross-spectrum between the corresponding complex POP coefficients are shown in Figure 15.2. In these spectra, a clockwise rotation (from real to imaginary part) is defined as positive. The POP frequencies f_1 and f_2 are therefore negative. As indicated by (15.20), the autospectrum of a complex POP coefficient is not symmetric about zero frequency, that is $\Gamma(+f) \neq \Gamma(-f)$ and a spectral peak is found around f_1 for POP1 (solid line in Figure 15.2a) and around f_2 for POP2 (dashed line in Figure 15.2a). Around the frequency range on which POP1 and POP2 oscillate, two modes are not related (Figure 15.2b).

This example demonstrates that the full spectral and spatial features of ENSO and QBO can be described by two POPs and their corresponding eigenvalues. The spectral relationship between the two modes is described by the cross-spectrum between the POP coefficients. Thus, we have considered here not only the relationship between one stage of the ENSO mode and one stage of the QBO mode, which is normally done by applying a cross-spectrum analysis to, say, the Southern Oscillation Index and the stratospheric wind at one level, but the relationship between two evolutions. One is characterized by an eastward propagation of surface wind anomalies and strengthening of SST in the central and eastern Pacific and the other by a downward propagation of stratospheric wind anomalies.

Since both ENSO and QBO modes display only one frequency peak, their spatial evolutions can certainly be also derived from a complex EOF analysis of filtered data. However, if the above combined time series is considered, the obtained modes have to satisfy the mathematical constraints discussed in Section 15.3.4. It is impossible to study the relationship between them, because by construction there is none.

15.4 Estimation of POPs and Interpretation Problems

In practical situations, when only a finite time series $\vec{\mathbf{x}}_t$ is available, $\mathcal{A}$ is estimated by first deriving the sample lag-1 covariance matrix $\hat{\boldsymbol{\Sigma}}(1) = \sum_t \vec{\mathbf{x}}(t+1)\vec{\mathbf{x}}_t^T$ and the sample covariance matrix $\hat{\boldsymbol{\Sigma}}(0) = \sum_t \vec{\mathbf{x}}_t\vec{\mathbf{x}}_t^T$ and then forming $\hat{\mathcal{A}} = \hat{\boldsymbol{\Sigma}}(1)\hat{\boldsymbol{\Sigma}}(0)^{-1}$. The eigenvectors of $\hat{\mathcal{A}}$ are the estimated POPs. The eigenvalues of this matrix satisfy $|\lambda| < 1$ for a stationary time series.

It can be shown that $\hat{\mathcal{A}} = \hat{\boldsymbol{\Sigma}}(1)\hat{\boldsymbol{\Sigma}}(0)^{-1}$ is the solution of the minimization problem:

$$\sum_{j=1}^{n} \left(\vec{\mathbf{x}}_{j,t} - \hat{\mathcal{A}} \vec{\mathbf{x}}_{j,t-1} \right)^2 = minimum \tag{15.28}$$

An estimation of multivariate spectral features of $\vec{\mathbf{x}}_t$ is given by the eigenvalues of $\hat{\mathcal{A}}$.

Let us assume that one (dynamical) process generates one POP and there are n (dynamical) processes which generate an m-dimensional time series. From the $m \times m$ system matrix, m POPs can be estimated. In the case $m < n$, a POP analysis of the m-dimensional time series can identify only m of n modes. One has to extend the spatial dimension in order to get the other n-m modes. However, in the case $m > n$, $m - n$ of the m POPs cannot be interpreted and are meaningless. Since n is unknown, there is no strict way to determine which m of n POPs are meaningless.

Although there is no complete solution to this problem, a rule-of-thumb is often used in practice. Since all processes are approximated by (15.1), the spectrum of the estimated POP coefficient must have the form of (15.19) and (15.20), which is a function of the estimated eigenvalue. In the case of a complex POP, the real and imaginary part of the POP coefficient should additionally be significantly coherent and 90^0-out-of phase on time scales near the estimated POP period. We therefore select only those modes whose coefficients satisfy these conditions.

15.5 POPs as Normal Modes

Equation (15.1) represents also the discrete form of a first order differential equation with $\mathcal{A}$ being the discrete version of a linear operator. The minimization in (15.28) suggests that the estimated system matrix $\hat{\mathcal{A}}$ describes optimally the approximated linear dynamics of the considered system in a least squares sense. Thus, the POPs are also referred to as the estimated normal modes of the system.

The ability of a POP analysis in identifying normal modes of a system is demonstrated by the example of medium-scale synoptic waves in the atmosphere (Schnur et al., 1993). In this study, the linear system matrix $\mathcal{A}$ has been obtained in two conceptually different ways. In one case, it was estimated from observational data via (15.28) and in the other it was derived theoretically from the quasigeostrophic equations, assuming small disturbances (linear stability analysis). Since the theoretical normal modes are known in this example, a comparison between the POPs provides insights into the interpretation of the POPs.

In the study of Schnur et al. (1993), the state vector $\vec{\mathbf{X}}$ was the same in both empirical and theoretical approaches and consists of Fourier expansion coefficients along each latitude and at each pressure level. This construction of $\vec{x}$ allows one to study the meridional-height structure of the waves. No zonal dependence of the waves was considered. The theoretical $\mathcal{A}$ was a function of a zonally symmetric basic state.

Schnur et al.'s theoretical $\mathcal{A}$ describes the linear mechanism of baroclinic instability which generates the synoptic waves, whereas their empirical $\hat{\mathcal{A}}$ contains in general linear approximations of all possible processes. Although they limited them by using filtered data, the empirical $\hat{\mathcal{A}}$ contains still more information than the theoretical $\mathcal{A}$. One candidate which operates on the same time scales is the one which causes nonlinear decaying of the waves. Thus, the empirical $\mathcal{A}$ may contain information about processes which generate and damp the waves.

Indeed, Schnur et al. found two categories of POPs. One is in good correspondence with the most unstable normal modes of the theoretical $\mathcal{A}$ in terms of both the frequency and wavelength of the waves and the meridional-height structure of the waves. The only inconsistency between this category of POPs and the baroclinic unstable waves lies in the damping and growth rates. For a stationary time series, a POP analysis always gets damped POP with corresponding eigenvalues $|\lambda| < 1$ whereas the most unstable modes have always a growth rate with $|\lambda| > 1$. This inconsistency is a consequence of the artificial set up of the theoretical approach in which only linear growth phase of the waves is considered. In reality, after waves grow out of noisy background, other processes take place. The growing modes appear in the data as stationary waves.

The second category of POPs describes the decaying phase of the baroclinically unstable modes. These modes were associated with patterns of Eliassen-Palm flux vectors and their divergence which characterize the decaying phase.

Schnur et al. demonstrated that at least in the case of baroclinic unstable waves, the empirical $\mathcal{A}$ is capable of approximating processes which are responsible for both the growing and decaying phase of the waves. Although the decaying phase itself is nonlinear, Simmons and Hoskins (1978) have shown that this nonlinear process is only important at the point where the linear growth of a wave comes to an end and barotropic decay begins. The evolution following this transition point displays linear character. That might be the reason why the linear POP analysis can describe also this phase.

Schnur et al.'s example shows the ability of empirical $\mathcal{A}$ in approximating the linear behavior of processes which generate the climate data. This implication can be extended to the situation where it is difficult to get the theoretical $\mathcal{A}$. An attempt was done by Xu (1993) in which a combined atmospheric and oceanic data set was used to approximate the processes in the coupled system.

References

The numbers given in parantheses at the end of the items in this bibliography refer to the sections in which the publication is quoted.

Alexandersson, H., 1986: A homogeneity test applied to precipitation data. J. Climatol. 6, 661-675 (4.2)

Anderson, T.W., 1971: The Statistical Analysis of Time Series. Wiley & Sons, 704 pp. (8.2)

Baillie, M.G.L., 1982: Tree-Ring Dating and Archaeology. Croom Helm, London, 274pp. (5.3)

Baker, D.G. 1975: Effect of observation time on mean temperature estimation. J. Applied Meteorology 14, 471-476 (4.2)

Bárdossy, A., and H. J. Caspary, 1990: Detection of climate change in Europe by analyzing European circulation patterns from 1881 to 1989. Theor. Appl. Climatol. 42, 155-167 (11.4)

Bárdossy, A., and E. J. Plate, 1991: Modeling daily rainfall using a semi-Markov representation of circulation pattern occurrence. J. Hydrol. 122, 33-47 (11.4)

Bárdossy, A., and E. J. Plate, 1992: Space-time model for daily rainfall using atmospheric circulation patterns. Water Resour. Res. 28, 1247-1259 (11.4)

Barnett, T.P., 1978: Estimating variability of surface air temperature in the Northern Hemisphere. Mon. Wea. Rev. 112, 303-312 (4.3)

Barnett, T.P., 1983: Interaction of the Monsoon and Pacific trade wind system at interannual time scale. Part I. Mon. Wea. Rev. 111, 756-773 (13.2)

Barnett, T.P., 1985: Variations in near-global sea level pressure. J. Atmos. Sci., 42, 478-501

Barnett, T.P., 1986: Detection of changes in the global troposphere field induced by greenhouse gases. J. Geophys. Res. 91, 6659-6667 (4.4)

Barnett, T.P., 1991: An attempt to detect the greenhouse-gas signal in a transient GCM simulation. In (M.E. Schlesinger, Ed.) Greenhouse-gas-induced Climatic Change: A critical appraisal of simulations and observations, Elsevier, 559-568 (4.4)

Barnett, T.P. and K. Hasselmann, 1979: Techniques of linear prediction with application to oceanic and atmospheric fields in the tropical Pacific. Revs. of Geophys. and Space Phys. 17, 949-955 (5.6)

Barnett, T.P. and R. Preisendorfer, 1987: Origins and levels of monthly and seasonal forecast skill for United States surface air temperature determined by canonical correlation analysis. Mon. Wea. Rev. 115, 1825-1850 (13.4)

Barnett, T.P. and M.E. Schlesinger, 1987: Detecting changes in global climate induced by greenhouse gases. J. Geophys. Res. 92, 14772-14780 (4.4)

Barnett, T.P., R.W. Preisendorfer, L.M. Goldstein and K. Hasselmann, 1981: Significance tests for regression model hierarchies. J. Phys. Oceanogr. 11, 1150-1154 (8.3)

Barnston, A.G., 1992: Correspondence among the correlation, RMSE, and Heidke forecast verification measures; refinement of the Heidke score. Wea. Forecasting, 7, 699-709 (10.3)

Barnston, A. G., 1994: Linear statistical short-term climate predictive skill in the Northern Hemisphere. J. Climate, 7, 699-709) (14.4)

Barnston, A.G. and R.E. Livezey, 1987: Classification, seasonality and persistence of low-frequency atmospheric circulation patterns. Mon. Wea. Rev. 115, 1825-1850 (12.1, 13.2)

Barnston, A. G. and H. M. van den Dool, 1993: A degeneracy in cross-validated skill in regression-based forecasts. J. Climate 6, 963-977 (10.5)

Barry, R.G. and Perry, A.H. 1973: Synoptic Climatology. Methuen, London. (7.2)

Barsugli, J.J. and D.S. Battisti, 1998: The basic effects of atmosphere-ocean thermal coupling on midlatitude variability. J. Atmos. Sci., 55, 477-493 (3.4)

Baur, F., P. Hess, and H. Nagel, 1944: Kalender der Grosswetterlagen Europas 1881-1939. Bad Homburg, 35 pp. (11.4)

Bjerknes, J., 1964: Atlantic air-sea interaction. Adv. in Geophysics 7, 1 - 82 (Academic Press, New York, London) (6.4)

Blackburn, T., 1983: A practical method of correcting monthly average temperature biases resulting from differing times of observations. J. Climate and Appl. Meteor. 22, 328-330 (4.2)

Blanke, B., J.D. Neelin, and D. Gutzler, 1997: Estimating the effect of stochastic wind stress forcing on ENSO irregularity. J. Climate, 10, 473-1486 (3.4)

Blasing, T.J., 1978: Time series and multivariate analysis in paleoclimatology. In. Time Series and Ecological Processes (H.H. Shuggart, Ed.), SIAM-SIMS Conference Series 5, Society for Industrial and Applied Mathematics, Philadelphia, 211-226 (5.6)

Blasing, T.J., A.M. Solomon and D.N. Duvick, 1984: Response functions revisited. Tree-Ring Bull. 44, 1-15 (5.6)

Bottomley, M., C.K. Folland, J. Hsiung, R.E. Newell and D.E. Parker, 1990: Global Ocean Surface Temperature Atlas (GOSTA). Joint Meteorological Office/Massachusetts Institute of Technology Project. Project supported by US Dept. of Energy, US National Science Foundation and US Office of Naval Research. Publication funded by UK Depts. of Energy and Environment. pp20 and 313 Plates. HMSO, London. (4.2, 4.3, 6.3)

Braconnot, P. and C. Frankignoul, 1993: Testing model simulations of the thermocline depth variability in the tropical Atlantic from 1982 through 1984. J. Phys. Oceanogr. 23, 1363-1388 (8.3)

Bradley, R.S. (Ed.), 1900: Global Changes of the Past. UCAR/OIES, Boulder, Colorado, 514pp. (5.1)

Bradley, R.S. and P.D. Jones (Eds.), 1992: Climate Since A.D. 1500, Routledge, London, 679pp. (5.7)

Bradley, R.S., P.M. Kelly, P.D. Jones, C.M. Goodess and H.F. Diaz, 1985: A climatic data bank for northern hemisphere land areas, 1851-1980. U.S. Dept. of Energy, Carbon Dioxide Research Division, Technical Report TRO17, 335pp. (4.2)

Bradley, R.S., H.F. Diaz, J.K. Eischeid, P.D. Jones, P.M. Kelly and C.M. Goodess, 1987: Precipitation fluctuations over Northern Hemisphere land areas since the mid-19th century. Science 237, 171-175 (4.2)

Bräker, O.U., 1981: Der Alterstrend bei Jahrringdichten und Jahrringbreiten von Nadelhölzern und sein Ausgleich. Mitt. Forstl. Bundes-Vers.-Anst. 142, 75-102 (5.2)

Breiman, L., J.H. Friedman, R.A. Olsen, and J.C. Stone, 1984: Classification and regression trees. Wadsworth, Monterey (11.4)

Bretherton, C.S. C. Smith and J.M. Wallace, 1992: An intercomparison of methods for finding coupled patterns in climate data. J. Climate 5, 541-560 (12.2, 12.3)

Briffa, K.R., 1984: Tree-climate relationships and dendroclimatological reconstruction in the British Isles. Unpublished Ph.D. Dissertation, University of East Anglia, Norwich, U.K. (5.4)

Briffa, K.R. and E.R. Cook, 1990: Methods of response function analysis. In E.R. Cook and L.A. Kairiukstis (Eds.) Methods of Dendrochronology. Applications in the Environmental Sciences. Kluwer / IIASA, Dordrecht, 240-247 (5.6)

Briffa, K.R. and P.D. Jones, 1990: Basic chronology statistics and assessment. In E.R. Cook and L.A. Kairiukstis (Eds.) Methods of Dendrochronology. Applications in the Environmental Sciences. Kluwer / IIASA, Dordrecht, 137-152 (5.4)

Briffa, K.R. and P.D. Jones, 1993: Global surface air temperature variation over the twentieth century: Part 2, Implications for large-scale palaeoclimatic studies of the Holocene. The Holocene 3, 165-179 (4.3, 5.7)

Briffa, K.R., P.D. Jones, T.M.L. Wigley, J.R. Pilcher and M.G.L. Baillie, 1983: Climate reconstruction from tree rings: Part 1, Basic methodology and preliminary results for England. J. Climatol. 3, 233-242 (5.6)

Briffa, K.R., P.D. Jones, T.M.L. Wigley, J.R. Pilcher and M.G.L. Baillie, 1986: Climate reconstruction from tree rings: Part 2, Spatial reconstruction of summer mean sea-level pressure patterns over Great Britain. J. Climatol. 6, 1-15 (5.6)

Briffa, K.R., T.M.L. Wigley and P.D. Jones, 1987: Towards an objective approach to standardization. In L. Kairiukstis, Z. Bednarz and E. Feliksik (Eds.) Methods of Dendrochronology. IIASA/Polish Academy of Sciences-Systems Research Institute, Warsaw, 239-253 (5.5)

Briffa, K.R., Jones, P.D. and Kelly, P.M. 1990: Principal component analysis of the Lamb catalogue of daily weather types: Part 2, seasonal frequencies and update to 1987. International J. of Climatology 10, 549-564 (7.1)

Briffa, K.R., P.D. Jones, T.S. Bartholin, D. Eckstein, F.H. Schweingruber, W. Karlén, P. Zetterberg and M. Eronen, 1992a: Fennoscandian summers from A.D.500: Temperature changes on short and long timescales. Climate Dynamics 7, 111-119 (5.5, 5.7)

Briffa, K.R., P.D. Jones and F.H. Schweingruber, 1992b: Tree-ring density reconstructions of summer temperature patterns across western North America since 1600. J. Climate 5, 735-754 (5.6)

Brink, K.H., 1989: Evidence for wind-driven current fluctuations in the western North Atlantic. J. Geophys. Res. 94, 2029-2044 (3.6)

Broomhead, D.S. and G. P. King, 1986a: Extracting qualitative dynamics from experimental data. Physica D 20, 217-236 (14.1)

Bryan, K., amd M. Cox, 1969: A nonlinear model of an ocean driven by wind and differential heating, Parts I and II. J. Atmos. Sci. 25, 945-978 (12.2)

Bürger, G., 1993: Complex Principal Oscillation Patterns. J. Climate, 6, 1972-1986 (15.1)

Cardone, V.J., Greenwood, J.G. and M.A. Cane, 1990: On trends in historical marine data. J. Climate 3, 113-127 (6.4)

Chave, A.D., D.S. Luther and J.H. Filloux, 1991: Variability of the wind stress curl over the North Pacific: implication for the oceanic response. J. Geophys. Res. 96, 18,361-18,379 (3.1, 3.2)

Chelliah, M. and P. Arkin, 1992: Large-scale interannual variability of monthly ourtgoing longwave radiation anomalies over global tropics. J. Climate 5, 371-389 (13.2)

Chen, J.M. and P.A. Harr, 1993: Interpretation of Extended Empirical Orthogonal Function (EEOF) analysis. Mon. Wea. Rev. 121, 2631-2636 (13.3)

Cheng, X. and J.M. Wallace, 1993: Cluster analysis of the Northern hemisphere wintertime 500-hPa Height field: Spatial patterns. J. Atmos. Sci., 50, 2674-2696 (14.4)

Cheng, X., G. Nitsche and J.M. Wallace, 1995: Robustness of low-frequency circulation patterns derived from EOF and rotated EOF analysis. J. Climate 8, 1709-1713 (13.2)

Chenoweth, M., 1992: A possible discontinuity in the U.S. Historical Temperature Record. J. Climate 5, 1172-1179 (4.2)

Cherry, S., 1996: Singular value decomposition and canonical correlation analysis, J. Climate, 9, 2003-2009 (12.3)

Cherry, S., 1997: Some comments on singular value decomposition analysis, J. Climate,10,1759-1761 (12.3)

Chervin, R.M. and S.H. Schneider, 1976: On determining the statistical significance of climate experiments with general circulation models. J. Atmos. Sci., 33, 405-412 (8.2)

Conrad, V. and L.D. Pollak, 1962: Methods in Climatology. Harvard University Press, Cambridge, Massachusetts (4.2)

Cook, E.R., 1990: Bootstrap confidence intervals for red spruce ring-width chronologies and an assessment of age-related bias in recent growth trends. Canadian J. Forest Res. 20, 1326-1331 (5.4)

Cook, E.R. and K.R. Briffa, 1990: A comparison of some tree-ring standardization methods. In E.R. Cook and L.A. Kairiukstis (Eds.) Methods of Dendrochronology. Applications in the Environmental Sciences. Kluwer/IIASA, Dordrecht, 153-162 (5.5)

Cook, E.R. and L.A. Kairiukstis (Eds.), 1990: Methods of Dendrochronology. Applications in the Environmental Sciences. Kluwer/IIASA, Dordrecht, 394pp. (5.1)

Cook, E.R., K.R. Briffa, S. Shiyatov and V. Mazepa, 1990a: Tree-ring standardization and growth trend estimation. In E.R. Cook and L.A. Kairiukstis (Eds.) Methods of Dendrochronology. Applications in the Environmental Sciences. Kluwer/IIASA, Dordrecht, 104-123 (5.5)

Cook, E.R., S. Shiyatov and V. Mazepa, 1990b: Estimation of the mean chronology. In E.R. Cook and L.A. Kairiukstis (Eds.) Methods of Dendrochronology. Applications in the Environmental Sciences. Kluwer/IIASA, Dordrecht, 123-132 (5.5)

Cook, E.R., T. Bird, M. Peterson, M. Barbetti, B. Buckley, R. D'Arrigo and F. Francey, 1992: Climatic change over the last millenium in Tasmania reconstructed from tree rings. The Holocene 2, 205-217 (5.7)

Cook, E.R., K.R. Briffa and P.D. Jones, 1994: Spatial regression methods in dendroclimatology: a review and comparison of two techniques. Int. J. Climatol. 14, 379-402 (5.6)

Cox, D.R., and P.A.W. Lewis, 1978: The Statistical Analysis of Series of Events. Metheun, London (11.4)

Craddock, J.M., 1977: A homogeneous record of monthly rainfall totals for Norwich for the years 1836-1976. Meteor. Magazine 106, 267-278 (4.2)

Cramer, J.S., 1987: Mean and variance of R^2 in small and moderate samples. J. Econometrics 35, 253-266 (5.6)

Cubasch, U., Hasselmann, K., Höck, H., Maier-Reimer, E., Mikolajewicz, U., Santer, B.D. and Sausen, R. 1992: Time-dependent greenhouse warming computations with a coupled ocean-atmosphere model. Climate Dynamics 8, 55-70 (7.4, 12.2)

Czaja, A. and C. Frankignoul, 1999: Decadal buoyancy forcing in a simple model of the subtropical gyre. J. Phys. Oceanogr., in press (3.6)

Daan, H., 1985: Sensitivity of verication scores to the classification of the predictand. Mon. Wea. Rev. 113, 1384-1392 (10.1)

Dansgaard, W., S.J. Johnsen, N. Reeh, N. Gundestrup, H.B. Clausen, and C.U. Hammer, 1975: Climatic changes, Norsemen and modern man. Nature 255, 24-28 (5.4)
Dansgaard, W., S.J. Johnsen, H.B. Clausen, D. Dahl-Jensen, N.S. Gundestrup, C.U. Hammer, C.S. Hvidberg, J.P. Steffensen, A.E. Sveinbjornsdottir, A.E. Jouzel and G. Bond, 1993: Evidence for general instability of past climate from a 250kyr ice core. Nature 364, 218-219 (5.4)
Delsol, F., K.Miyakoda and R. H. Clarke, 1971: Parameterized processes in the surface boundary layer of an atmopsheric circulation model, Quart. J. Roy. Met. Soc. 97, 181-208 (1.2)
Delworth, T.L. and S. Manabe, 1988: The infuence of potential evaporation on the variabilities of simulated soil wetness and climate. J. Climate 1, 523-547 (3.5)
Deser, C. and M.L. Blackmon, 1993: Surface climate variations over the North Atlantic ocean during winter. J. Climate 6, 1743-1753 (3.4)
Devijver and J. Kittler, 1982: Pattern Recognition: A Statistical Approach. Prentice-Hall, New York, 367pp. (14.3)
Diaz, H.F., R.S. Bradley and J.K. Eischeid, 1989: Precipitation fluctuations over global land areas since the late 1800s. J. Geophys. Res. 94, 1195-1210 (4.2)
Donnell, C.A., 1912: The effect of the time of observation on mean temperatures. Mon. Wea. Rev. 40, 708 (4.2)
Douglas, A.V., D.R. Cayan and J. Namias, 1982: Large-scale changes in North Pacific and North American weather patterns in recent decades. Mon. Wea. Rev. 110, 1851-1862 (4.3)
Draper, N.R. and H. Smith, 1981: Applied Regression Analysis. Wiley, New York. (5.6)
Dunbar, R.B. and J.E. Cole, 1993: Coral Records of Ocean-Atmosphere Variability. NOAA Climate and Global Change Program. Special Report No.10, UCAR, Boulder, 37pp. (5.3)

Easterling D.R., B. Horton, P.D. Jones, T.C. Peterson, T.R. Karl, D.E. Parker, M.J. Salinger, V. Razuvayev, N. Plummer, P.Jamason and C.K. Folland, 1997: A new look at maximum and minimum temperature trends for the globe. Science, 277, 364-367 (4.3)
Ebisuzaki, W., 1997: A method to estimate the statistical significance of a correlation when the data are serially correlated. J. Climate, 10, 2147-2153 (9.1, 9.4)
Eddy, J.A., 1992: The PAGES project: proposed implementation plans for research activities. Global IGBP Change Report No.19, Stockholm: IGBP. (5.2)
Efron, B., 1979: Bootstrap methods: another look at the Jackknife. Annals of Statistics 7, 1-26 (5.4)

Ellis, W., 1890: On the difference produced in the mean temperature derived from daily Maxima and Minima as dependent on the time at which the thermometers are read. Quart. J. Royal Meteor. Soc. 16, 213-218 (4.2)
Elsaesser, H.W., M.C. MacCracken, J.J. Walton and S.L. Grotch, 1986: Global climatic trends as revealed by the recorded data. Rev. Geophys. 24, 745-792 (4.3)
Elsner, J.B. and A.A. Tsonis, 1991: Comparisons of observed Northern Hemisphere surface air temperature records. Geophys. Res. Lett. 18, 1229-1232 (4.3)
Esbensen, S.K., 1984: A comparison of intermonthly and interannual teleconnections in the 700mb geopotential height field during the Northern Hemisphere winter. Mon. Wea. Rev. 112, 2016-2032. (12.1)

Farge, M., E. Goirand and T. Philipovitch, 1993: Wavelets and wavelet packets to analyse, filter and compress two-dimensional turbulent flows. In P. Müller and D. Henderson (eds): Statistical Methods in Physical Oceanography. Proceedings of the 'Aha Huliko'a Hawaiin Winter Workshop, January 12-15, 1993 (13.2)
Flury, B.N., 1989: Common Principal Components and Related Multivariate Models. Wiley and Sons, 258 pp. (8.3)
Folland, C.K., 1991: Sea temperature bucket models used to correct historical SST data in the Meteorological Office. Climate Research Tech. Note No. 14. Available from the National Meteorological Library, Meteorological Office, Bracknell, Berkshire, UK. (6.3)
Folland C.K. and D.E. Parker, 1991: Worldwide surface temperature trends since the mid 19th century. In M.E. Schlesinger (Ed.): Greenhouse-gas-induced Climatic Change: A critical appraisal of simulations and observations. Elsevier Scientific Publishers, 173-194 (4.2)
Folland, C.K. and D.E. Parker, 1995: Correction of instrumental biases in historical sea surface temperature data. Quart. J. Roy. Meteor. Soc. 121, 319-367 (4.2)
Folland, C.K., T.N. Palmer and D.E. Parker, 1986a: Sahel rainfall and worldwide sea temperatures, 1901-85. Nature 320, 602-607 (6.2, 6.5)
Folland, C.K., A. Woodcock and L.D. Varah, 1986b: Experimental monthly long-range forecasts for the United Kingdom. Part III. Skill of the monthly forecasts. Meteor. Mag. 115, 1373, 377-393 (14.4)
Folland, C.K., T.R. Karl and K. Y. Vinnikov, 1990: Observed climate variations and change. In J.T. Houghton, G.J. Jenkins and J.J. Ephraums, (Eds.) Climate Change: The IPCC Scientific Assessment. Cambridge University Press, 195-238 (4.2, 4.3)
Folland, C.K., J.A. Owen, M.N. Ward, and A.W. Colman, 1991: Prediction of seasonal rainfall in the Sahel region of Africa using empirical and dynamical methods. J. Forecasting 10, 21-56 (6.3)
Folland, C.K., T.R. Karl, N. Nicholls, B.S. Nyenzi, D.E. Parker and

K. Y. Vinnikov, 1992: Observed climate variability and change. In J.T. Houghton, B.A. Callander and S.K. Varney (Eds.) Climate Change 1992, The Supplementary Report to the IPCC Scientific Assessment. Cambridge University Press, 135-170 (4.2, 4.3)

Foufoula-Georgiou, E., 1985: Discrete-time point process models for daily rainfall. Water Resources Technical Report No. 93, Univ. of Washington, Seattle (11.2)

Foufoula-Georgiou, E., and D. P. Lettenmaier, 1987: A Markov renewal model for rainfall occurrences. Water Resour. Res. 23, 875-884 (11.3)

Frankignoul, C., 1979: Stochastic forcing models of climate variability. Dyn. Atmos. Oc. 3, 465-479 (3.4)

Frankignoul, C., 1981: Low-frequency temperature fluctuations off Bermuda. J.Geophys. Res. 86, 6522-6528 (3.6)

Frankignoul, C., 1985: Sea surface temperature anomalies, planetary waves, and air-sea feedback in the middle latitudes. Rev. Geophys. 23, 357-390 (3.4)

Frankignoul, C., C. Duchêne and M. Cane, 1989: A statistical approach to testing equatorial ocean models with observed data. J. Phys. Oceanogr. 19, 1191-1208 (8.3)

Frankignoul, C., S. Février, N. Sennéchael, J. Verbeek and P. Braconnot, 1995: An intercomparison between four tropical ocean models. Part 1: Thermocline variability. Tellus 47A, 351-364 (8.3)

Frankignoul, C. and K. Hasselmann, 1977: Stochastic climate models, Part 2. Application to sea-surface temperature anomalies and thermocline variability. Tellus 29, 289-305 (3.4)

Frankignoul, C. and A. Molin, 1988a: Response of the GISS general circulation model to a midlatitude sea surface temperature anomaly in the North Pacific. J. Atmos. Sci. 45, 95-108 (8.3)

Frankignoul, C. and A. Molin, 1988b: Analysis of the GISS GCM to a subtropical sea surface temperature anomaly using a linear model. J. Atmos. Sci. 45, 3834-3845 (8.3)

Frankignoul, C. and P. Müller, 1979a: Quasi-geostrophic response of an infinite beta-plane ocean to stochastic forcing by the atmosphere. J. Phys. Oceanogr. 9, 104-127 (3.2, 3.6)

Frankignoul, C. and P. Müller, 1979b: On the generation of geostrophic eddies by surface buoyancy flux anomalies. J. Phys. Oceanogr. 9, 1207-1213 (3.6)

Frankignoul, C., P. Müller and E. Zorita, 1997: A simple model of the decadal response of the ocean to stochastic wind forcing. J. Phys. Oceanogr., 1997, 1533-1546 (3.6)

Frankignoul, C. and R.W. Reynolds, 1983: Testing a dynamical model for mid-latitude sea surface temperature anomalies. J. Phys. Oceanogr. 13, 1131- 1145 (3.4)

Frankignoul, C., A. Czaja and B. L'Heveder, 1998: Air-sea feedback in the North Atlantic and surface boundary conditions for ocean models. J. Climate, 11, 2310-2324 (3.4)

Freilich, M.H. and D.B. Chelton, 1986: Wavenumber spectra of Pacific winds measured by the Seasat scatterometer. J. Phys. Oceanogr. 16, 741-757 (3.2)

Fritts, H.C., 1976: Tree Rings and Climate. Academic Press, London, 567pp. (5.1, 5.5, 5.6)

Fritts, H.C., 1991: Reconstructing large-scale climate patterns from tree-ring data. The University of Arizona Press, Tucson. (5.6)

Fritts, H.C. and T.W. Swetnam, 1986: Dendroecology: A tool for evaluating variations in past and present forest environments. Laboratory of Tree-Ring Research, University of Arizona, Tucson. (5.5)

Fritts, H.C., T.J. Blasing, B.P. Hayden and J.E. Kutzbach, 1971: Multivariate techniques for specifying tree-growth and climate relationships and for reconstructing anomalies in palaeoclimate. J. Appl. Meteorol. 10, 845-864 (5.6)

Fritts, H.C., J. Guiot and G.A. Gordon, 1990: Verification. In E.R. Cook and L.A. Kairiukstis (Eds.) Methods of Dendrochronology. Applications in the Environmental Sciences. Kluwer/IIASA, Dordrecht, 178-185 (5.6)

Fu, G., H. Diaz, J.O. Fletcher, 1986: Characteristics of the response of sea surface temperature in the Central Pacific associated with warm episodes of the Southern Oscillation. Mon. Wea. Rev. 114, 1716-1738 (6.4)

Gabriel, K. R., and J. Neumann, 1957: On a distribution of weather cycles by lengths. Qaurt. J. Roy Meteor. Soc. 83, 375-380 (11.2)

Gabriel, K. R., and J. Neumann, 1962: A Markov chain model for daily rainfall occurrences at Tel Aviv. Quart. J. Roy. Meteor. Soc. 88, 90-95 (11.2)

Gandin, L. S. and A. H. Murphy, 1992: Equitable skill scores for categorical forecasts. Mon. Wea. Rev., 120, 361-370 (10.3)

Georgakakos, K. P., and M. L. Kavvas, 1987: Precipitation analysis, modeling, and prediction in hydrology. Rev. Geophys., 25, 163-178 (11.3)

Gibson, J.K., P. Kallberg, S. Uppala, A. Hernandez, A. Nomura and E. Serrano, 1997: ECMWF Reanalyses, Project Report Series 1: ERA Description, 72pp. (4.2)

Gill, A.E., 1982: Atmosphere-ocean dynamics. International Geophysics Series, Vol. 30, Academic Press, Orlando, London, pp666 (6.2, 6.4)

Ghil, M. and K.C. Mo, 1991: Intraseasonal oscillations in the global atmosphere-Part I: Northern Hemisphere and tropics. J. Atmos. Sci. 48, 752-779 (14.4)

Ghil, M. and R. Vautard, 1991: Interdecadal oscillations and the warming trend in global temperature time series. Nature 350, 324-327 (14.4)

Giorgi, F., and L.O. Mearns, 1991: Approaches to the simulation of regional climate change: A review. Rev. Geophys. 29, 191-216 (11.3)

Glynn, W.J. and R.J. Muirhead, 1978: Inference in canonical correlation analysis. J. Multivariate Analysis 8, 468-478 (13.3)

Gohberg, I.C. and M.G. Krein, 1969: Introduction to the theory of nonself-adjoint operators, AMS Monograph, 18, 378pp. (12.2)

Golub, G.H. and C. Reinsch, 1970: Singular value decomposition and least square solutions. Numerical Math. 14, 403-420 (12.2)

Golub, G. H. and C. F. Van Loan, 1996: Matrix Computations, 3rd edition, Johns Hopkins University Press, 694pp (12.2)

Gordon, G.A., 1982: Verification of dendroclimatic reconstructions. In M.K. Hughes, P.M. Kelly, J.R. Pilcher and V.C. La Marche, Jr. (Eds.) Climate from Tree Rings. Cambridge University Press, Cambridge, 58-61 (5.6)

Graumlich, L. and L.B. Brubaker, 1986: Reconstruction of annual temperatures (1590-1979) for Longmire, Washington, derived from tree rings. Quaternary Res. 25, 223-234 (5.6)

Graybill, D.A. and S.G. Shiyatov, 1992: Dendroclimatic evidence from the northern Soviet Union. In R.S. Bradley and P.D. Jones (Eds.) Climate Since A.D. 1500. Routledge, London, 393-414 (5.7)

Grootes, P.M., M. Stuiver, J.W.C. White, S. Johnsen and J. Jouzel, 1993: Comparison of oxygen isotope records from the GISP2 and GRIP Greenland ice cores. Nature 366, 552-554 (5.4)

Gregory, J.M., Jones, P.D., Wigley, T.M.L. 1991: Precipitation in Britain: an analysis of area-average data updated to 1989. Intern. J. Climatol. 11, 331-345 (7.5)

Groisman, P.Y., V.V. Koknaeva, T.A. Belokrylova and T.R. Karl, 1991: Overcoming biases of precipitation measurement: a history of the USSR experience. Bull. Amer. Met. Soc. 72, 1725-1733 (4.3)

Grotch, S.L., 1988: Regional intercomparisons of general circulation model predictions and historical climate data. U.S. Department of Energy Report DOE/NBB-0084, Atmospheric and Geophysical Sciences Group, Lawrence Livermore National Laboratory, Livermore, CA (11.3)

Guiot, J., 1990: Comparison of methods. In E.R. Cook and L.A. Kairiukstis (Eds.) Methods of Dendrochronology. Applications in the Environmental Sciences. Kluwer/IIASA, Dordrecht, 185-193 (5.4, 5.6)

Guiot, J., A.L. Berger and Munaut, A.V., 1982: Response functions. In M.K. Hughes, P.M. Kelly, J.R. Pilcher and V.C. La Marche, Jr. (Eds.) Climate from Tree Rings. Cambridge University Press, Cambridge, 38-45 (5.6)

Haan, T.N., D.M. Allen, and J.O. Street, 1976: A Markov chain model of daily rainfall. Water Resour. Res. 12, 443-449 (11.3)

Halliwell, G.R., Jr. and D.A. Mayer, 1996: Frequency response properties of forced climatic SST anomaly variability in the North Atlantic. J. Climate, 9, 3575-3587 (3.4)

Hannoschöck, G. and C. Frankignoul, 1985: Multivariate statistical analysis of sea surface temperature anomaly experiments with the GISS general circulation model. J.Atmos.Sci. 42, 1430-1450 (8.2, 8.3, 13.2)
Hansen, J.E. and S. Lebedeff, 1987: Global trends of measured surface air temperature. J. Geophys. Res. 92, 13345-13372 (4.3)
Hansen, J.E. and S. Lebedeff, 1988: Global surface temperatures: update through 1987. Geophys. Res. Letters 15, 323-326 (4.3)
Hansen, J., I. Fung, A. Lacis, D. Rind, S. Lebedeff, R. Ruedy and G. Russell, 1988: Global climate changes as forecast by Goddard Institute for Space Studies three-dimensional model. J. Geophys. Res. 93, 9341-9364 (4.4)
Hansen, A.R. and A. Sutera, 1986: On the probability density function of planetary-scale atmospheric wave amplitude. J. Atmos. Sci. 43, 3250-3265 (2.5)
Hartzell, F.Z., 1919: Comparison of methods for computing daily mean temperatures. Mon. Wea. Rev. 47, 799-801 (4.2)
Hastenrath, S., 1984: Interannual variability and annual cycle: Mechanisms of circulation and climate in the tropical Atlantic sector. Mon. Wea. Rev. 112, 1097-1107 (6.5)
Hasselmann, K., 1976: Stochastic climate models. Part 1. Theory. Tellus, 28, 473-485 (3.1, 3.3)
Hasselmann, K., 1979: On the signal to noise problem in atmospheric response studies. Meteorology over the Tropical Oceans, D.B. Shaw, Ed., Bracknell, U.K., Roy. Meteor. Soc., pp. 251-259 (8.3)
Hasselmann, K., 1988: PIPs and POPs: The reduction of complex dynamical systems using Principal Interaction and Oscillation Patterns. J. Geophys. Res. 93, 11015-11021 (8.3, 15.1)
Hasselmann, K., 1993: Optimal fingerprints for the detection of time-dependent climate change. J. Climate 6, 1957-1971 (8.3)
Hay, L. E., G. J. McCabe, Jr., D. M. Wolock, and M. A. Ayers, 1991: Simulation of precipitation by weather type analysis. Water Resour. Res. 27, 493-501 (11.4, 11.5)
Hegerl, G.C., H. von Storch, K. Hasselmann, B.D. Santer. U. Cubasch and P. Jones, 1996: Detecting anthropogenic climate change with an optimal fingerprint method, J. Climate 9, 2281-2306 (4.4, 13.2)
Helland, I.S., 1987: On the interpretation and use of R^2 in regression analysis. Biometrics 43, 61-69 (5.6)
Hense, A., R. Glowienka-Hense, H. von Storch and U. Stähler, 1990: Northern Hemisphere atmospheric response to changes of Atlantic Ocean SST on decadal time scales: a GCM experiment. Climate Dyn. 4, 157-174 (13.2)
Herterich, K. and K. Hasselmann, 1987: Extraction of mixed layer advection velocities, diffusion coefficients, feedback factors and atmospheric forcing parameters from the statistical analysis of North Pacific SST anomaly

fields. J. Phys. Oceanogr. 17, 2145-2156 (3.4)
Hildebrandson, H.H., 1897: Quelque recherches sure les centres d'action de l'atmosphère. K. Sven. Vetenskaps akad. Handl. 29, 1-33 (1.1)
Holton, J.R., 1983: The stratosphere and its links to the troposphere. In B. Hoskins and R. Pearce (Eds.) Large-Scale Dynamical Processes in the Atmosphere. Academic Press, 277-303 (15.2)
Hotelling, H., 1936: Relations between two sets of variants. Biometrika 28, 321-377 (13.4)
Horel, J.D. and J.M. Wallace, 1981: Planetary scale phenomena associated with the Southern Oscillation. Mon. Wea. Rev. 109, 813-829 (12.1)
Hoskins, B.J. and D.J. Karoly, 1981: The steady linear response of a spherical atmosphere to thermal and orographic forcing. J. Atmos. Sci. 38, 1179-1196 (6.2)
Houghton, J.T., G.J. Jenkins and J.J. Ephraums, (Eds.) 1990: Climate Change: The IPCC Scientific Assessment. Cambridge University Press, 365pp. (4.4)
Hughes, J.P., 1993: A class of stochastic models for relating synoptic atmospheric patterns to local hydrologic phenomena. Ph.D. Dissertation, Department of Statistics, University of Washington (11.4)
Hughes, J.P. and P. Guttorp, 1994: A class of stochastic models for relating synoptic atmospheric patterns to regional hydrologic phenomena. Water Resour. Res. 30, 1535-1546 (11.4)
Hughes, J.P., D.P. Lettenmaier, and P. Guttorp, 1993: A stochastic approach for assessing the effects of changes in synoptic circulation patterns on gauge precipitation. Water Resour. Res. 29, 3303-3315 (11.4)
Hughes, M.K., 1990: The tree ring record. In R.S. Bradley (Ed.) Global Changes of the Past. UCAR/OIES, Boulder, Colorado, 117-137 (5.1)
Hughes, M.K. and H.F. Diaz, 1994: Was there a "medieval warm period", and if so, where and when? Climatic Change 26, 109-142 (5.7)
Hughes, M.K., P.M. Kelly, J.R. Pilcher and V.C. La Marche, Jr., Eds., 1982: Climate from Tree Rings. Cambridge University Press, Cambridge, 223pp. (5.1)
Hulme, M., 1992: A 1951-80 global land precipitation climatology for the evaluation of general circulation models. Climate Dyn. 7, 57-72 (6.5)
Hulme, M. and Jones, P.D., 1991: Temperatures and windiness over the UK during the winters of 1988/89 and 1989/90 compared to previous years. Weather 46, 126-135 (7.5)
Hulme, M., Briffa, K.R., Jones, P.D. and Senior, C.A. 1993: Validation of GCM control simulations using indices of daily airflow types over the British Isles. Climate Dynamics 9, 95-105 (7.1, 7.5)
Hurrell, J.W., 1995: Decadal trends in the North Atlantic Oscillation and relationships to regional temperature and precipitation. Science, 269, 676-679 (4.3)

James, R.W. and P.T. Fox, 1972: Comparative sea-surface temperature measurements. Marine Science Affairs Report No. 5, WMO 336, 27pp. (4.2)

Janicot, S., 1992a: Spatiotemporal variability of West African rainfall. Part I: Regionalizations and typings. J. Climate 5, 489- 497 (6.5)

Janicot, S., 1992b: Spatiotemporal variability of West African rainfall. Part II: Associated surface and airmass characteristics. J. Climate 5, 499-511 (6.5)

Janowiak, J.E., 1988: An investigation of interannual rainfall variability in Africa. J. Climate 1, 240-255 (6.5)

Jenkinson, A.F. and Collison, F.P. 1977: An initial climatology of gales over the North Sea. Synoptic Climatology Branch Memorandum No. 62. Meteorological Office, Bracknell (7.3, 7.5)

Jones, P.D., 1987: The early twentieth century Arctic high - fact or fiction? Climate Dynamics 1, 63-75 (4.2)

Jones, P.D., 1988: Hemispheric surface air temperature variations: Recent trends and an update to 1987. J. Climate 1, 654-660 (4.3)

Jones, P.D., 1991: Southern Hemisphere sea-level pressure data: an analysis and reconstruction back to 1951 and 1911. Int. J. Climatol. 11, 585-607 (4.2)

Jones, P.D. and K.R. Briffa, 1992: Global surface air temperature variations over the twentieth century: Part 1 Spatial, Temporal and seasonal details. The Holocene 2, 174-188 (4.3, 4.5)

Jones, P.D. and M. Hulme, 1996: Calculating regional climatic time series for temperature and precipitation: methods and illustrations. Intern. J. Climatol. 16, 361-377 (4.2)

Jones, P.D., T. Jónsson and D. Wheeler, 1997b: Extension to the North Atlantic Oscillation using early instrumental pressure oberservations from Gibraltar and SW Iceland. Intern. J. Climatol. 17, 1433-1450 (4.3)

Jones, P.D. and P.M. Kelly, 1982: Principal component analysis of the Lamb catalogue of daily weather types: Part 1, annual frequencies. J. Climate 2, 147-157 (7.1)

Jones, P.D. and P.M. Kelly, 1983: The spatial and temporal characteristics of Northern Hemisphere surface air temperature variations. Intern. J. Climatol. 3, 243-252 (4.3)

Jones, P.D., M. New, D.E. Parker, S. Martin and I.G. Rigor, 1999a: Surface air temperature and its changes over the past 150 years. Rev. Geophys. (in press) (4.4)

Jones, P.D., T.J. Osborn and K.R. Briffa, 1997a: Estimating sampling errors in large-scale temperature averages. J. Climate 10, 2548-2568 (4.3)

Jones, P.D., M.J. Salinger and A.B. Mullan, 1999b: Extratropical circulation indices in the Southern Hemisphere based on station data. Intern. J. Climatol. (in press) (4.3)

Jones, P.D. and T.M.L. Wigley, 1990: Global warming trends. Scientific American 263, 84-91 (4.1, 4.3)

Jones, P.D., S.C.B. Raper, B.D. Santer, B.S.G. Cherry, C.M. Goodess, P.M. Kelly, T.M.L. Wigley, R.S. Bradley and H.F. Diaz, 1985: A Grid Point Surface Air Temperature Data Set for the Northern Hemisphere. US. Dept. of Energy, Carbon Dioxide Research Division, Technical Report TR022, 251pp. (4.3)

Jones, P.D., S.C.B. Raper, R.S. Bradley, H.F. Diaz, P.M. Kelly and T.M.L. Wigley, 1986a: Northern Hemisphere surface air temperature variations: 1851-1984. J. Climate Appl. Meteor. 25, 161-179 (4.2, 4.3)

Jones, P.D., S.C.B. Raper and T.M.L. Wigley, 1986b: Southern Hemisphere surface air temperature variations: 1851-1984. J. Climate and Applied Meteorology 25, 1213-1230 (4.3)

Jones, P.D., S.C.B. Raper, B.S.G. Cherry, C.M. Goodess and T.M.L. Wigley, 1986c: A Grid Point Surface Air Temperature Data Set for the Southern Hemisphere. U.S. Dept. of Energy, Carbon Dioxide Research Division, Technical Report TRO27, 73pp. (4.2)

Jones, P.D., P. Y. Groisman, M. Coughlan, N. Plummer, W-C. Wang and T.R. Karl, 1990: Assessment of urbanization effects in time series of surface air temperatures over land. Nature 347, 169-172 (4.2)

Jones, P.D., T.M.L. Wigley and G. Farmer, 1991: Marine and Land temperature data sets: A Comparison and a look at recent trends. In M.E. Schlesinger (Ed.) Greenhouse-gas-induced climatic change: A critical appraisal of simulations and observations. Elsevier Scientific Publishers, 153-172 (4.3, 4.4)

Jones, P.D., Hulme, M. and Briffa, K.R. 1993: A comparison of Lamb circulation types with an objective classification derived from gridpoint mean-sea-level pressure data. Intern. J. Climatol. 13, 655-663 (7.1)

Jones, R.H., 1976: On estimating the variance of time averages. J. Appl. Meteor. 15, 514-515 (8.2)

Joyce, T.M. and P.Robbins, 1996: The long-term hydrographic record at Bermuda. J. Climate, 9, 3122-3131 (3.6)

Kalnay, E., M. Kanamitsu, R. Kistler, W. Collins, D. Deavon, L. Gandin, M. Iredell, S. Saha, G. White, J. Woollen, Y. Zhu, M. Chelliah, W. Ebisuzaki, W. Higgins, J. Janowiak, K.C. Mo, C. Ropelewski, J. Wang, A. Leetmaa, R. Reynolds, R. Jenne and D. Joseph, 1996: The NMC/NCAR 40-year Reanalysis project. Bull. Amer. Meteor. Soc., 77, 437-471 (4.2)

Khanal, N.N., and R.L. Hamrick, 1974: A stochastic model for daily rainfall data synthesis. Proceedings, Symposium on Statistical Hydrology, Tucson, AZ, U.S. Dept. of Agric. Publication No. 1275, 197-210 (11.3)

Kalkstein, L.S., G. Tan, and J.A. Skindlov, 1987: An evaluation of three clustering procedures for use in synoptic climatological classification. J. of Climate and Applied Meteorology 26(6), 717-730 (11.4)

Karl, T.R. and C.N. Williams, Jr., 1987: An approach to adjusting climatological time series for discontinuous inhomogeneities. J. Clim. Appl. Met. 26, 1744-1763 (4.2)

Karl, T.R., C.N. Williams Jr. and P.J. Young, 1986: A model to estimate the time of observation bias associated with monthly mean maximum, minimum and mean temperatures for the United States. J. Clim. Appl. Met. 23, 1489-1504 (4.2)

Karl, T.R., H.F. Diaz and G. Kukla, 1988: Urbanization: its detection and effect in the United States climate record. J. Climate 1, 1099-1123 (4.2)

Karl, T.R., P.D. Jones, R.W. Knight, G. Kukla, N. Plummer, V. Razuvayev, K.P. Gallo, J. Lindesay, R.J. Charlson and T.C. Peterson, 1993: Asymmetric trends of daily maximum and minimum temperature: empirical evidence and possible causes. Bull. Am. Meteor. Soc. 74, 1007-1023 (4.3, 4.5)

Karoly, D.J., 1984: Typicalness of the FGGE year in the Southern Hemisphere. In, GARP Special Report No. 42, World Meteor. Organization, WMO/TD - No. 22, Geneva, 129-143 (4.2)

Karoly, D.J., 1987: Southern Hemisphere temperature trends: A possible greenhouse gas effect? Geophys. Res. Lett. 14, 1139-1141 (4.4)

Karoly, D.J., 1989: Northern Hemisphere temperature trends: A possible greenhouse gas effect? Geophys. Res. Lett. 16, 465-468 (4.4)

Karoly, D.J., 1995: Observed variability of the Southern Hemisphere atmospheric criculation. In D.G. Martinson, K.Bryan, M. Ghil, M.M. Hall, T.R. Karl, E.S. Sarachik, S. Sorooshian and L.D. Talley (Ed.) Natural Climate Variability on Decade-to-Century Time Scales. National Research Council/National Academy of Sciences, Washington D.C. (4.4)

Karoly, D.J., J.A. Cohen, G.A. Meehl, J.F.B. Mitchell, A.H. Oort, R.J. Stouffer and R.T. Wetherald, 1994: An example of fingerprint detection of greenhouse climate change. Climate Dynamics 10, 97-105 (8.2)

Katz, R.W., 1982: Statistical evaluation of climate experiments with general circulation models: A parametric time series modeling approach. J. Atmos. Sci. 39, 1446-1455 (2.4)

Kavvas, M. L., and J. W. Delleur, 1981: A stochastic cluster model of daily rainfall sequences. Water Resour. Res., 17(4), 1151-1160 (11.3)

Kimoto, M. and M. Ghil, 1993: Multiple flow regimes in the Northern hemisphere winter. Part II: Sectorial regimes and preferred transitions. J. Atmos. Sci. 50, 2645-2673 (14.4)

Können, G.P., P.D. Jones, M.H. Kaltofen and R.J. Allan, 1998: Pre-1866 extensions of the Southern Oscillation Index using early Indonesian and Tahitian meteorological readings. J. Climate 11, 2325-2339 (4.3)

Kulkarni, A. and H. von Storch, 1995: Monte Carlo experiments on the effect of serial correlation on the Mann-Kendall test of trend. Meteorol. Z. (NF) 4, 82-85 (2.3)

Kushnir, Y. and N.C. Lau, 1992: The general circulation model response to a North Pacific SST anomaly: Dependence on time scale and pattern po-

larity. J. Climate 5, 271-283 (3.4)
Kushnir, Y. and J.M. Wallace, 1989: Low frequency variability in the Northern Hemisphere winter. Geographical distribution, structure and time dependence. J. Atmos. Sci. 46, 3122-3142. (12.1)
Kutzbach, G., 1979: The thermal theory of cyclones. AMS Monograph, Boston, 254pp. (1.1)
Kutzbach, J.E. and R.A. Bryson, 1974: Variance spectrum of holocene climatic fluctuations in the North Atlantic sector. J. Atmos. Sci. 31, 1958-1963 (3.7)

Labitzke, K., 1987: Sunspot, the QBO, and the stratospheric temperature in the north polar region. Geophys. Res. Lett. 14, 535-537 (2.2)
Labitzke, K. and H. van Loon, 1988: Association between the 11-year solar cycle, the QBO, and the atmosphere. Part I: the troposphere and the stratosphere in the northern hemisphere in winter. J. Atmos. Terr. Phys. 50, 197-206 (2.2)
La Marche, V.C., Jr., 1982: Sampling strategies. In M.K. Hughes, P.M. Kelly, J.R. Pilcher and V.C. La Marche, Jr. (Eds.) Climate from Tree Rings. Cambridge University Press, Cambridge, 2-6(5.6)
Lamb, H.H. 1950: Types and spells of weather around the year in the British Isles. Quarterly J. of the Royal Meteorological Society 76, 393-438 (7.1, 7.2)
Lamb, H.H., 1972: British Isles weather types and a register of daily sequence of circulation patterns, 1861-1971. Geophysical Memoir 116, HMSO, London, 85pp. (4.3, 7.1, 7.2, 11.4)
Lamb, H.H., 1977: Climate: Present, Past and Future. Volume 2, Climatic History and the Future. Methuen, London, 835pp. (5.4)
Lamb, P.J., 1978a: Case studies of tropical Atlantic surface circulation patterns during recent sub-Saharan weather anomalies: 1967 and 1968. Mon. Wea. Rev. 106, 482-491 (6.5)
Lamb, P.J., 1978b: Large-scale tropical Atlantic surface circulation patterns associated with Subsaharan weather anomalies. Tellus A30, 240-251 (6.5)
Lanzante, J. R., 1984: Strategies for assessing skill and signficance of screening regression models with emphasis on Monte Carlo techniques. J. Climate Appl. Meteor. 24, 952-966 (10.5, 12.1)
Lanzante, J.R., 1984: A rotated eigenanalysis of the correlation between 700-mb heights and sea surface temperatures in the Pacific and Atlantic. Mon. Wea. Rev. 112, 2270-2280
Lanzante, J. R., 1986: Reply. J. Climate Appl. Meteor. 25, 1485-1486 (10.5)
Lara, A. and R. Villalba, 1993: A 3620-year temperature record from Fitzroya cupressoides tree rings in southern South America. Science 260, 1104-1106 (5.7)

Latif, M., 1998: Dynamics of interdecadal variability in coupled ocean-atmosphere models. J. Climate, 11, 602-624 (3.6)

Lau, N.-C., 1985: Modeling the seasonal dependence of the atmospheric response to observed El Niños in 1962-76. Mon. Wea. Rev. 113, 1970-1996 (1.2)

LeCam, L., 1961: A stochastic description of precipitation. Paper presented at the 4th Berkeley Symposium on Mathematics, Statistics, and Probability, University of California, Berkeley, CA (11.3)

Lemke, P., 1977: Stochastic climate models. Part 3. Application to zonally averaged energy models. Tellus 29, 385-392 (3.7)

Lemke, P., E.W. Trinkl and K. Hasselmann, 1980: Stochastic dynamic analysis of polar sea ice variability. J. Phys. Oceanogr. 10, 2100-2120 (3.5)

Lindzen, R.S. and Nigam, S., 1987: On the role of sea surface temperature gradients in forcing low-level winds and convergence in the tropics. J. Atmos. Sci. 44, 2418-2436 (6.2)

Liu, R.Y., and K. Singh, 1992: Moving blocks bootstrap captures weak dependence. In: R. Lepage and L. Billard (Eds.), Exploring the Limits of Bootstrap. John Wiley, 225-248 (9.4)

Livezey, R. E., 1985: Statistical analysis of general circulation model climate simulation sensitivity and prediction experiments. J. Atmos. Sci. 42, 1139-1149 (9.1)

Livezey, R. E., 1987: Caveat emptor! The evaluation of skill in climate predictions. Toward Understanding Climate Change. The J.O. Fletcher Lectures on Problems and Prospects of Climate Analysis and Forecasting. Editor U. Radok, Westview Press, Boulder, CO, 149-178 (10.1, 10.2)

Livezey, R. E. and W. Y. Chen, 1983: Statistical field significance and its determination by Monte Carlo techniques. Mon. Wea. Rev. 111, 46-59 (6.5, 8.2, 9.1, 9.2, 9.3, 9.4)

Livezey, R. E., and J. D. Hoopingarner, 1995: Verification of official monthly mean 700-hPa height forecasts: An update. Wea. Forecasting, 10, 512-527 (10.4)

Livezey, R. E., T.M. Smith, 1999: Covariability of Aspects of North American Climate with Global Sea Surface Temperatures on Interannual to Interdecadal Timescales. Journal of Climate, 12, 289-302,1999 (12.1)

Lorenz, E. N., 1963: Deterministic nonperiodic flow. J. Atmos. Sci. 20, 130-141 (14.2)

Lough, J.M., 1986: Tropical Atlantic sea surface temperatures and rainfall variations in Subsaharan Africa. Mon. Wea. Rev. 114, 561-570 (6.5)

Liu, Q. and C.J.E. Schuurmanns, 1990: The correlation of tropospheric and stratospheric temperatures and its effects on the detection of climate changes. Geophys. Res. Lett. 17, 1085-1088 (4.4)

Lofgren, G.R. and J.H. Hunt, 1982: Transfer functions. In M.K. Hughes, P.M. Kelly, J.R. Pilcher and V.C. La Marche, Jr. (Eds.) Climate

from Tree Rings. Cambridge University Press, Cambridge, 50-56 (5.6)
Luksch, U. and H. von Storch, 1992: Modeling the low-frequency sea surface temperature variability in the North Pacific. J. Climate 5, 893-906 (3.4)
Luther, D.S., A.D. Chave, J.H. Filloux, and P.F. Spain, 1990: Evidence for local and nonlocal barotropic responses to atmospheric forcing during BEMPEX. Geophys. Res. Letters, 17, 949-952 (3.6)

MacCracken, M.C. and H. Moses, 1982: The first detection of carbon dioxide effects: Workshop Summary, 8-10 June 1981, Harpers Ferry, West Virginia. Bull. Am. Met. Soc. 63, 1164-1178 (4.4)
Madden, R.A. and P.R. Julian, 1971: Detection of a 40-50 day oscillation in the zonal wind in the tropical Pacific. J. Atmos. Sci. 28, 702-708 (8.2)
Madden, R.A. and V. Ramanathan, 1980: Detecting climate change due to increasing carbon dioxide. Science 209, 763-768 (4.4)
Madden, R.A., D.J. Shea, G.W. Branstator, J.J. Tribbia and R. Weber, 1993: The effects of imperfect spatial and temporal sampling on estimates of the global mean temperature: Experiments with model and satellite data. J. Climate 6, 1057-1066 (4.3)
Manabe, S. and K. Bryan, 1969: Climate calculations with a combined ocean-atmosphere model. J. Atmos. Sci. 26, 786-789 (1.2)
Manabe, S. and R.J. Stouffer, 1994: Multiple-century response of a coupled ocean-atmopsphere model to an increase of atmopsheric carbon dioxide. J. Climate 7, 5-23 (1.2)
Manley, G. 1974: Central England temperatures: monthly means 1659 to 1973. Quart. J. Roy. Meteor. Soc. 100, 389-405 (7.5)
McCabe, G. J., Jr., 1990: A conceptual weather-type classification procedure for the Philadelphia, Pennsylvania, area. Water-Resources Investigations Report 89-4183, U.S. Geological Survey, West Trenton (11.4, 11.5)
Mjelde, J.W., D.S. Peel, S.T. Sonku, and P.J. Lamb, 1993: Characteristics of climate forecast quality: Implications for economic value of midwestern corn producers. J. Climate 6, 2175-2187 (10.1)
Meyers, S.D., B.G. Kelly and J.J. O'Brien. 1993: An introduction to wavelet analysis in ocenoagrpahy and meteorology: With applications for the dispersion of Yanai waves. Mon. Wea. Rev. 121, 2858-2866 (13.2)
Michaelson, J., 1987: Cross-validation in statistical climate forecast models. J. Climate Appl. Meteor. 26, 1589-1600 (10.1, 10.5)
Mielke, P. W., K. J. Berry and G. W. Brier, 1981: Application of the multi-response permutation procedure for examining seasonal changes in monthly mean sea level pressure patterns. Mon. Wea. Rev. 109, 120-126 (9.1)
Mikolajewicz, U. and E. Maier-Reimer, 1990: Internal secular variability in an ocean general circulation model. Climate Dyn. 4, 145-156 (3.6)

Mitchell, J.F.B., Senior, C.A. and Ingram, W.J. 1989: CO_2 and climate: a missing feedback. Nature 341, 132-134 (7.4)

Mitchell J.M., Jr., 1961: The Measurment of Secular Temperature Changes in the Eastern United States, Research Paper No. 43. U.S. Department of Commerce, Weather Bureau, Washington, D.C. (4.2)

Miyakoda, K., and J. Sirutis, 1977: Comparative global prediction experiments on parameterized subgrid-scale vertical eddy transports. Beitr. Phys. Atmos. 50, 445-487 (1.2)

Molchanov, S.A., O.I. Piterbarg and D.D. Sokolov, 1987: Generation of large-scale ocean temperature anomalies by short-periodic atmospheric processes. Izvestiya, Atmos. Ocean. Phys. 23, 405-409 (3.4)

Molteni, F., S. Tibaldi and T. N. Palmer, 1990: Regimes in the wintertime circulation over the Northern extratropics. I: Observational evidence. Quart. J. Roy. Meteorol. Soc. 116, 31-67 (14.4)

Moron, V., A. Navarra, N. Ward and E. Roeckner, 1998: Skill and reproducibility of regional seasonal anomalies, Climate Dynamics, 14,83-100 (12.4)

Moron, V., R. Vautard and M. Ghil, 1998: Trends, interdecadal and interannual oscillations in global sea-surface temperatures. Climate Dynamics, 14, 545-569 (14.4)

Morrison, D.F., 1976: Multivariate Statistical Methods. McGraw-Hill, 414 pp. (8.3)

Müller, P. and C. Frankignoul, 1981: Direct atmospheric forcing of geostrophic eddies. J. Phys. Oceanogr. 11, 287-308 (3.3, 3.6)

Munk, W.H., G.W. Groves and G.F. Carrier, 1950: Note on the dynamics of the Gulf Stream. Journal of Mar. Research 9, 218-238 (1.2)

Munro, M.A.R., 1984: An improved algorithm for cross-dating tree-ring series. Tree-Ring Bull. 44, 17-27 (5.5)

Murphy, A. H., 1996: General decompositions of MSE-based skill scores: Measures of some basic aspects of forecast quality. Mon. Wea. Rev., 124, 2353-2369

Murphy A. H. and E. S. Epstein, 1989: Skill scores and correlation coefficients in model verification. Mon. Wea. Rev. 117, 572-581 (10.4)

Murphy A. H. and R. L. Winkler, 1987: A general framework for forecast verification. Mon. Wea. Rev. 115, 1330-1338 (10.3)

Murray, R. and Lewis, R.P.W. 1966: Some aspects of the synoptic climatology of the British Isles as measured by simple indices. Meteor. Mag. 95, 192-203 (7.1)

Mysak, L.A., D.K. Manak and R.F. Marsden, 1990: Sea-ice anomalies observed in the Greenland and Labrador seas during 1901-1984 and their relation to an interdecadal Arctic climate cycle. Climate. Dyn. 5, 111-133 (3.5)

Namias, J., 1981: Teleconnections of 700mb anomalies for the Northern

hemisphere. CalCOFI Atlas No. 29, Scripps Institute, La Jolla, Calif. (12.1)
Namias, J., X. Yuan and D.R. Cayan, 1988: Persistence of North Pacific sea surface temperatures and atmospheric flow patterns. J. Climate, 1, 682-703 (12.1)
Neelin, J.D., 1989: On the interpretation of the Gill Model. J. Atmos. Sci. 46, 2466-2468 (6.2)
Newman, M., and P.D. Sardeshmukh, 1995: A caveat concerning singular value decomposition, J. Climate, 8, 352-360 (12.3)
Ng, C.N. and Young, P.C., 1990: Recursive estimation and forecasting of non-stationary time-series. J. Forecasting 9, 173-204 (6.4)
Nicholls, N., G.V. Gruza, J. Jouzel, T.R Karl, L.A. Ogallo and D.E. Parker, 1996: Observed climate variability and change. In J.T. Houghton, L.G. Meira Filho, B.A. Callander, N. Harris, A. Kattenberg and K. Maskell (Eds.) The IPCC Second Scientific Assessment. Cambridge University Press, 133-192 (4.2, 4.3)
Nicholson, S.E., 1980: The nature of rainfall fluctuations in subtropical West Africa. Mon. Wea. Rev., 108, 473-487 (6.5)
Nicholson, S.E., 1985: Sub-Saharan rainfall 1981-84. J. Climate Appl. Meteor. 24, 1388-1391 (6.5)
Nicholson, S.E., 1988: Land surface atmosphere interaction: Physical processes and surface changes and their impact. Progr. Phys. Geogr. 12, 36-65 (6.6)
Nicholson, S.E., 1989: African drought: characteristics, causal theories and global teleconnections. Geophysical Monographs 52, 79-100
Nicholson, S.E. and I.M. Palao, 1993: A re-evaluation of rainfall variability in the Sahel. Part I. Characteristics of rainfall fluctuations. Int. J. Climatol. 4, 371-389 (6.5)
Nitsche, G., J.M. Wallace and C. Kooperberg, 1994: Is there evidence of multiple equilibria in the planetary-wave amplitude? J. Atmos. Sci.51, 314-322 (2.5)
Nitta, T. and S. Yamada, 1989: Recent warming of tropical sea surface temperature and its relationship to the Northern Hemisphere circulation. J. Met. Soc. Japan 67, 375-383 (4.3)
North, G.R., 1984: Empirical Orthogonal Functions and Normal Modes. J. Atmos. Sci. 41, 879-887 (13.3)
North, G.R., T.L. Bell, R.F. Cahalan and F.J. Moeng, 1982: Sampling errors in the estimation of empirical orthogonal functions, Mon. Wea. Rev. 110, 699-706 (13.2)

Oke, T.R., 1974: Review of urban climatology, 1968-1973. WMO Tech. Note No. 169, WMO No. 539, World Meteor. Organization, Geneva (4.2)
Oke, T.R., (Ed.) 1986: Urban Climatology and its applications with special regard to tropical areas. Proc. Tech. Conf. organized by WMO, World Meteor. Organization No. 652 Geneva, 534pp. (4.2)

O'Lenic, E.A. and R.E. Livezey, 1988: Practical considerations in the use of rotated principal component analysis (RPCA) in diagnostic studies of upper-air height fields. Mon. Wea. Rev., 116, 1682-1689 (6.3)
Ortiz, M.J. and A. Ruiz de Elvira, 1985: A cyclo-stationary model of sea surface temperature in the Pacific ocean. Tellus 37A, 14-23 (3.4)

Palmer, T.N., 1986: Influence of the Atlantic, Pacific and Indian Oceans on Sahel rainfall. Nature 322, 251-253 (6.5)
Palmer, T.N. and Sun, 1985: A modelling and observational study of the relationship between sea surface temperature anomalies in the north-west Atlantic and the atmospheric general circulation. Quart. J. Roy. Meteor. Soc. 111, 947-975 (6.2)
Palmer, T.N., Brankovic, C., Viterbo, P. and M.J. Miller, 1992: Modeling interannual variations of summer monsoons. J. Climate 5, 399-417 (6.5)
Parker, D.E., Legg, T.P. and C.K. Folland, 1992: A new daily Central England Temperature series, 1772-1991. Intern. J. Climatol. 12, 317-342 (7.5)
Pedlosky, J., 1990: Baroclinic Instability, the Charney paradigm. In R.S. Lindzen, E.N. Lorenz and G.W. Platzman (Eds) The Atmosphere - A challenge: the Science of J. G. Charney. AMS Monograph, Boston 321pp. (1.2)
Pedlosky, J., 1987: Geophysical Fluid Dynamics. Springer, 710pp.
Penland, M. C., M. Ghil and K. Weickmann, 1991: Adaptative filtering and maximum entropy spectra with application to changes in atmospheric angular momentum. J. Geophys. Res. D96, 22659-22671 (14.4)
Peterson, T.C., D.R. Easterling, T.R. Karl, P. Groisman, N. Nicholls, N. Plummer, S. Torok, I. Auer, R. Böhm, D. Gullett, L. Vincent, R. Heino, H. Tuomenvirta, O. Mestre, T. Szentimrey, M.J. Salinger, E.J. Førland, I Hanssen-Bauer, H. Alexandersson, P.D. Jones and D.E. Parker, 1998: Homogeneity adjustments of in situ atmospheric climate data: A review. Intern. J. Climatol., 18, 1493-1517 (4.2)
Peterson, T.C., K.P. Gallo, J. Lawrimore, T.W. Owen, A. Huang and D.A. McKittrick, 1999: Global rural temperature trends. Geophys. Res. Letts. 26, 329-332 (4.2)
Petterson, G., I. Renberg, P. Geladi, A. Lindberg and F. Lindren, 1993: Spatial uniformity of sediment accumulation in varved lake sediments in northern Sweden. J. Paleolimnol. 9, 195-208 (5.3)
Philander, S.G., 1990: El Niño, La Niña and the Southern Oscillation. International Geophysics Series, Vol. 46, Academic Press, San Diego, London, pp293. (1.1, 6.2)
Pittock, A.B., and M.J. Salinger, 1991: Southern hemisphere climate scenarios. Climate Change 8, 205-222 (11.3)
Plaut, G. and R. Vautard, 1994: Spells of low-frequency oscillations and weather regimes over the Northern Hemisphere. J. Atmos. Sci., 51, 210-236

(14.1, 14.3, 14.4)

Pollard, D., 1982: The performance of an upper-ocean model coupled to an atmospheric GCM: Preliminary results. Climatic Research Institute Report 31, Corvallis, Oregon (4.4)

Potter, K.W., 1981: Illustration of a new test for detecting a shift in mean in precipitation series. Mon. Wea. Rev. 109, 2040-2045 (4.2)

Potts, J. M.., C. K. Folland, I. T. Jollife, and D. Sexton, 1996: Revised LEPS scores for assessing climate model simulations and long-range forecasts. J. Climate, 9, 34-53.

Preisendorfer, R. W. and T. P. Barnett, 1983: Numerical model reality intercomparison tests using small-sample statistics. J. Atmos. Sci. 40, 1884-1896 (9.1)

Preisendorfer, R.W. and J.E. Overland, 1982: A significance test for principal components applied to a cyclone climatology. Mon. Wea. Rev. 110, 1-4 (13.3)

Preisendorfer, R.W., F.W. Zwiers and T.P. Barnett, 1981: Foundations of Principal Component Selection Rules. SIO Reference Series 81-4, Scripps Institute of Oceanography, La Jolla, California (5.6)

Radok, U., 1988: Chance behavior of skill scores. Mon. Wea. Rev. 116, 489-494 (10.3)

Rasmusson, E. and T. Carpenter, 1981: Variations in tropical SST and surface wind fields associated with the Southern Oscillation / El Niño. MOn. Wea. rev. 110, 354-384 (8.3)

Rasmusson, E.M., X. Wang and C.F. Ropelewski, 1990: The biennial component of ENSO variability. J. Mar. Syst., 1, 71-96

Reverdin, G., P. Delecluse, C. Levi, A. Morliére and J. M. Verstraete, 1991: The near surface tropical Atlantic in 1982-1984. Results from a numerical simulation and data analysis, Progress in Oceanogr. 27, 273-340 (8.3)

Rencher, A.C. and F.C. Pun, 1980: Inflation of R2 in best subset regression. Technometrics 22, 49-53 (5.6)

Reynolds, R.W., 1978: Sea surface temperature anomalies in the North Pacific ocean. Tellus, 30, 97-103 (3.4)

Richman, M.B., 1986: Rotation of principal components. Internat. J. Climatol. 6, 293-335 (13.2)

Robock, A., 1978: Internally and externally caused climate change. J. Atmos. Sci. 35, 1111-1122 (3.7)

Rodda, J.C., 1969: The Rainfall Measurement Problem. International Association of Scientific Hydrology Publication No. 78. Wallingford, United Kingdom, 215-231 (4.2)

Rodriguez-Iturbe, I., B. Febres de Power, and J. B. Valdes, 1987: Rectangular pulses point process models for rainfall: Analysis of empirical data. J. Geophys. Res. 92, 9645-9656 (11.3)

Roeckner,E, K.Arpe, L. Bengtsson, L. Dümenil, M. Esch, E. Kirk, F. Lunkeit, W. Ponater, B. Rockel, R. Sausen, U. Schlese, S. Schubert, M. Windelband, 1992: Simulation of the present-day climate with the ECHAM model: impact of model physics and resolution. Report 93, Max-Planck-Institut für Meteorologie, Hamburg, Germany (1.2)

Roeckner E., K. Arpe, L. Bengtsson, M. Christoph, M. Claussen, L. Dümenil, M. Esch, M. Giorgetta, U. Schlese, U. Schulzweida, 1996: The atmospheric general circulation model ECHAM-4: model description and simulation of present-day climate. Report Number 218, MPI, Hamburg (1.2)

Ropelewski, C.F. and M.S. Halpert, 1987: Global and regional scale precipitation patterns associated with the El Niño / Southern Oscillation. Mon. Wea. Rev. 115, 1606-1626 (6.5)

Ropelewski, C.F. and M.S. Halpert, 1989: Precipitation patterns associated with the high index phase of the Southern Oscillation. J. Climate 2, 268-284 (6.5)

Rowell, D.P., 1991: The impact of SST forcing on simulations of the Indian and African monsoons. Extended abstracts of papers presented at TOGA / WGNE MONEG workshop, 21-23 October 1991, WCRP-68, WMO/TD No. 470, pp. 145-151 (6.4)

Rowell, D.P., 1998: Assessing potential seasonal predictability with an ensemble of multidecadal GCM simulations. J. Climate, 11, 109-120 (8.2)

Rowell, D.P., Folland, C.K., Maskell, K., Owen, J.A. and Ward, M.N., 1992: Modelling the influence of global sea surface temperatures on the variability and predictability of seasonal Sahel rainfall. Geophys. Res. Lett. 19, 905-908 (6.5)

Rowntree, P.R., 1972: The influence of tropical east Pacific Ocean temperatures on the atmosphere, Quart. J. Roy. Meteor. Soc. 98, 290- 321 (6.5)

Ruiz de Elvira, A. and P. Lemke, 1982: A Langevin equation for stochastic climate models with periodic feedback and forcing variance. Tellus 34, 313-320 (3.4)

Rumbaugh, W.F., 1934: The effect of time of observations on mean temperature. Mon. Wea. Rev. 62, 375-376 (4.2)

Saltzman, B. and K.A. Maasch, 1991: A first-order global model of late cenozoic climatic change. 2. Further analysis based on a simplification of CO_2 dynamics. Climate Dyn. 6, 201-210 (3.7)

Santer, B.D., K.E. Taylor, T.M.L. Wigley, J.E. Penner, P.D. Jones and U. Cubasch, 1995: Towards the detection and attribution of an anthropogenic effect on climate. Climate Dynamics, 12, 77-100. (4.4)

Santer, B.D., K.E. Taylor, T.M.L. Wigley, T.C. Johns, P.D. Jones, D.J. Karoly, J.F.B. Mitchell, A.H. Oort, J.E. Penner, V. Ramaswamy, M.D. Schwarzkopf, R.J. Stouffer and S.F.B. Tett, 1996a: A search for human influences on the thermal structure of the atmosphere.

Nature, 382, 39-46 (4.4)
Santer, B.D., T.M.L. Wigley, T.P. Barnett and E. Anyamba, 1996: Detection of climate change and attribution of causes. In J.T. Houghton, L.G. Meira Filho, B.A. Callander, N. Harris, A. Kattenberg and K. Maskell (Eds.) The IPCC Second Scientific Assessment. Cambridge University Press, 406-443 (4.4)
Santer, B. D. and T. M. L. Wigley, 1990: Regional validation of means, variances and spatial patterns in general circulation model control runs. J. Geophys. Res. 95 (D), 829-850 (9.1)
Santer, B., T.M.L. Wigley and P.D. Jones, 1993: Correlation methods in fingerprint detection studies. Climate Dyn. 8, 265-276 (4.5, 13.2)
Santer, B.D., T.M.L. Wigley, P.D. Jones and M.E. Schlesinger, 1991: Multivariate methods for the detection of greenhouse-gas-induced climate change. In (M.E. Schlesinger Ed.) Greenhouse-gas-induced climate change: A critical appraisal of simulations and observations. Elsevier, 511-536. (4.4)
Schlesinger, M.E. and Z.-C. Zhao, 1989: Seasonal climate changes induced by doubled CO_2 as simulated by OSU atmospheric GCM mixed-layer ocean model. J. Climate 2, 459-495 (4.4)
Schnur, R., G. Schmitz, N. Grieger and H. von Storch, 1993: Normal Modes of the atmosphere as estimated by principal oscillation patterns and derived from quasi-geostrophic theory. J. Atmos. Sci. 50, 2386-2400 (15.5)
Schweingruber, F.H., 1988: Tree Rings: Basics and Applications of Dendrochronology. Reidel, Dordrecht, 276pp. (5.1)
Schweingruber, F.H., T.S. Bartholin, E. Schär and K.R. Briffa, 1988: Radiodensitometric-dendroclimatological conifer chronologies from Lapland (Scandinavia) and the Alps (Switzerland). Boreas 17, 559-566 (5.7)
Seber, G.A.F., 1984: Multivariate Observations. Wiley and Sons, 686pp. (4.4)
Serre-Bachet, F. and L. Tessier, 1990: Response function analysis for ecological study. In. Methods of Dendrochronology. Applications in the Environmental Sciences (E.R. Cook and L.A. Kairiukstis, Eds.), Kluwer/IIASA, Dordrecht, 247-258 (5.6)
Severuk, B., 1982: Methods for correction for systematic error in point precipitation measurement for operational use. WMO Operational Hydrology Report 21 (WMO-NO. 589)
Shackleton, N.J. and J. Imbrie, 1990: The $\delta^{18}O$ spectrum of oceanic deep water over a five-decade band. Climatic Change 16, 217-230 (3.7)
Shapiro, L. J., 1984: Sampling errors in statistical models of tropical cyclone motion: A comparison of predictor screening and EOF techniques. Mon. Wea. Rev. 112, 1378-1388 (10.5)
Shapiro, L. and D. Chelton, 1986: Comments on "Strategies for assessing skill and significance of screening regression models with emphasis on

Monte Carlo techniques." J. Climate Appl. Meteor. 25, 1295-1298 (10.5)

Shiyatov, S., V. Mazepa and E.R. Cook, 1990: Correcting for trend in variance due to changing sample size. In E.R. Cook and L.A. Kairiukstis (Eds.) Methods of Dendrochronology. Applications in the Environmental Sciences. Kluwer/IIASA, Dordrecht, 133-137 (5.5)

Shukla, J., 1984: Predictability of time averages: Part I: Dynamical predictability of monthly means. In D.M. Burridge and E. Källen (Eds) Problems and Prospects in Long and Medium Range Weather Forecasting. Springer-Verlag, Berlin, pp. 109 (14.4)

Simmons, A.J. and B.J. Hoskins, 1976: Baroclinic instability on the sphere: Normal modes of the primitive and quasigeostrophic equations. J. Atmos. Sci. 33, 1454-1477 (15.5)

Sirutis, J., and K. Miyakoda, 1990: Subgrid scale physics in 1-month forecasts. Part I: Experiment with four parameterization packages. Mon. Wea. Rev. 118, 1043-1064 (1.2)

Smagorinsky, J., 1963: General circulation experiments with the primitive equations. Part I: the basic experiment. Mon. Wea. Rev. 91, 99-164 (1.2)

Smagorinsky, J., S. Manabe, and L. Holloway, 1965: Numerical results from a nine-level general circulation model of the atmosphere. Mon. Wea. Rev. 93, 727-768 (1.2)

Smith, J. A., and A. F. Karr, 1985: Statistical inference for point process models of rainfall. Water Resour. Res. 21, 73-80 (11.3)

Smithies, F., 1970: Integral Equations. Cambridge University Press, 172pp. (12.2)

Stahle, D.W., M.K. Cleaveland and J.G. Hehr, 1988: North Carolina climate changes reconstructed from tree rings: A.D. 372 to 1985. Science 240, 1517-1519 (5.7)

Stern, R.D., and R. Coe, 1984: A model fitting analysis of daily rainfall data. J. R. Statist. Soc. A 147, 1-34 (11.3)

Stocker, F.T. and L.A. Mysak, 1992: Climatic fluctuations on the century time scale: a review of high-resolution proxy data and possible mechanisms. Climatic Changes 20, 227-250 (3.5, 3.7)

Stommel, H., 1948: The westward intensification of wind-driven ocean currents. Trans. Amer. Geo. Union 29, 202-206 (1.2)

Street-Perrott, F. and Perrott, R.A., 1990: Abrupt climate fluctuations in the tropics: the influence of Atlantic Ocean circulation. Nature 343, 607-612 (6.2)

Sutera, A., 1981: On stochastic perturbation and long term climate behavior. Quart. J. R. Met. Soc 107, 137-153 (3.7)

Sverdrup, H.U., 1947: Wind-driven currents in a baroclinic ocean; with applications to the equatorial currents of the eastern Pacific. Proc. Nat. Acad. Sciences USA 33, 318-326 (1.2)

Takens, F., 1981: Detecting strange attractors in turbulence. In D.A.

Rand and L.-S. Young (Eds.) Dynamical Systems and Turbulence. Lecture Notes in Mathematics 898. Springer, Berlin, pp. 366-381 (14.2)

Taylor, K., R.B. Alley, J. Fiacco, P.M. Grootes, G.W. Lamorey, P.A. Mayewski and M.J. Spencer, 1992: Ice core dating and chemistry by direct-current electrical conductivity. J. Glaciol. 38, 325-332 (5.4)

Tett, S.F.B., J.F.B. Mitchell, D.E. Parker and M.R. Allen, 1997: Human influence on the atmospheric vertical temperature structure: Detection and observations. Science, 274, 1170-1173 (4.4)

Thiébeaux, H. J. and F. W. Zwiers, 1984: The interpretation and estimation of effective sample size. J. Climate Appl. Meteor. 23, 800-811 (2.4, 8.2, 9.1, 9.4)

Thompson, L.G., 1990: Ice-core records with emphasis on the global record of the last 2000 years. In R.S. Bradley (Ed.) Global Changes of the Past. UCAR/OIES, Boulder, Colorado, 201-224 (5.3)

Tiedtke, M., 1986: Parameterizations of cumulus convection in large scale models. Physically-based modelling and simulation of climate and climate change. M.E. Schlesinger, Reidel, 375-431 (1.2)

Timmermann, A., M. Latif, R. Voss, and A. Grötzner, 1998: Northern hemisphere interdecadal variability: a coupled air-sea mode. J. Climate, 11, 1906-1931 (3.6)

Trenberth, K. E., 1984: Some effects of finite sample size and persistence on meteorological statistics. Part I: Autocorrelations. Mon. Wea. Rev. 112, 2359-2368 (9.1, 9.4)

Trenberth, K.E., 1990: Recent observed interdecadal climate changes in the Northern Hemisphere. Bull. Amer. Met. Soc. 71, 988-993 (4.3)

Trenberth, K.E. and J.W. Hurrell, 1994: Decadal atmosphere-ocean variations in the Pacific. Climate Dyn. 9, 303-318 (3.2)

Trenberth, K.A. and D.A. Paolino, 1980: The Northern Hemisphere Sea Level Pressure data set: trends, errors and discontinuities. Mon. Wea. Rev. 108, 855-872 (4.2)

Trenberth, K.E., J.R. Christy and J.W. Hurrell, 1992: Monitoring global monthly mean surface temperatures. J. Climate 5, 1405 - 1423 (4.2, 4.3)

Troup, A.J., 1965: The Southern Oscillation. Q. J. R. Meteor, Soc. 91, 490-506 (1.1)

Tyler, D.E., 1982: On the optimality of the simultaneous redundancy transformations. Psychometrika 47, 77-86 (13.2)

Ulrych, T.J. and T.N. Bishop, 1975: Maximum entropy spectral analysis and autoregressive decomposition, Rev. Geophys. Space Phys. 13, 1, 183-200 (14.3)

van den Dool, H. M., 1987: A bias in skill in forecasts based on analogues and antilogues. J. Climate Appl. Meteor. 26, 1278-1281 (10.5)

van den Dool, H. M., S. Saha and A. Johansson, 2000: Empirical Orthogonal teleconnections. J. Climate (submitted) (13.5)

van Loon, H., 1972: Pressure in the Southern Hemisphere. In Meteorology of the Southern Hemisphere. Meteor. Monograph Vol, 13. American Meteor. Soc., Boston, 59-86 (4.2)

van Loon, H. and H. J. Rogers, 1978: The seesaw in winter temperature between Greenland and northern Europe. Part 1: General description. Mon. Wea. Rev. 106, 296-310 (1.1, 4.3)

Vautard, R., 1990: Multiple weather regimes over the North Atlantic: Analysis of precursors and successors. Mon. Wea. Rev. 118, 2056-2081 (14.4)

Vautard, R. and M. Ghil, 1989: Singular spectrum analysis in nonlinear dynamics with applications to paleoclimatic time series, Physica D 35, 395-424 (14.3)

Vautard, R., C. Pires and G. Plaut, 1996: Long-range atmospheric predictability using space-time principal components. Mon Wea Rev., 124, 288-307 (14.4)

Vautard, R., P. Yiou and M. Ghil, 1992: Singular spectrum analysis: A toolkit for short, noisy chaotic signals. Physica D 58, 95-126 (14.1)

Vautard, R., G. Plaut, R. Wang and G. Brunet, 1999: Seasonal prediction of North American surface air temperatures using space-time principal components. J. Climate, 12, 380-294 (1.4)

Venzke, S., M.A. Allen, R.T. Sutton, and D.P. Rowell, 1999: The atmospheric response over the North Atlantic to decadal changes in sea surface temperature. J. Climate, in press (8.3)

Veronis, G., 1981: Dynamics of large-scale ocean circulation. In B.A Warren and C. Wunsch (Eds.) Evolution of Physical Oceanography. MIT Press, 620pp. (1.2)

Villalba, R., 1990: Climatic fluctuations in northern Patagonia during the last 1000 years inferred from tree-ring records. Quaternary Res. 34, 346-360 (5.7)

Vinnikov, K. Y., P. Y. Groisman and K.M. Lugina, 1990: The empirical data on modern global climate changes (temperature and precipitation). J. Climate 3, 662-677 (4.3)

Vinnikov, K.Y. and I.B. Yeserkepova, 1991: Soil moisture: empirical data and model results. J. Climate 4, 66-79 (3.5)

von Storch, H., 1982: A remark on Chervin-Schneider's algorithm to test significance. J. Atmos. Sci. 39, 187-189 (8.2)

von Storch, H., 1987: A statistical comparison with observations of control and El Niño simulations using the NCAR CCM. Beitr. Phys. Atmos. 60, 464-477 (13.2)

von Storch, H. and G. Hannoschöck, 1984: Comment on "Empirical Orthogonal Function analysis of wind vectors over the tropical Pacific region". Bull. Am. Met. Soc. 65, 162 (13.3)

von Storch, H. and G. Hannoschöck, 1986: Statistical aspects of estimated principal vectors (EOFs) based on small sample sizes. J. Clim. Appl. Meteor. 24, 716-724 (13.3)
von Storch, H. and H.A. Kruse, 1985: The extratropical atmospheric response to El Niño events - a multivariate significance analysis. Tellus 37A, 361-377 (8.3)
von Storch, H. and E. Roeckner, 1983: Methods for the verification of general circulation models applied to the Hamburg University GCM. Part 1: Test of individual climate states. Mon. Wea. Rev. 111, 1965-1976 (8.3)
von Storch, H. and F.W. Zwiers, 1988: Recurrence analysis of climate sensitivity experiments. J. Climate, 1, 157-171 (8.2)
von Storch, H. and F. W. Zwiers, 1999: Statistical Analysis in Climate Research. Cambridge University Press, 494 pp (10.1, 10.3, 10.4, 13.2, 15.1)
von Storch, H., T. Bruns, I. Fischer-Bruns and K. Hasselmann, 1988: Principal Oscillation Pattern analysis of the 30- to 60-day oscillation in a General Circulation Model equatorial troposphere. J. Geophys. Res. 93, 11022-11036 (15.1)
von Storch, H., E. Zorita, and U. Cubasch, 1993: Downscaling of climate change estimates to regional scales: Application to winter rainfall in the Iberian Peninsula. J. Climate, 6, 1161-1171 (11.3)
von Storch, H., G. Bürger, R. Schnur and J.S. von Storch, 1995: Principal Oscillation Patterns. J. Climate 8, 377-400 (15.1)

Walker, G.T. and Bliss, E.W., 1932: World Weather V, Mem. R. Meteorol. Soc. 4, 53-84 (1.1)
Wallace, J. M. and R.E. Dickinson, 1972: Empirical orthogonal representation of time series in the frequency domain. Part I: Theoretical considerations. J. Appl. Meteor. 11, 887-892 (13.2)
Wallace, J. M. and D. S. Gutzler, 1981: Teleconnection in the geopotential height field during the northern hemisphere winter. Mon. Wea. Rev. 109, 784-812 (12.1, 12.3)
Wallace, J.M. and Q. Jiang, 1987: On the observed structure of the interannual variability of the atmosphere/ocean system. Variability of the atmosphere/ocean system, Roy. Meteor. Soc., 182pp (12.1)
Wallace, J.M., C. Smith and Q. Jiang, 1990: Spatial patterns of atmosphere-ocean interaction in northern winter. J. Climate 3, 990-998 (12.1)
Wallace, J.M., C. Smith and C.S. Bretherton, 1992: Singular value decomposition of wintertime sea surface temperature and 500-mb height anomalies. J. Climate. 5, 561-576 (12.3)
Walsh, J.E. and W.L. Chapman, 1990: Arctic contribution to upper-ocean variability in the North Atlantic. J. Climate 3, 1462-1473 (3.5)
Ward, M.N., 1992: Provisionally corrected surface wind data, worldwide ocean-atmosphere surface fields and Sahelian rainfall variability. J. Climate 5, 454-475 (6.4, 12.1)

Ward, M.N., 1994: Tropical North African rainfall and worldwide monthly to multi-decadal climate variations. PhD thesis, Reading University, 313p (6.4, 6.5)

Ward, M.N. and C.K. Folland, 1991: Prediction of seasonal rainfall in the North Nordeste of Brazil using eigenvectors of sea-surface temperature. Int. J. Climatol. 11, 711-743 (10.3)

Ward, M.N., K. Maskel, C.K. Folland, D.P. Rowell, and R. Washington, 1994: A tropic-wide oscillation of boreal summer rainfall and patterns of sea-surface temperature. Hadley Centre Climate Research Technical Note 48, available from the National Meteorological Library, Meteorological Office, London Road, Bracknell, Berkshire, England, pp 29.(6.4)

Ward, N., and A. Navarra, 1997: Pattern Analysis of ensemble GCM simulations with prescribed SST: boreal summer examples over Europe and the tropical Pacific. Journal of Climate, 10, 2210-2220, 1997 (12.4)

WASA, 1998: Changing waves and storms in the Northeast Atlantic? - Bull. Amer. Met. Soc. 79, 741-760 (13.5)

Weare, B.C. and J.N. Nasstrom, 1982: Examples of extended empirical orthogonal function analyses. Mon. Wea. Rev. 110, 481-485 (13.2, 14.1)

Weaver, A.J., J. Marotzke, P.F. Cummins and E.S. Sarachik, 1993: Stability and variability of the thermohaline circulation. J. Phys. Oceanogr. 23, 39-60 (3.6)

Weiss, L.L., 1964: Sequences of wet and dry days described by a Markov chain model. Mon. Wea. Rev. 92, 169-176 (11.3)

Weisse, R., U. Mikolajewicz and E. Maier-Reimer, 1994: Decadal variability of the North Atlantic in an ocean general circulation model. J. Geophys. Res. 99C6, 14411-14421 (3.6)

Werner, P. and H. von Storch, 1993: Interannual variability of Central European mean temperature in January / February and its relation to the large-scale circulation. Clim. Res. 3, 195-207 (13.3)

Wetherald, R.T. and S. Manabe, 1990: Cited by Cubasch, U. and Cess, R.D., In J.T. Houghton, G.J. Jenkins and J.J. Ephraums (Eds.) Climate Change: The IPCC Scientific Assessment. Cambridge University Press, p.81. (4.4)

Wigley, T.M.L. and T.P. Barnett, 1990: Detection of the Greenhouse Effect in the Observations. In J.T. Houghton, G.J. Jenkins and J.J. Ephraums (Eds.) Climate Change: The IPCC Scientific Assessment. Cambridge University Press, 239-255 (4.4)

Wigley, T.M.L. and P.D. Jones, 1981: Detecting CO_2-induced climate change. Nature 292, 205-208 (4.4)

Wigley, T. M. L. and B. D. Santer, 1990: Statistical comparison of spatial fields in model validation, perturbation and predictability experiments. J. Geophys. Res. 95 (D), 51-866 (9.1, 11.3)

Wigley, T.M.L., K.R. Briffa and P.D. Jones, 1984: On the average value of correlated time series, with applications in dendroclimatology and

hydrometeorology. J. Climate Appl. Meteor. 23, 201-213 (5.4)
Wigley, T.M.L., J.M. Lough and P.D. Jones, 1984: Spatial patterns of precipitation in England and Wales and a revised homogeneous England and Wales precipitation series. Intern. J. Climatol. 4, 1-26 (5.4, 7.2)
Wigley, T.M.L., J.K. Angell and P.D. Jones, 1985: Analysis of the temperature record. In M.C. MacCracken and F.M. Luther (Eds.) Detecting the Climatic Effects of Increasing Carbon Dioxide. U.S. Dept. of Energy, Carbon Dioxide Research Division, Washington D.C., 55-90 (4.4)
Wigley, T.M.L., P.D. Jones and P.M. Kelly, 1986: Empirical climate studies: warm world scenarios and the detection of climatic change induced by radiatively active gases. In B. Bolin, B.R. Ds, J. Jäger and R.A. Warrick (Eds.) The Greenhouse Effect, Climatic Change and Ecosystems. SCOPE series, Wiley, 271-323 (4.3)
Wigley, T.M.L., P.D. Jones and K.R. Briffa, 1987: Cross-dating methods in dendrochronology. J. Archaeological Sci. 14, 51-64 (5.5)
Wigley, T.M.L., P.D. Jones, K.R. Briffa, and G. Smith, 1990: Obtaining sub-grid-scale information from coarse-resolution general circulation model output. J. Geophys. Res. 95, 1943-1953
Wilks, D. S., 1995: Statistical Methods in the Atmospheric Sciences. San Diego, Academic Press. 464 pp. (10.1, 10.3, 10.4)
Wilks, D. S., 1997: Resampling hypothesis tests for autocorrelated fields. J. Climate, 10, 65-82 (9.1, 9.4)
Willebrand, J., 1978: Temporal and spatial scales of the wind field over the North Pacific and North Atlantic. J. Phys. Oceanogr., 8, 1080-1094 (3.1)
Williams, J. and H. van Loon, 1976: An examination of the Northern Hemisphere Sea Level Pressure data set. Mon. Wea. Rev. 104, 1354-1361 (4.2)
Wilson, C.A. and J.F.B. Mitchell, 1987: A doubled CO_2 climate sensitivity experiment with a GCM including a simple ocean. J. Geophys. Res. 92, 13315-13343 (4.4)
Wilson, M.A.G., 1975: A wavenumber-frequency analysis of large-scale tropospheric motions in the extratropical northern hemisphere. J. Atmos. Sci. 32, 478-488 (3.2)
Wilson, L. L., D. P. Lettenmaier, and E. F. Wood, 1991: Simulation of daily precipitation in the Pacific Northwest using a weather classification scheme. In E. F. Wood (Ed.) Land Surface-Atmosphere Interactions for Climate Modeling: Observations, Models, and Analysis. Surv. Geophys., 12(1-3), 127-142, Kluwer, Dordrecht, The Netherlands (11.4, 11.5)
Wilson, L.L., D.P. Lettenmaier, and E. Skyllingstad, 1992: A hierarchical stochastic model of large-scale atmospheric circulation patterns and multiple station daily precipitation. J. Geophys. Res. 97, 2791-2809 (11.4, 11.5)
Wiser, E. H., 1965: Modified Markov probability models of sequences of precipitation events. Mon. Weath. Rev. 93, 511-516 (11.4)

Woolhiser, D.A., and G.G.S. Pegram, 1979: Maximum likelihood estimation of Fourier coefficients to describe seasonal variations of parameters in stochastic daily precipitation models. J. Appl. Meteorol. 8, 34-42 (11.3)
Woolhiser, D.A., and J. Roldan, 1982: Stochastic daily precipitation models, 2. A comparison of distributions of amounts. Water Resour. Res. 18, 1461-1468 (11.2)
Woodruff, S.D., R.J. Slutz, R.J. Jenne and P.M. Steurer, 1987: A Comprehensive Ocean-Atmosphere Data Set. Bulletin American Meteor. Soc. 68, 1239-1250 (4.2, 6.3)
World Meteorological Organization (WMO), 1966: Climatic Change. (Technical Note No. 79). Geneva, Switzerland (4.2)

Xu, J.-S., 1992: On the relationship between the stratospheric QBO and the tropospheric SO. J. Atmos. Sci. 49, 725-734 (15.3)
Xu, J.-S., 1993: The Joint Modes of the Coupled Atmosphere-Ocean System Observed from 1967 to 1986. J. Climate 6, 816-838 (15.5)

Yarnal, B. 1993: Synoptic Climatology in Environmental Analysis. Belhaven Press, London. 195pp. (7.2)
Young, K.C., 1993: Detecting and removing inhomogeneities from long-term monthly sea level pressure time series. J. Climate 6, 1205-1220 (4.2)
Young, P.C., Ng, C.N., Lane, K. and Parker, D.E., 1991: Recursive forecasting, smoothing and seasonal adjustment of non-stationary environmental data. J. Forecasting 10, 57-89. (6.4)

Zhang, Y., J.M. Wallace and N. Iwasaka, 1996: Is climate variability over the North Pacific a linear response to ENSO? J. Climate 9, 1468-1478 (3.2)
Zorita, E., V. Kharin and H. von Storch, 1992: The atmospheric circulation and sea surface temperature in the North Atlantic in winter. Their interaction and relevance for Iberian rainfall. J. Climate 5, 1097-1108 (11.5, 13.3, 13.4)
Zorita, E., J.P. Hughes, D.P. Lettenmaier, and H. von Storch, 1995: Stochastic characterization of regional circulation patterns for climate model diagnosis and estimation of local precipitation. J. Climate 8, 1023-1042 (11.4, 11.5)
Zucchini, W., and P. Guttorp, 1991: A hidden Markov model for space-time precipitation. Water Resour. Res. 27, 1917-1923 (11.4)
Zwiers, F. W., 1987: Statistical consideration for climate experiments. Part II: Multivariate tests. J. Climate Appl. Meteor. 26, 477-487 (9.1, 9.3, 9.4)
Zwiers, F. W., 1990: The effect of serial correlation on statistical inferences made with resampling procedures. J. Climate 3, 1452-1461 (9.1, 9.4)
Zwiers, F.W., 1996: Interannual variability and predictability in an ensem-

ble of AMIP climate simulations conducted with the CCC GCM2. Climate Dyn., 12, 825-847 (8.2)

Zwiers, F.W., 1993: Simulation of the Asian Summer Monsoon with the CCC GCM-1. J. Climate 6, 470-486 (13.3)

Zwiers, F. W. and G. J. Boer, 1987: A comparison of climates simulated by a General Circulation Model when run in the annual cycle and perpetual modes. Mon. Wea. Rev. 115, 2626-2644 (9.1)

Zwiers, F.W. and H.J. Thiébaux, 1987: Statistical considerations for climate experiments. Part I. Scalar tests. J. Climate Appl. Meteor. 26, 464-476 (8.2)

Zwiers, F. W. and H. von Storch, 1995: Taking serial correlation into account in tests of the mean. J. Climate 8, 336-351 (2.2, 8.2, 9.1, 9.4)

Abbreviations

AGCM	atmospheric General Circulation Model
BP	bootstrap procedure
CART	Classification and Regression Trees
CCA	Canonical Correlation Analysis
COADS	Comprehensive Atmosphere Ocean Data Set
ECMWF	European Centre for Medium Range Forecast
ENSO	El Niño Southern Oscillation
EOF	Empirical Orthogonal Function
GCM	General Circulation Model
GISS	Goddard Institut for Space Studies
LODYC	Laboratoire d'Océanographie Dynamique et de Climatologie
MPI	Max-Planck-Institut für Meteorologie
MSE	mean square error
MSSA	Multichannel Singular Spectrum Analysis
NAO	North Atlantic Oscillation
NHMM	Nonhomogeneous Hidden Markov Model
NMC	National Meteorological Centre, USA
PIP	Principal Interaction Pattern
PNA	Pacific North-America Pattern
POP	Principal Oscillation Pattern
PP	permutation procedure
SNR	signal-to-noise ratio
SO	Southern Oscillation
SSA	Singular Spectrum Analysis
SST	sea-surface temperature
SVD	Singular Value Decomposition
T-EOF	Time EOF
ST-EOF	Space-Time EOFs

Index

Made in United States
Orlando, FL
04 November 2026

90201249R00199